Handbook of Geospatial Approaches to Sustainable Cities

This comprehensive handbook presents the current state of knowledge on geospatial technologies, techniques, and methods that are imperative for providing solutions to sustainable cities. It addresses the role of geospatial big data and AI techniques and how they are applied when analyzing the sustainability of urban development, land use, urban planning, and resource management, as well as monitoring the impact urbanization has on the environment and the ecosystem. Taking an interdisciplinary approach to sustainable cities, and with contributions from renowned experts around the world, this holistic handbook is a toolbox for geospatial, urban, and sustainability professionals, the artificial intelligence community, and those who work in related fields.

Features:

- Explores cutting-edge geospatial and AI techniques in support of efficient, resilient, digital, and smart cities
- Bridges urban science and sustainability science via geospatial methods
- Contributes to the efforts of GEO by addressing and exemplifying pertinent societal benefit areas and engagement priorities
- Includes 16 case studies with a broad geographic scope that integrate societal needs with technological advances
- Draws expertise in geospatial technology, big data, and artificial intelligence from leading experts in the world

This book is intended for researchers and scientists interested in learning techniques in GeoAI, including the technologies for collecting, analyzing, managing, processing, and visualizing geospatial datasets.

Imaging Science Series

Qihao Weng
Hong Kong Polytechnic University

Imaging science is a multidisciplinary field concerned with the generation, collection, duplication, analysis, modification, and visualization of images, including imaging things that the human eye cannot detect. As an evolving field, it includes research and researchers from physics, mathematics, electrical engineering, computer vision, computer science, remote sensing, and perceptual psychology. The book series will review the subfields within imaging science that include: image processing, computer vision, 3D computer graphics, animations, atmospheric optics, astronomical imaging, digital image restoration, digital imaging, color science, digital photography, holography, magnetic resonance imaging, medical imaging, microdensitometry, optics, photography, remote sensing, radar imaging, radiometry, silver halide, ultrasound imaging, photoacoustic imaging, thermal imaging, visual perception, and various printing technologies. The books in this series assume that readers have a basic knowledge of remote sensing and imaging science, and therefore will not cover very basic materials but concentrate on those aspects of imaging science and the application discipline(s) that are pertinent for a 'how to' knowledge. This provides academic authors and leading professionals an opportunity to combine their research, consultancy, and interesting project work into a book that is then relevant to professional lines of work or teaching modules focusing on imaging science applications.

High Spatial Resolution Remote Sensing: Data, Analysis, and Applications
Edited by Yuhong He and Qihao Weng

Remote Sensing Time Series Image Processing
Qihao Weng

Handbook of Geospatial Approaches to Sustainable Cities
Edited by Qihao Weng, in collaboration with Cheolhee Yoo

https://www.routledge.com/Imaging-Science/book-series/CRCIS

Handbook of Geospatial Approaches to Sustainable Cities

Edited by
Qihao Weng
in collaboration with Cheolhee Yoo

The Hong Kong Polytechnic University

The Hong Kong Institute of Surveyors

CRC Press
Taylor & Francis Group
Boca Raton London New York

CRC Press is an imprint of the
Taylor & Francis Group, an **informa** business

Designed cover image: © Tavarius, Shutterstock

First edition published 2024
by CRC Press
2385 NW Executive Center Drive, Suite 320, Boca Raton FL 33431

and by CRC Press
4 Park Square, Milton Park, Abingdon, Oxon, OX14 4RN

CRC Press is an imprint of Taylor & Francis Group, LLC

ISBN: 978-1-032-15481-7 (hbk)
ISBN: 978-1-032-15534-0 (pbk)
ISBN: 978-1-003-24456-1 (ebk)

DOI: 10.1201/9781003244561

Typeset in Times
by MPS Limited, Dehradun

Contents

PART III Geospatial Techniques for Resilient Cities

PART IV Digital Cities

Preface

Geospatial technology is one of the three most important, evolving fields in the 21st century, along with nanotechnology and biotechnology. Earth Observation (EO) technologies, in conjunction with field surveys, have been used to observe, monitor, measure, and model many of the components that comprise the ecosystems, contributing to the sustainability and productivity of natural and human systems. Driven by societal needs and improvement in geospatial technology, we have witnessed a great increase in research and development, technology transfer, and engineering activities worldwide. Their integration with the Internet and communications technologies, as well as Geographic Information System, have led to the emergence of geospatial big data over the web and on hands. These geo-referenced big data and imagery, in conjunction with GPS, mobile mapping, and modern telecommunication technologies, have sparked increasing interest in governments, industries, and the public.

Artificial Intelligence (AI), especially deep learning, has transformed the field of computer vision, and now rivals human-level performance in certain tasks, such as image recognition and semantic segmentation. In this context, there is a strong need for research and development of new approaches to multi-sensory and multi-modal deep learning within the new processing and recognition frameworks of remotely sensed imagery. Since the 21st century, with the improvement of spatial resolution of remote sensing images, image classification methods have gradually developed into three parallel branches at different levels: pixel-level, object-level, and scene-level classification. Many AI-based methods employ scene classification, which uses the layout of objects within the scene, in addition to the ambient context, for classification. It requires solving the problems in data training and validation, model generalization across different modalities of EO images, and supercomputing power. In recent years, AI has shown great promise to fulfil the challenging need for image processing. AI can aid in the collecting and processing of image data using neural networks and deep learning models and through Computer Vision models to better understand and interpret data more efficiently. Since circa 2016, we have witnessed the convergence of the improved capacities in geospatial and AI technology, leading to a new field of Geospatial AI (abbr. GeoAI).

This book explores the convergence of geospatial technology, GeoAI, big data, and sustainable urban development, providing insights into the nexus of these fields, the current state, and future potentials. It consists of four sections, each addressing a unique aspect of these synergies. This book would be of great value for researchers, students, and professionals in the fields of geospatial technologies, AI, urban science, and sustainability science.

PART I – ARTIFICIAL INTELLIGENCE AND BIG DATA ANALYTICS: NEEDS AND REQUIREMENTS

This section covers three chapters discussing the current state of AI and big data analytics in understanding and managing urban environments. Chapter 1 provides an overview of urban physical background sensing via GeoAI and street-level imagery, discussing the available datasets, processing methodologies, current applications, and potential future trajectories of urban remote sensing. Chapter 2 explores using geospatial big data to tackle urban sustainability issues, providing case studies on agglomeration economies, urban freight logistics, and urban functionalities. Chapter 3 reviews the recent advancements in geospatial data, techniques, and platforms dedicated to urban flood monitoring and management. It covers mitigation, preparation, response, and recovery strategies, underlining the crucial role of multi-sensor data and integrating geospatial techniques with advanced AI. This chapter also points out existing challenges and prospects in the field.

PART II – GEOSPATIAL TECHNIQUES FOR RENEWABLE CITIES

This section explores using geospatial techniques to create sustainable, renewable urban environments, emphasizing their potential to improve urban living. Chapter 4 investigates the challenges resource-based cities in China face in achieving the UN's Sustainable Development Goals (SDGs) and the role of geospatial techniques' remote sensing data in assessing environment-related SDG indicators. The findings suggest that resource-based cities typically struggle more with environmental challenges and sustainable development than non-resource-based cities. Chapter 5 presents a GeoAI model that leverages crowdsourced data for mapping urban green spaces from Sentinel-2A satellite images, demonstrating how integrating remote sensing with crowdsourced geospatial big data can boost urban green space mapping. Chapter 6 uses nighttime light time series data to analyze changes in urban lands across 34 global mega-cities, categorizing them into five groups: urban growth, deurbanization, urban land use intensification, land use degradation, and no significant change. This chapter offers new insights into global urbanization, highlighting the broader extent and greater magnitude of changes in developing countries' mega-cities. Chapter 7 studies the impact of nature-based solutions, such as green roofs, permeable pavements, and street trees, on reducing urban heat islands, improving thermal comfort, and lessening building energy consumption. The analysis shows that green roofs can significantly reduce energy demand in hot, sunny regions and potentially eliminate annual energy demands for cooling or heating in low-rise urban buildings.

PART III – GEOSPATIAL TECHNIQUES FOR RESILIENT CITIES

This section provides an in-depth overview of how geospatial techniques contribute to identifying, assessing, and managing various urban risks, encompassing both natural and anthropogenic threats, towards creating resilient cities. Chapter 8 unveils a two-step GeoAI algorithm for mapping hourly urban air temperature with a detailed 250 m spatial resolution in Seoul, South Korea. This algorithm addresses the limitations inherent in satellite imagery and ground-based observation stations, producing air temperature maps that effectively capture intricate urban temperature patterns and their temporal variations, aiding in monitoring urban heat island (UHI) effects. Chapter 9 introduces various satellite land surface temperature (LST) products, LST retrieval algorithms, and the unique biases associated with urban surfaces. It further discusses different methods for quantifying UHI and outlines the contexts where these methods are best employed. Chapter 10 outlines strategies for mitigating urban air pollution, encompassing observation, modeling, and control aspects. Chapter 11 assesses emerging drought risks through geospatial analysis, unveiling underlying mechanisms from a land-atmosphere coupling perspective. Chapter 12 presents methodologies to enhance urban air temperature forecasting performance by merging numerical weather prediction (NWP) models with machine learning, featuring case studies in Seoul. This chapter underscores how machine learning can augment urban air temperature forecasting accuracy for practical applications.

PART IV – DIGITAL CITIES

This section emphasizes the transformative potential of geospatial techniques, big data, and AI in sculpting the future of urban landscapes into digitally driven cities. Chapter 13 conducts a comprehensive review and analysis of existing datasets and spatial modeling methods to estimate and map carbon emissions from urban areas. This effort underpins policy decisions concerning urban energy resilience and promoting low-carbon development. Chapter 14 formulates visual indicators, namely greenness, openness, and enclosure, derived from street view data using a GeoAI-based semantic segmentation approach. Through multilevel regression analysis, the study investigates the relationship between these indicators and depression and anxiety levels among the elderly. Chapter 15 delves into applying deep learning AI models for mapping Local Climate Zones (LCZ), presenting an innovative method to evaluate urban form across varied landscapes objectively. This chapter highlights the potential of deploying GeoAI in LCZ mapping, paving the way for future breakthroughs in urban climate studies. Chapter 16 introduces a spatial agent-based model developed to simulate the diffusion of private autonomous vehicles (AVs) across time and space. It investigates their potential impact on housing expenses saved via residential relocation and in-vehicle work. The findings from this chapter provide valuable insights for AV-related stakeholders, including local authorities, urban planners, and vehicle manufacturers.

Qihao Weng and Cheolhee Yoo
Hong Kong
October 20, 2023

Acknowledgments

We wish to thank all the contributors for making this book possible. Furthermore, we offer our sincere appreciation to all the reviewers, who have taken precious time from their busy schedules to review the chapters submitted to this book. Further, we appreciate the members at the JC STEM Lab of Earth Observations and Research Centre for Artificial Intelligence in Geomatics, The Hong Kong Polytechnic University, for their support. Last but not least, we acknowledge the financial support from The Hong Kong Polytechnic University and The Hong Kong Institute of Surveyors.

We hope that the publication of this book will provide much stimulation to students, researchers, and practitioners to conduct more in-depth studies on geospatial technology, GeoAI, urban science, and sustainability science, and will open up a new way of thinking for years ahead in these communities.

The reviewers of this book are listed below in alphabetical order: Yinxia Cao, Yue Chang, Dongjin Cho, Siqi Jia, Zhiwei Li, Xiaoyan Lu, Pir Mohammad, Seonyoung Park, Huijuan Xiao, Cheolhee Yoo, Qiming Zheng, and Qingwei Zhong.

Editors

Qihao Weng, a Foreign Member of The Academy of Europe (Academia Europaea) and a Fellow of IEEE, AAAS, AAG, ASPRS, and AAIA, is currently a Chair Professor at the Hong Kong Polytechnic University and has worked as the Director of the Center for Urban and Environmental Change and a Professor of Geography at Indiana State University, 2001–2021, and a Senior Fellow at the National Aeronautics and Space Administration from 2008 to 2009. He received his Ph.D. in geography from the University of Georgia in 1999. Weng is the Lead of GEO Global Urban Observation and Information Initiative, 2012–2022, and an Editor-in-Chief of *ISPRS Journal of Photogrammetry and Remote Sensing*. Additionally, he serves as the Series Editor of Taylor & Francis Series in Remote Sensing Applications, and Taylor & Francis Series in Imaging Science. Weng has been the Organizer and Program Committee Chair of the biennial IEEE/ISPRS/GEO sponsored International Workshop on Earth Observation and Remote Sensing Applications conference series since 2008; a National Director of American Society for Photogrammetry and Remote Sensing from 2007 to 2010; and a panelist of U.S. DOE's Cool Roofs Roadmap and Strategy in 2010.

Cheolhee Yoo is currently a Research Assistant Professor at the Hong Kong Polytechnic University. He earned his Bachelor of Engineering and Ph.D. from the Department of Urban and Environmental Engineering at the Ulsan National Institute of Science and Technology (UNIST), South Korea, in 2017 and 2022, respectively. In 2019, he further expanded his research experience as a research intern at the RIKEN Center for Advanced Intelligence Project (AIP) in Japan. Following the completion of his doctoral studies, Dr. Yoo served as a Postdoctoral Fellow at the Hong Kong Polytechnic University from April to September 2022. He is currently an Editorial Board Member of the *ISPRS Journal of Photogrammetry and Remote Sensing* and *GIScience & Remote Sensing*.

Contributors

Ren Chao
The University of Hong Kong
Hong Kong S.A.R., China

Haojie Chen
The Hong Kong Polytechnic University
Hung Hom, Kowloon, Hong Kong

Yang Chen
Wuhan University
Wuhan, China

Dongjin Cho
Ulsan National Institute of Science and Technology
Ulsan, South Korea

Hai Guo
The Hong Kong Polytechnic University
Hung Hom, Kowloon, Hong Kong

Ce Hou
The Hong Kong University of Science and Technology
Hong Kong, China

Jungho Im
Ulsan National Institute of Science and Technology
Ulsan, South Korea

Siqi Jia
The Hong Kong Polytechnic University
Hung Hom, Kowloon, Hong Kong

Xiaoyu Jiao
The Hong Kong Polytechnic University
Hung Hom, Kowloon, Hong Kong

Siwoo Lee
Ulsan National Institute of Science and Technology
Ulsan, South Korea

Yong Li
The Hong Kong University of Science and Technology
Hong Kong, China

Zhiwei Li
The Hong Kong Polytechnic University
Hung Hom, Kowloon, Hong Kong

Xintao Liu
The Hong Kong Polytechnic University
Hung Hom, Kowloon, Hong Kong

Cai Meng
Wuhan University
Wuhan, China

Yamin Qing
The Hong Kong Polytechnic University
Hung Hom, Kowloon, Hong Kong

Justin Hayse Chiwing G. Tang
The Hong Kong Polytechnic University
Hung Hom, Kowloon, Hong Kong

Shuo Wang
The Hong Kong Polytechnic University
Hung Hom, Kowloon, Hong Kong

Yu Wang
The Hong Kong Polytechnic University
Hung Hom, Kowloon, Hong Kong

Huijuan Xiao
The Hong Kong Polytechnic University
Hung Hom, Kowloon, Hong Kong

Song Xue
The Hong Kong Polytechnic University
Hung Hom, Kowloon, Hong Kong

Yuhei Yamamoto
Chiba University
Chiba-shi, Japan

Jiang Ying
The Hong Kong Polytechnic University
Hung Hom, Kowloon, Hong Kong

Zidong Yu
The Hong Kong Polytechnic University
Hung Hom, Kowloon, Hong Kong

Yangzong Zeren
The Hong Kong Polytechnic University
Hung Hom, Kowloon, Hong Kong

Fan Zhang
Peking University
Beijing, China

Qiming Zheng
The Hong Kong Polytechnic University
Hung Hom, Kowloon, Hong Kong

Chengxiang Zhuge
The Hong Kong Polytechnic University
Hung Hom, Kowloon, Hong Kong

Part I

Artificial Intelligence and Big Data Analytics

Needs and Requirements

1 Sensing Urban Physical Environment with GeoAI and Street-Level Imagery

Ce Hou, Yong Li, and Fan Zhang

INTRODUCTION

Sensing the urban physical environment is crucial to understanding, managing, planning, and designing cities. The paramount significance of the urban physical environment has been repeatedly emphasized by landscape architects, urban planners, and researchers. The "City Beautiful Movement", which flourished from the 1890s to the 1920s, was a dedicated endeavor aimed at enhancing the quality of the urban living environment and elevating people's overall quality of life through the improvement of cities' physical surroundings. This movement sought to augment the aesthetic appeal of urban areas, creating a more visually pleasing and comfortable perception of the city (Wilson, 1989). In Kevin Lynch's masterpiece The Image of the City, he provides a comprehensive exploration of how urban residents perceive and construct their own perception of the physical environment within the city (Lynch, 1964). The concept of investigating the human-centered perception of the urban physical environment has exerted a significant influence on numerous urban planning theorists, including William H. Whyte, Jane Jacobs, and Jan Gehl (Jacobs, 2016; Whyte et al., 1980; Gehl, 2013). By sensing the physical environment of a city, urban planners and designers gain more profound insights into its distinctive attributes and intricacies. This understanding enables them to devise effective strategies and solutions aimed at bolstering the sustainability, livability, and overall quality of life for the city's inhabitants (Zhang et al., 2019).

In the past few decades, various methodologies for perceiving and analyzing the physical urban environment have been developed, refined, and employed. Before the flourishing of information technology, traditional work of sensing the urban physical environment primarily relied on manual observation methods, such as questionnaires, census data, and geographic measurements (Hoehner et al., 2005; Galea et al., 2005; Handy et al., 2002). For instance, urban researchers physically visited cities, meticulously observed buildings, roads, and traffic patterns, and collected valuable insights from residents through questionnaires and interviews (Davison and Lawson, 2006; Frank et al., 2005). Traditional surveying tools and maps were utilized to gather geographical data pertaining to the city, including topographical features and architectural structures

DOI: 10.1201/9781003244561-2

(Harley, 1990; DeMers, 2008). These manual-based urban sensing methods are often labor-intensive, time and resource-consuming, and sometimes difficult to implement (Fleischhacker et al., 2013).

The advancement of information technology has led to the emergence of urban sensing methods utilizing remote sensing (RS), global positioning systems (GPS), and geographic information systems (GIS), which enables deeper sensing of the urban physical environment (Liang and Weng, 2010; Miller and Small, 2003). RS technology provides large-scale imagery data with a high spatio-temporal resolution, facilitating the acquisition of large-scale information on cities, such as city sprawl, land use, and vegetation cover (Ji et al., 2006; Alqurashi and Kumar, 2013; Xie et al., 2008). Moreover, a wide range of sensor data enables real-time sensing of various physical environmental parameters in cities, including climate conditions, air quality, and urban noise (Dutta et al., 2009). GPS enables the sensing of the urban physical environment through the collection of travel tracks from city dwellers (Van der Spek et al., 2009). By leveraging GPS, large volumes of location data can be gathered and utilized to analyze parameters such as traffic flow and travel patterns and to assess challenges such as road congestion and traffic safety (Kamran and Haas, 2007; Zheng et al., 2009). Subsequently, the integration of multiple data sources within a GIS platform allows for the overlay analysis of geographic data and the identification of spatial relationships, leading to a deeper understanding of the characteristics, challenges, and potential solutions related to the urban physical environment (Jat et al., 2008; Liang and Weng, 2010). The emergence of multi-source big data and GIS technologies enables sensing the urban physical environment from various perspectives and scales, extracting valuable information and patterns that unveil trends and problematic areas within the urban landscape. In this context, using street-level Imagery (SI) and geospatial artificial intelligence (GeoAI) are valuable complements, providing fine-grained spatial visual information and advanced spatial analysis techniques. These advancements further enhance urban physical sensing methodologies, enabling scholars to uncover more comprehensive information previously hidden within the city (Zhang 2023a, 2023b, 2023c).

Notably, utilizing SI and GeoAI to conduct urban sensing has demonstrated remarkable potential in recent years. In contrast to the aforementioned urban sensing methods, SI effectively supplements urban analysis data sources, particularly from a human perspective (Zhang et al., 2018; Dubey et al., 2016b). Unlike satellite imagery and aerial imagery, which capture information from vertical or oblique viewpoints, SIs provide more detailed visual information at the neighborhood level (shown in Figure 1.1), which is aligned with the human perspective of observing the urban physical environment (Kang et al., 2020; Biljecki and Ito, 2021). Consequently, SIs enable virtual urban audits for governmental purposes and assist urban planners in investigating people's objective perceptions of places, fostering a better understanding of human-land interaction within the city (Kang et al., 2023). The rapid development of GeoAI technology further amplifies SI's potential for information mining (Zhang et al., 2023a, 2023b, 2023c). By leveraging SIs and GeoAI, we possess the capacity to not merely evaluate the condition of urban infrastructure by employing semantic segmentation methodologies (Xiao and

(a) Satellite Imagery (b) Aerial Imagery (c) Street-view Imagery

FIGURE 1.1 Examples of (a) satellite imagery, (b) aerial imagery, and (c) street view imagery.

Quan, 2009; Rundle et al., 2011; Li et al., 2015b; Gong et al., 2018), but also to distill more profound insights concealed within the visual data, including safety, architectural age and style, as well as the urban socio-economic environment (Zhang et al., 2021; Sun et al., 2022; Suel et al., 2019). The combination of SI and GeoAI enhances the existing research paradigm established since the inception of geographic information science and technology, thus broadening the spectrum of urban physical environment sensing.

Within the scope of this article, we conduct an overview of prevailing studies centered around the integration of SI and GeoAI for the purpose of sensing the urban physical environment, concurrently discussing the applications and achievements within the discipline of urban science. Furthermore, we have addressed the shortcomings inherent in current research endeavors and have projected potential avenues for future exploration.

STREET-LEVEL IMAGERY (SI)

SI refers to the photographs taken along street networks, depicting the profile view of the urban streetscape from a similar view of human vision and describing the urban physical environment comprehensively (Zhang et al., 2019, 2023a, 2023b, 2023c). To perceive the urban physical environment using SI, it is important to collect relevant data that aligns with the research area and apply it to the intended tasks. However, it's worth noting that different SI data sources may vary in terms of data quality (Hou and Biljecki, 2022). Additionally, the sampling method employed for SI can also impact the performance of the model on specific tasks (Zhang et al., 2023a, 2023b, 2023c). In this section, we will explore various SI data sources and discuss the quality aspects associated with them.

Street-Level Imagery Sources

Based on the providers, SI can be divided into the following two categories:

Commercial SI is provided by commercial companies or organizations and is typically only accessible through payment (Anguelov et al., 2010). Some well-known providers of commercial SI include Google Street View,

Bing Maps Streetside, Here Street View, Baidu Street View, AutoNavi, and Tencent Street View.

One of the main advantages of commercial SI is its high data quality. Commercial providers invest significant resources in data collection, processing, and maintenance, resulting in consistent image quality during shooting and processing. As a result, commercial SI often has higher standards for image quality, resolution, coverage, and update frequency.

In addition, commercial SI typically has extensive global coverage, including in rural and urban areas, and across multiple countries. This is due to the greater resources available to commercial providers, which allows them to ensure that street-view imagery is available in more areas. As such, commercial SI has become an invaluable resource for applications such as urban planning, transportation, tourism, and GIS.

Crowdsourcing SI, such as those provided by Mapillary and OpenStreetView, is a freely accessible resource contributed and maintained by various entities, including academic institutions, government departments, and community projects (Juhasz and Hochmair, 2016). These sources of SI are often supplemented by data collected from self-driving cars and other methods.

However, the quality of crowdsourced SI may vary depending on the contributor, including aspects such as image resolution, coverage, and update frequency. As such, the data quality of crowdsourced SI may not be as consistent or high-precision as that of commercial SI. Additionally, crowdsourced SI has limited coverage, especially in remote areas (Yan et al., 2020; Goodchild, 2007; Goodchild and Li, 2012). Despite these limitations, crowdsourced SI remains a valuable resource for various applications, particularly in areas where commercial SI is unavailable or insufficient.

Street-Level Imagery Attributes

Here, we examine how various attributes of SI impact its use in different applications and analyses. SI attributes refer to the characteristics and properties associated with SI. These attributes provide information about image quality, coverage, timeliness, and metadata availability. By understanding the impact of these attributes, researchers can assess the reliability, suitability, and limitations of street view data for the specific application scenario.

Image quality is one such attribute that assesses the visual quality of SI, including resolution, sharpness, color accuracy, and absence of distortions or artifacts (Lauko et al., 2020). Poor image quality issues, such as low-light conditions, blurriness, and variable weather, can impact the accuracy of object detection and recognition algorithms. For example, computer vision algorithms may struggle to correctly identify and classify objects in low-light conditions or blurry images, leading to lower accuracy in tasks like object detection, tracking, or scene understanding (Gao et al., 2022). Similarly, the presence of artifacts or color inaccuracies can affect the performance of machine learning models trained on SI, leading to reduced accuracy in categorizing and segmenting different objects and regions in the images (Choi et al., 2022).

Coverage is another important attribute that refers to the spatial distribution of SI. Unbalanced spatial coverage means that some regions or locations have more comprehensive SI coverage compared to others. This can lead to insufficient data for assessing and analyzing specific areas, potentially leading to inaccurate understanding of the existing environment or infrastructure needs. For example, the significant gap in SI coverage between Asia and Africa can result in inaccurate assessments of urban greening and potentially mislead policymakers and researchers (Liang et al., 2023; O'Regan et al., 2022).

Timeliness is an attribute that refers to the temporal coverage of SI. It often exhibits heterogeneity with sporadic updates, resulting in inconsistent coverage across different locations. This can introduce biases and inconsistencies in analyses that involve multiple locations within a city or region. Comparisons or assessments that assume consistent coverage across all areas may be inaccurate or incomplete, potentially leading to flawed conclusions or misleading interpretations (Yu et al., 2022).

Metadata availability is an attribute that refers to the extent of metadata provided with SI. Metadata provides useful information about the properties and characteristics of SI, and its availability can vary across different platforms and data sources. The lack of complete metadata in SI can introduce uncertainties, limit data selection, hinder spatial analysis and visualization, impede time-based analysis, challenge data quality assessment, and restrict data integration and interaction capabilities (Vo et al., 2023). Commercial providers, such as Google Street View, often provide comprehensive metadata, while crowdsourced platforms may have more limited or inconsistent metadata availability.

METHODS FOR SENSING THE URBAN PHYSICAL ENVIRONMENT

Methods of sensing the urban physical environment through SI involve using computer vision, deep learning, data analysis, and other technical means to perceive and understand the city's physical environment. This method has various applications in city-related tasks, such as urban planning, traffic management, environmental monitoring, and smart cities.

To categorize perception methods based on their tasks in the context of urban physical environment perception, we can further partition the underlying techniques based on their learning approaches.

Sensing Tasks

Sensing the urban physical environment with SI can be categorized into the following tasks (Table 1.1).

Scene Element Extraction

Scene element extraction is the most direct and common way to use street-level images in representing built environments. Scene elements can be derived either by object detection DCNN models, such as SSD (Liu et al., 2016) and

TABLE 1.1
Approaches Based on Perception Tasks

Category	Work	Method
Element extraction	Object detection Object segmentation Relation detection	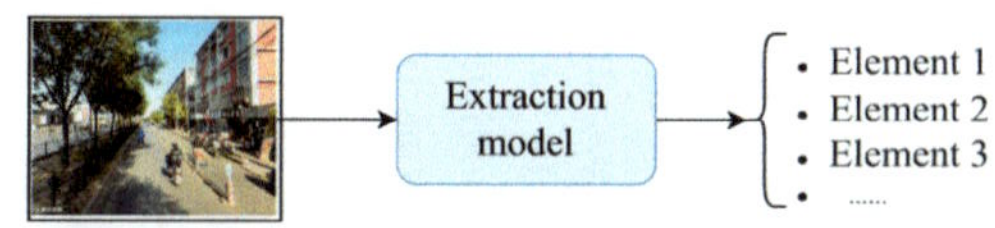
Perception	Scene category Perception model	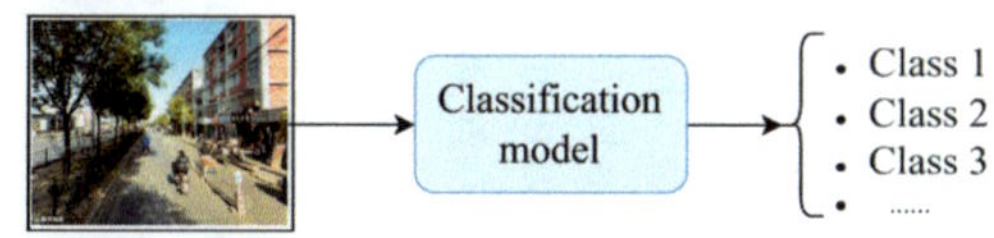
Inference	En2End learning Crime rates House prices Human dynamics	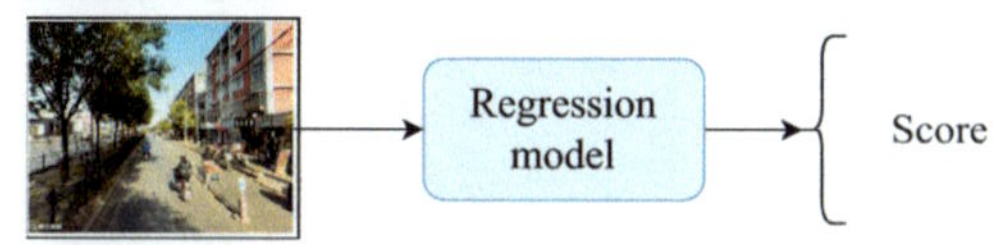
Embedding	Representation learning Image deep feature	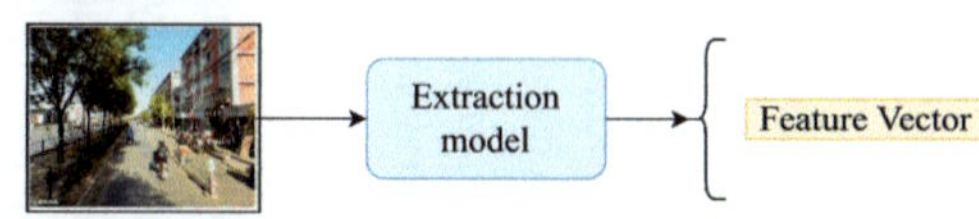
Generation	Scene generation Scene reconstruction	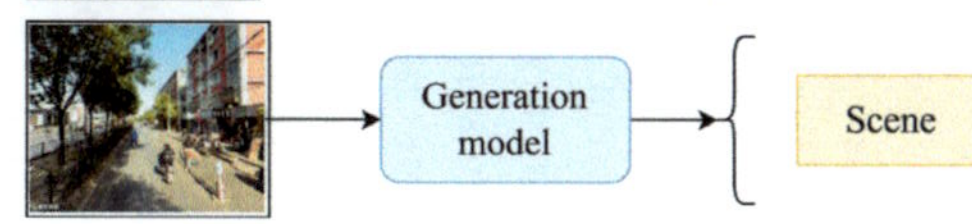

YOLO (Bochkovskiy et al., 2020), or object segmentation models, e.g., PSPNet (Zhao et al., 2017), Mask RCNN (He et al., 2017), and HRNet (Wang et al., 2020). The former models output the detected objects with bounding boxes, so the number of different objects in a scene can be counted. The latter model predicts the object categories of every pixel in an image, which can be further calculated as the object proportions of a scene. Both models provide an approach to measuring a scene with quantitative metrics.

Scene Perception

Scene perception is based on typical image classification models. Scene perception models are mainly based on the Place Pulse 2.0 database (Dubey et al., 2016a), which contains one million human ratings on street-level images along six dimensions: "safe", "lively", "beautiful", "wealthy", "boring", and "depressing". The models trained on Place Pulse have been proven to be effective in estimating these perceptual dimensions of new scenes in most cities.

Scene Inference

Beyond information from visible scenes, scene inference models aim at inferring the scene information that cannot be directly observed from the visible scenes, such as crime (Khosla et al., 2014), real estate values (Law et al., 2019), and long-term human dynamics (Zhang et al., 2019), etc. These kinds of models are trained through an end-to-end learning process, in which the model automatically learns efficient features between the initial input image and the final output labels (crime rate, house price value, etc.), without manually indicating what visual cues matter to the model.

Scene Embedding

Scene embedding aims to learn and represent a scene with a generic feature, which is typically a vector of continuous numbers that capture the semantics of a scene. As a computational representation, this vector makes scenes computable, which can be used either to measure similarity among scenes or as the input of other machine learning tasks (Ye et al., 2020; Zhu et al., 2020a). In practice, Zhou et al. (2017) proposed to use the last activation layer from the trained model by Places dataset as a generic scene embedding, also termed deep features.

Scene Generation

Scene generation is enabled by a special architecture of DCNN called Generative Adversarial Nets (GAN) (Goodfellow et al., 2014), which can generate entirely new predicted images after learning what realistic scenes look like from hundreds of thousands of real street scenes. GAN creates computer-generated urban scenes based on user-generated inputs, such as objective characteristics extracted from images (e.g., buildings, roads, and vehicles) or encoded perceptions (e.g., attractiveness, safety, and liveliness of a place). Urban scenes can be also generated and edited based on some attributes of the built environment (Bau et al., 2020; Zhu et al., 2020b; Richter et al., 2022). These types of models are useful for scenario planning and urban design applications which frequently require that participants envision images of cities that don't exist. Urban designers are also using GAN to create vivid high-resolution imagery. A pioneering approach looking at this was introduced by Noyman and Larson (2020), who designed a physical platform that allows users to generate street scenes by combining a wide range of street elements based on their preferences, including different land uses, types of roads, density of buildings, and presence of sidewalks.

Approaches Based on Techniques

To gain a comprehensive understanding of the methods used in perceiving the urban physical environment using SI, we can categorize the models into the following categories based on their learning approaches (Figure 1.2).

Supervised Learning

Supervised learning methods use labeled training data to train a model. In SI urban physical environment perception, it is possible to use labeled SI (Ye et al., 2020; Zhu et al., 2020a), point cloud data (Micusik and Kosecka, 2009), or other annotated datasets, these methods enable the training of models to perform various

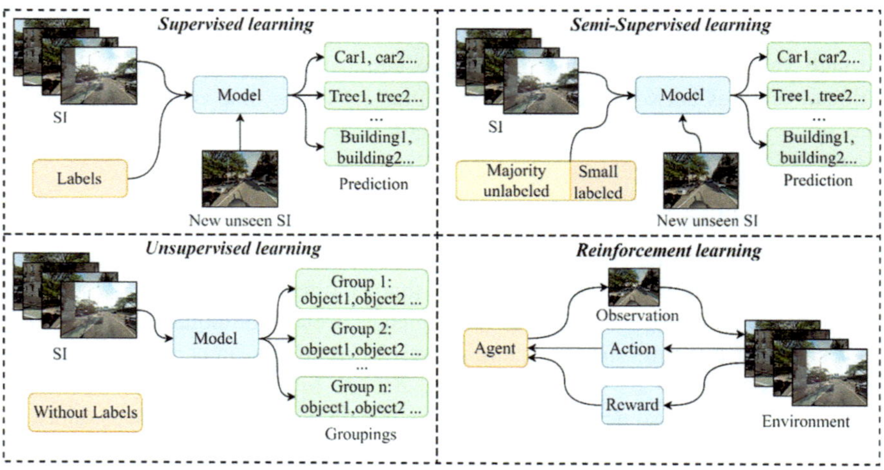

FIGURE 1.2 Approaches based on techniques. The supervised learning model utilizes known street-level imagery along with its corresponding labels during the training phase. The learning algorithm generates an inference function, which is eventually used to predict unseen street-level imagery. Unsupervised learning describes hidden structures from unlabeled data. In the context of street views, it involves learning to partition or classify street view images into different groups or categories without any prior labels or annotations. Semi-supervised models use both labeled and unlabeled data for training. Reinforcement learning is when an agent interacts with an environment to learn and improve its decision-making. The agent takes action in the environment and receives feedback through rewards or penalties. The environment represents the system or context in which the agent operates, providing states and dynamics for interaction. The agent's goal is to learn the best actions to maximize cumulative rewards over time in different environmental states.

tasks, including scene classification, object detection, semantic segmentation, and 3D reconstruction.

Unsupervised Learning

Unsupervised learning, including self-supervised learning, is an important approach to perceiving the urban physical environment using SI. Unlike supervised learning that relies on labeled data, unsupervised learning methods utilize unlabeled data for training models and aim to discover hidden structures and patterns in the data (Wang and Biljecki, 2022). In the context of SI urban physical environment perception, unsupervised learning methods can be applied to various tasks:

Clustering: Unsupervised learning algorithms can group similar SI features together based on their intrinsic properties. Clustering can help identify common patterns or distinct groups within the data, which can be useful for understanding the diversity of urban environments or discovering specific types of scenes or objects (Ke et al., 2022).

Dimensionality reduction: Unsupervised learning techniques such as principal component analysis (PCA) (Yuan et al., 2023) or autoencoders (Kamal and Bae, 2022) can reduce the dimensionality of the street view data while retaining important information. This can help visualize and analyze the data in a lower-dimensional space, facilitate data exploration, and improve the efficiency of subsequent processing tasks.

Anomaly detection: Unsupervised learning methods can identify anomalous or outlier SIs that deviate from the expected patterns. This can be valuable for detecting unusual events, identifying infrastructure issues, or discovering uncommon urban conditions that require further investigation (Zhao et al., 2022).

Representation learning: Unsupervised learning can also be used to learn meaningful representations or embeddings from unlabeled street view data. By training models to capture the underlying structure of the data, these representations can be leveraged for subsequent tasks such as similarity comparison (Mohamed et al., 2022), retrieval (Lin et al., 2022), or contrastive learning (Liu et al., 2023).

Semi-Supervised Learning

Semi-Supervised learning methods play a significant role in SI and physical environment perception by leveraging labeled and unlabeled training data. These methods can effectively utilize limited labeled data in conjunction with a large amount of unlabeled data to enhance model performance and generalization capabilities (Oliveira et al., 2022; Yang et al., 2022). In the context of SI physical environment perception, semi-supervised learning methods can be applied in various ways:

Pseudo-labeling: Initially, a small portion of the SI data is manually labeled with high-quality annotations. These labeled samples are then used to train a model. The trained model can generate pseudo-labels for the remaining unlabeled data based on its predictions. The unlabeled data, along with their pseudo-labels, are then incorporated into the training process to refine the model iteratively (Li and Ma, 2022).

Consistency regularization: Semi-supervised learning methods often involve the use of consistency regularization techniques. These techniques aim to ensure that similar inputs produce similar outputs, even when small perturbations are applied to the input data. By enforcing consistency across different augmentations or views of the same street view image, the model can learn robust representations and improve its ability to generalize to unseen data (Bortoloti et al., 2022).

Reinforcement Learning

Reinforcement learning methods learn the optimal behavioral strategy through the interaction between the agent and the environment. In SI urban physical environment perception, reinforcement learning methods can be applied to intelligent navigation, traffic control, and other scenarios to optimize urban environment perception and decision-making process (Wang et al., 2022b; Grabaskas and Wang, 2022).

APPLICATIONS OF URBAN SENSING

According to the depth of urban physical environment sensing, the application can be divided into three levels hierarchically (Figure 1.3). Observational sensing

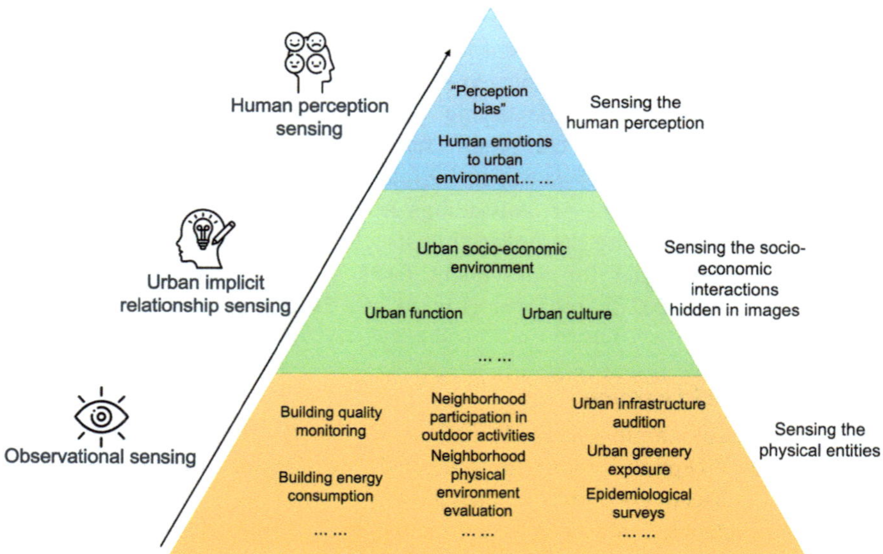

FIGURE 1.3 An overview of the application of urban sensing based on GeoAI and SI. Based on the depth of urban physical environment sensing, the application can be hierarchically stratified into three levels. Each sensing level offers a distinct perspective and an understanding of the urban physical environment. The first level, observational sensing, primarily centers on the physical entities within urban settings. The second level, urban implicit relationship sensing, delves into the intangible conceptions that extend beyond mere physical entities. Lastly, human perception sensing is devoted to capturing and analyzing human responses and perceptions of the environment.

focuses on the physical entities described in the cities; urban implicit relationship perception aims to capture the more abstract information hidden in the images, such as urban socio-economic environments and urban culture; and human perception sensing emphasizes the emotional experience when urban dwellers participate in the places and the difference between the objective measure and subjective human perception.

Observational Sensing: Sensing the Physical Entity

Observational sensing primarily encompasses the detection of tangible entities present within the SIs, such as buildings, plants, traffic lights, and other elements scattered in the city. Given that SIs offer extensive and meticulous data regarding the urban physical environment, encompassing virtually all places within the city, GeoAI models demonstrate the capability to effectively acquire information ranging from individual units to the entirety of the urban landscape through semantic segmentation, classification, and other feature extraction techniques.

Instance-Level Sensing

The combined approach of SI and GeoAI is proving invaluable for municipal engineering by facilitating automated monitoring and streamlining of environmental audits for various aspects of urban infrastructure (Xiao and Quan, 2009; Kang et al., 2018). This virtual auditing capability enables efficient analysis and evaluation of diverse environmental factors, such as urban infrastructure and green spaces, among other critical elements that contribute to the overall built environment.

One of the key advantages of this approach is the ability to supplement traditional maps with fine-grained details often overlooked. Through semantic segmentation of street images, elements such as traffic lights (Campbell et al., 2019), sideroads (Ning et al., 2022), poles (Ao, 2019; Krylov et al., 2018), traffic signs (Balali et al., 2015), and zebra crossings (Ahmetovic et al., 2015) can be accurately identified and incorporated into urban data. This supplementation enhances GIS and drives the development of three-dimensional GIS and digital twins (Li and Yao, 2020; Lehtola et al., 2022; Pang and Biljecki, 2022).

Additionally, SIs and GeoAI can be employed to monitor the status of urban infrastructure to assist with construction and transportation in cities. In this context, SI serves as a readily accessible data source, enabling numerous studies to extract relevant information regarding building energy consumption using deep learning models. These models utilize SIs to infer various factors, including building age (Sun et al., 2021), window-to-wall ratios (Szczesniak et al., 2022), and image-based energy efficiency estimation (Mayer et al., 2023). In addition, SIs aid in road quality monitoring, building damage detection (He et al., 2022), and landslide detection (Voumard et al., 2017). They also enable the observation of pedestrian and vehicle flow on roads, providing valuable insights for traffic planning (Chen et al., 2020b; Yin et al., 2015). The strategic utilization of the extensive data aggregated by SIs considerably enhances the protection and surveillance of urban construction and transportation systems and fosters the facilitation of sustainable urban development (Hak et al., 2016).

Street-Level Sensing

Streets, as a vital component of urban morphology, serve as the primary interface for social interactions between urban residents and the built environment (Li and Ratti, 2019). The layout of streets has a complex impact on the urban climate, encompassing factors such as airflow patterns, solar radiation variations, and precipitation distribution within the city. These factors, in turn, have indirect consequences for human comfort, ambient air quality, and even the broader climate cycle (Masson et al., 2020; Shi et al., 2016; Carrasco-Hernandez et al., 2015).

The utilization of SI in combination with GeoAI methods enables the modeling of key indicators of street metrics, such as sky view factor (SVF) (Xia et al., 2021), solar radiation (Gong et al., 2019), and light pollution (Li et al., 2021). These indicators serve a wide range of purposes, ranging from microclimate studies to guiding urban planning decisions. By utilizing these indicators, cities can optimize their resource allocation and promote more efficient use of resources. For instance, Deng et al. (2021) employ semantic segmentation of SIs to assess the potential for collecting street-level solar radiation, thus offering promising prospects for improving the efficiency of solar radiation utilization. Furthermore, Li et al. (2019) utilize image segmentation techniques to detect buildings that may block sunlight, enabling the prediction of locations prone to sun glare on the streets and facilitating better guidance for road travel. Due to the important role of greenery in urban street spaces, a large number of articles focus on the use of GeoAI methods to generate an indicator to measure urban greenery and its visibility in the city (which is called green view index, GVI) from SIs (Yang et al., 2009; Li et al., 2015b; Cai et al., 2018). The impact of urban greenery extends to considerations of thermal comfort, as green spaces can provide shading and cooling effects, improving the microclimate and mitigating heat-related issues in urban environments (Li et al., 2018b). Moreover, the inclusion of green spaces is recognized as an essential component of sustainable urban development, with studies exploring the relationship between greenery and urban growth, livability, and ecological balance (Chen et al., 2020a).

The combination of SIs and GeoAI can be effectively integrated into spatial syntax analysis frameworks (which is a quantitative method for perceiving the interaction of the urban spatial environment) (Abusaada and Elshater, 2021; Hidalgo and Hernandez, 2001), either by expanding the scope of investigation from to objective spatial configuration to subjective urban sensing, or by modeling key parameters (e.g., enclosure, connectivity, accessibility, visibility) within the context of space syntax (Jiang et al., 2000; van Nes et al., 2021). Other studies explore the re-assessment of enclosure (Yin and Wang, 2016), employ isovist analysis (Takizawa and Kinugawa, 2020; Xiang et al., 2021), and investigate the relationship between human perception of street accessibility (Wang et al., 2022a) to further understand the relationship between visual perception and spatial syntax.

The utilization of SIs and GeoAI provides valuable insights for street metrics, empowering informed decision-making in streets such as urban planning, energy efficiency, and resource management; and integrating SIs and GeoAI into spatial syntactic analysis enhances our understanding of the interaction between subjective perception and objective spatial configuration, leading to a more comprehensive assessment of urban physical environments and the potential for improved urban design, planning, and user experience.

Neighborhood-Level Sensing

When we focus our perspective on the neighborhood level, the integration of SI and GeoAI enables us to observe the harmonious atmosphere in neighborhoods and the interaction between humans and their surroundings. This capability proves highly beneficial in the context of public health. Extensive research has consistently demonstrated a significant relationship between the built environment and the physical and mental well-being of community residents. Communities that boast well-maintained surroundings and abundant green spaces tend to foster greater outdoor activities and promote better mental health among their residents (Barton and Pretty, 2010). In contrast, communities characterized by disorderly environments and inadequate infrastructure are often associated with the spread of epidemics and higher rates of psychological depression among local residents (Wilson and Kelling, 2015; Sampson and Raudenbush, 1999).

Scientists can employ semantic segmentation techniques to identify various elements within neighborhoods and evaluate their influence on public health issues (Rzotkiewicz et al., 2018; Kang et al., 2020; Chen et al., 2023). The analysis of urban greenery is involved across different scales of urban sensing. Research at the neighborhood level focuses on the relationship between green spaces and neighborhood engagement in physical exercise, highlighting the positive influence of greenery in promoting physical activity and healthy lifestyles (Ki and Lee, 2021). The mental health benefits of urban green spaces have also been examined, emphasizing their role in reducing stress, improving mood, and enhancing overall mental well-being (Soga et al., 2021). Furthermore, the presence of greenery has been studied in relation to enhanced safety perceptions and reduced crime rates in neighborhoods (Li et al., 2015a). Additionally, numerous studies have investigated specific and detailed information relevant to public health within neighborhoods, such as the distribution of retail alcohol stores (Less et al., 2015), the presence of no smoking signs around schools and hospitals (Wilson and Thomson, 2015), and the evaluation of community accessibility levels (Seekins et al., 2014). Research in this field also extends to transportation safety, including the analysis of road traffic accidents and pedestrian injuries (Johnson and Gabler, 2015; Mooney et al., 2016).

Another prominent area of research in neighborhood-level investigations revolves around public participation in physical exercises. The integration of SIs and GeoAI provides a human-centered perspective on the built environment, thereby highlighting the importance of walkability or bikeability for individuals (Bartzokas-Tsiompras et al., 2020; Nagata et al., 2020; Chow et al., 2014; Zhou et al., 2019). SIs and GeoAI also play a crucial role in measuring the physical and mental health of young and elderly individuals in urban settings (Wang et al., 2019; Li et al., 2018a). Furthermore, they assist in the detection of environmental factors associated with healthy or unhealthy lifestyles, such as obesity, alcohol abuse, and diabetes, among residents (Clews et al., 2016; Yang et al., 2020; Egli et al., 2019). SIs provide valuable insights into the community places frequently visited by younger and older individuals, shedding light on the development of strategies for building healthier cities and promoting active lifestyles (Kang et al., 2020).

City-Level Sensing

We can also conduct a comprehensive analysis encompassing the entire city due to the extensive accessibility of SIs within the urban area. By integrating RS imagery and SIs, GeoAI methodologies are capable of producing multi-perspective portrayals of an identical urban physical environment, thereby augmenting the performance of urban sensing (Najmi et al., 2022; Zou and Wang, 2022). City-level urban sensing proves invaluable for examining places characterized by "unusual" conditions on a large scale, which is meaningful for improving the overall quality of life for urban dwellers and reducing socio-economic disparities. By identifying urban slums (Najmi et al., 2022; Chen et al., 2022; Fan et al., 2023) and areas with abundant housing (Zou and Wang, 2022), we can detect the physical segregation between the affluent and the underprivileged in the city as early as possible, enabling city managers to proactively address this issue. Moreover, this approach serves the purpose of uncovering overlooked locales within urban settings (Zhang et al., 2020a). The identification of so-called "ghost cities" also contributes to understanding inter-city interactions, which is helpful in guiding broader urban development strategies (Jin et al., 2017).

By leveraging SIs and GeoAI, we are empowered to assess the vulnerability of cities in the face of natural disasters, such as flooding and hurricanes (Xing et al., 2023). The intricate link between the spread of infectious diseases and the community environment is widely recognized (Perdue et al., 2003). Through comparing SIs at different time stamps, it is also possible to evaluate the urban regeneration process (Thackway et al., 2023) and guide the process of post-disaster recovery (Meyer and Hendricks, 2018). The integration of SIs and GeoAI offers a synergistic approach enabling environmental monitoring to identify potential factors contributing to disease transmission. Furthermore, this approach facilitates the inference of potential transmission trajectories based on the spatial distribution of infectious disease outbreaks (Nguyen et al., 2020; Delicado and Rowland, 2021; Haddawy et al., 2019).

Urban Implicit Relationship Sensing: Sensing Beyond the Image

The information provided by SIs extends far beyond the visual representation itself. The application of GeoAI enables us to uncover deeper information pertaining to the intangible characteristics of cities, including their socio-economic environment, urban function, and cultural aspects.

Numerous studies have demonstrated the link between the physical environment and the socio-economic conditions of communities. For instance, neighborhoods characterized by cleanliness and visual unity often indicate higher socio-economic status, while areas with higher levels of graffiti and disarray tend to be associated with poorer socio-economic conditions. By combining SI and GeoAI, we can uncover hidden socio-economic indicators and gain a deeper understanding of the socio-economic environment of a city, in addition to analyzing the physical environment depicted in SIs (Arietta et al., 2014). SIs offer a novel perspective for measuring the socio-economic composition of cities. For example, Gebru et al. (2017) utilized SIs and GeoAI to measure the demographic composition of the United States. Additionally, Arietta et al. (2014) proposed a GeoAI method to extract predictive relationships between the visual appearance of cities captured in SIs and non-physical attributes such as crime rates, house prices, and population density. The topic of social inequality is also a significant focus in urban research. Through the analysis of SIs, researchers have identified various forms of inequality, including disparities in green space exposure (Wang et al., 2021b), walkability (Zhou et al., 2019), health (Suel et al., 2019), environment (Wang et al., 2021a), and racial disparities (Hwang and Sampson, 2014). Furthermore, the integration of SI and GeoAI enables a reconsideration of phenomena such as social segregation and gentrification from a new perspective (Ilic et al., 2019; Thackway et al., 2023; Fan et al., 2023).

The discrimination of urban functions is also one of the primary applications of SI and GeoAI. Many studies have focused on discriminating urban points of interest (POIs) and analyzing land use and cover change (LUCC) (Fan et al., 2022; Fang et al., 2021; Hu et al., 2020; Ye et al., 2021; Cao et al., 2018). Specifically, SIs provide the ability to capture the changes and details in POI and land use analysis with their rich visual information (Fang et al., 2021). Moreover, the combination of Street View and GeoAI enables the exploration of various social phenomena, including commuting patterns, work-life balance, and crime (Zhang et al., 2019; Yao et al., 2021; Khorshidi et al., 2021). It has also been utilized to enhance the accuracy of real estate value estimates by incorporating neighborhood appearance into the analysis, a previously underexplored factor (Law et al., 2019; Kang et al., 2021). By utilizing SI and GeoAI, researchers can gain insights into urban functions, analyze spatial dynamics, and explore various social phenomena related to urban environments.

Even though only a limited number of articles specifically focus on urban culture, the integration of SI and GeoAI facilitates the utilization of sensing technology to extract valuable insights about urban culture. The style of buildings, for instance, can be easily identified from SIs (Sun et al., 2022; Zhang et al., 2020b), and landmarks that represent the unique identity of cities can be detected through the analysis of multiple SIs (Lander et al., 2017). These visual distinctions allow us to perceive the humanistic accumulation of cities. The diverse civilizations that have shaped different cities are reflected in their architectural styles and ultimately captured in the streetscape.

By harnessing SI and GeoAI, researchers can enhance their understanding of the socio-economic environment of cities, uncover patterns of social inequality, and investigate phenomena such as social segregation and gentrification, enabling evidence-based policy-making, urban planning, and interventions to address socio-economic disparities and promote equitable and inclusive urban environments. Moreover, the utilization of SI and GeoAI empowers researchers to explore urban functions, analyze spatial dynamics, and investigate various social phenomena, facilitating informed decision-making in urban planning, land management, and policy development, thereby promoting efficient resource allocation and the creation of sustainable and livable cities. The integration of SI and GeoAI also enables the extraction of valuable insights about urban culture, reflecting the diverse civilizations that have shaped them.

Human Perception Sensing: Sensing with Human-Centric Perspective

Understanding how people perceive streets is crucial for urbanists aiming to create more livable and thriving urban environments. Streets that are perceived as comfortable and safe foster emotional attachment among residents, leading to improved mental and physical well-being (Hidalgo and Hernandez, 2001).

SIs, captured from a human perspective, provide a valuable opportunity for researchers to model urban perceptions and support human-centered urban planning (Kang et al., 2020). One notable project in this realm is the "Place Pulse" project developed by MIT's Senseable City Lab. This initiative invites individuals to rate the safety of street images from around the world, resulting in a collection of over one million street images with associated safety ratings. These data have been used to train GeoAI models to accurately predict perceptions of safety (Dubey et al., 2016b; Naik et al., 2014). The outcomes of this study have inspired numerous urban researchers and facilitated various related investigations (Naik et al., 2017; Zhang et al., 2018; Li et al., 2015a).

Furthermore, studies have focused on uncovering people's emotional perception of community environments and roads, examining the emotions evoked by the built environment (Zhang et al., 2022; Ye et al., 2019; Xiang et al., 2021). Researchers have also explored simulating residents' subjective perceptions of the urban environment using visual information (Verma et al., 2020).

It is worth noting that the objective perception of urban phenomena may not always align with the subjective perception. For instance, an environment that appears safe may not necessarily be safe in reality, and vice versa. By combining SIs with questionnaires, researchers can compare these differences, known as perception biases (Zhang et al., 2021). For example, Kang et al. (2023) conducted a study in Stockholm, examining residents' perceptions of safety on streets using GeoAI and questionnaires concurrently. The research explores the discrepancies between modeling results and residents' subjective perceptions, utilizing the GeoAI approach.

Apart from the diverse applications described earlier, SIs and GeoAI can be applied to augmented reality (AR) and virtual reality (VR) to augment human sensing for the built environment. One such application is the utilization of SIs to generate virtual environments within AR or VR platforms. This approach enhances the realism of the virtual environment, thereby improving users' perception and experience of the virtual built environment (Kostakos et al., 2019; Johnson, 2019). This technology can be valuable for architectural and urban design, allowing designers and planners to visualize and explore different design options and their potential impact on the built environment.

Through the integration of SIs, GeoAI, and questionnaire-based assessments, urbanists gain insights into the subjective perceptions of streets and urban environments. This knowledge helps bridge the gap between objective and subjective perceptions, informing urban planning decisions and interventions to create safer, more enjoyable, and people-centric cities.

FUTURE TRENDS

We currently reside in an era where sensing is ubiquitous, with people's approach to sensing the world expanding rapidly from the utilization of our natural sense organs to various smart sensing devices. SI effectively broadens the fine-grained urban sensing methods at the fundamental level of the urban environment. Simultaneously, we find ourselves amidst an era dominated by the advancements of artificial intelligence (AI), wherein diverse AI models are employed to accomplish various objectives with exceptional performance. This newfound efficacy enables human beings to undertake tasks previously deemed inconceivable. The remarkable achievements attained by large-sized deep learning models, such as ChatGPT and Segment Anything Model (SAM), demonstrate the immense potential of AI in accomplishing multipurpose tasks and analyzing multi-modal data. These models surpass the capabilities of most pre-existing models, thereby reshaping the research paradigm of deep learning. Sensor technology and AI systems, distinguished by SI and GeoAI, continue to offer the profound potential for advancements in every step of urban studies, including the detecting, handling, and interpreting of urban challenges.

First, the innovation of SIs and GeoAI can provide a more profound and comprehensible interpretation of the underlying principles behind urban issues. Future studies should place greater emphasis on examining the augmentation effect of SIs through the integration of other geospatial data sources, such as social media and GPS trajectories. By harnessing a diverse range of multi-modal data, we can expand our urban sensing capabilities and render previously abstract information in SIs visible. From a public management perspective, it is paramount to understand the rationale behind AI-based decision-making rather than solely focusing on the outcomes. Incorporating interpretable AI techniques enables effective analysis of the reasoning behind the approaches taken by AI models in addressing urban problems, as well as understanding the disparities between AI and human policy-making. Consequently, this fosters policymakers' awareness of previously over-looked phenomena.

Furthermore, harnessing the power of SI and GeoAI offers a wide array of diverse and precise methodologies to enrich urban studies. Artificial Intelligence-Generated Content (AIGC) empowers us to generate practical solutions tailored to address specific problems in a variety of scenarios. Moreover, it facilitates the creation of urban designs that are customized to align with the expectations and objectives of urban development set forth by governments or policymakers. By leveraging SI and GeoAI, we can introduce a range of alternative approaches to urban planning and design processes, thereby yielding more targeted and effective solutions for the challenges faced in urban development.

Finally, the promising prospect of artificial general intelligence (AGI), which possesses heightened intelligence and extraordinary information-gathering capabilities, holds the potential to serve as a powerful agent for the autonomous identification and resolution of urban issues by human urban governments. This further augments the overall efficiency and effectiveness of urban management. Additionally, AGI can play a crucial role in unveiling the universal laws underlying complex urban phenomena through data mining and analysis. By leveraging AGI's capabilities, we can uncover hidden patterns and causal relationships that contribute to a deeper understanding of urban dynamics.

Above all, these advancements in SI, and GeoAI hold significant potential to revolutionize our understanding and approach to urban challenges. By leveraging these technologies, we can delve deeper into the intricate dynamics of cities, promote evidence-based decision-making, and foster sustainable and livable urban environments.

REFERENCES

Abusaada, H. and Elshater, A. (2021). Effect of people on placemaking and affective atmospheres in city streets. *Ain Shams Engineering Journal*, 12(3):3389–3403.

Ahmetovic, D., Manduchi, R., Coughlan, J. M., and Mascetti, S. (2015). Zebra crossing spotter: Automatic population of spatial databases for increased safety of blind travelers. In Proceedings of the 17th International ACM SIGACCESS Conference on Computers & Accessibility, pages 251–258.

Alqurashi, A. and Kumar, L. (2013). Investigating the use of remote sensing and GIS techniques to detect land use and land cover change: A review. *Advances in Remote Sensing*, 2: 193–204.

Anguelov, D., Dulong, C., Filip, D., Frueh, C., Lafon, S., Lyon, R., Ogale, A., Vincent, L., and Weaver, J. (2010). Google street view: Capturing the world at street level. *Computer*, 43(6):32–38.

Ao, Y. (2019). Fully Convolutional Networks for Street Furniture Identification in Panorama Images. Master's thesis, University of Twente.

Arietta, S. M., Efros, A. A., Ramamoorthi, R., and Agrawala, M. (2014). City forensics: Using visual elements to predict non-visual city attributes. *IEEE Transactions on Visualization and Computer Graphics*, 20(12):2624–2633.

Balali, V., Ashouri Rad, A., and Golparvar-Fard, M. (2015). Detection, classification, and mapping of US traffic signs using Google street view images for roadway inventory management. *Visualization in Engineering*, 3:1–18.

Barton, J. and Pretty, J. (2010). What is the best dose of nature and green exercise for improving mental health? a multi-study analysis. *Environmental Science & Technology*, 44(10):3947–3955.

Bartzokas-Tsiompras, A., Tampouraki, E. M., and Photis, Y. N. (2020). Is walkability equally distributed among downtowners? evaluating the pedestrian streetscapes of eight European capitals using a micro-scale audit approach. *City Sustainability and Regeneration*, 4, page 135.

Bau, D., Zhu, J.-Y., Strobelt, H., Lapedriza, A., Zhou, B., and Torralba, A. (2020). Understanding the role of individual units in a deep neural network. *Proceedings of the National Academy of Sciences*, 117(48):30071–30078.

Biljecki, F. and Ito, K. (2021). Street view imagery in urban analytics and GIS: A review. *Landscape and Urban Planning*, 215:104217.

Bochkovskiy, A., Wang, C.-Y., and Liao, H.-Y. M. (2020). Yolov4: Optimal speed and accuracy of object detection. arXiv preprint arXiv:2004.10934.

Bortoloti, F. D., Tavares, J., Rauber, T. W., Ciarelli, P. M., and Botelho, R. C. G. (2022). An annotated image database of building facades categorized into land uses for object detection using deep learning: Case study for the city of Vila Velha-ES, Brazil. *Machine Vision and Applications*, 33(5):80.

Cai, B. Y., Li, X., Seiferling, I., and Ratti, C. (2018). Treepedia 2.0: applying deep learning for large-scale quantification of urban tree cover. In 2018 IEEE International Congress on Big Data, pages 49–56. IEEE.

Campbell, A., Both, A., and Sun, Q. C. (2019). Detecting and mapping traffic signs from Google street view images using deep learning and GIS. *Computers, Environment and Urban Systems*, 77:101350.

Cao, R., Zhu, J., Tu, W., Li, Q., Cao, J., Liu, B., Zhang, Q., and Qiu, G. (2018). Integrating aerial and street view images for urban land use classification. *Remote Sensing*, 10(10):1553.

Carrasco-Hernandez, R., Smedley, A. R., and Webb, A. R. (2015). Using urban canyon geometries obtained from Google street view for atmospheric studies: Potential applications in the calculation of street level total shortwave irradiances. *Energy and Buildings*, 86:340–348.

Chen, B., Feng, Q., Niu, B., Yan, F., Gao, B., Yang, J., Gong, J., and Liu, J. (2022). Multimodal fusion of satellite and street-view images for urban village classification based on a dual-branch deep neural network. *International Journal of Applied Earth Observation and Geoinformation*, 109:102794.

Chen, J., Chen, L., Li, Y., Zhang, W., and Long, Y. (2023). Measuring physical disorder in urban street spaces: A large-scale analysis using street view images and deep learning. *Annals of the American Association of Geographers*, 113(2):469–487.

Chen, J., Zhou, C., and Li, F. (2020a). Quantifying the green view indicator for assessing urban greening quality: An analysis based on internet-crawling street view data. *Ecological Indicators*, 113:106192.

Chen, L., Lu, Y., Sheng, Q., Ye, Y., Wang, R., and Liu, Y. (2020b). Estimating pedestrian volume using street view images: A large-scale validation test. *Computers, Environment and Urban Systems*, 81:101481.

Choi, K., Lim, W., Chang, B., Jeong, J., Kim, I., Park, C.-R., and Ko, D. W. (2022). An automatic approach for tree species detection and profile estimation of urban street trees using deep learning and Google street view images. *ISPRS Journal of Photogrammetry and Remote Sensing*, 190:165–180.

Chow, C. K., Corsi, D. J., Lock, K., Madhavan, M., Mackie, P., Li, W., Yi, S., Wang, Y., Swaminathan, S., Lopez-Jaramillo, P., et al. (2014). A novel method to evaluate the community built environment using photographs–environmental profile of a community health (epoch) photo neighbourhood evaluation tool. *PLoS One*, 9(11): e110042.

Clews, C., Brajkovich-Payne, R., Dwight, E., Ahmad Fauzul, A., Burton, M., Carleton, O., Cook, J., Deroles, C., Faulkner, R., Furniss, M., et al. (2016). Alcohol in urban streetscapes: A comparison of the use of Google street view and on-street observation. *BMC Public Health*, 16:1–8.

Davison, K. K. and Lawson, C. T. (2006). Do attributes in the physical environment influence children's physical activity? A review of the literature. *International Journal of Behavioral Nutrition and Physical Activity*, 3(1):1–17.

Delicado, A. and Rowland, J. (2021). Visual representations of science in a pandemic: Covid-19 in images. *Frontiers in Communication*, 6:645725.

DeMers, M. N. (2008). *Fundamentals of Geographic Information Systems*. John Wiley & Sons.

Deng, M., Yang, W., Chen, C., Wu, Z., Liu, Y., and Xiang, C. (2021). Street-level solar radiation mapping and patterns profiling using Baidu street view images. *Sustainable Cities and Society*, 75:103289.

Dubey, A., Naik, N., Parikh, D., Raskar, R., and Hidalgo, C. A. (2016a). Deep learning the city: Quantifying urban perception at a global scale, *In Computer Vision–ECCV 2016: 14th European Conference*, Amsterdam, The Netherlands, October 11–14, 2016, Proceedings, Part I 14 (pp. 196–212).

Dubey, A., Naik, N., Parikh, D., Raskar, R., and Hidalgo, C. A. (2016b). Deep learning the city: Quantifying urban perception at a global scale. In Computer Vision–ECCV 2016: 14th European Conference, Amsterdam, The Netherlands, October 11–14, 2016, Proceedings, Part I 14, pages 196–212. Springer.

Dutta, P., Aoki, P. M., Kumar, N., Mainwaring, A., Myers, C., Willett, W., and Woodruff, A. (2009). Common sense: Participatory urban sensing using a network of handheld air quality monitors. In Proceedings of the 7th ACM Conference on Embedded Networked Sensor Systems, pages 349–350.

Egli, V., Zinn, C., Mackay, L., Donnellan, N., Villanueva, K., Mavoa, S., Exeter, D. J., Vandevijvere, S., and Smith, M. (2019). Viewing obesogenic advertising in children's neighbourhoods using Google street view. *Geographical Research*, 57(1): 84–97.

Fan, M., Huang, J., and Wang, H. (2022). Dumapper: Towards automatic verification of largescale pois with street views at Baidu maps. In Proceedings of the 31st ACM International Conference on Information & Knowledge Management, pages 3063–3071.

Fan, Z., Su, T., Sun, M., Noyman, A., Zhang, F., Pentland, A. S., and Moro, E. (2023). Diversity beyond density: Experienced social mixing of urban streets. *PNAS Nexus*, 2(4): PNAS Nexus, 2023, 1–11.

Fang, F., Yu, Y., Li, S., Zuo, Z., Liu, Y., Wan, B., and Luo, Z. (2021). Synthesizing location semantics from street view images to improve urban land-use classification. *International Journal of Geographical Information Science*, 35(9):1802–1825.

Fleischhacker, S. E., Evenson, K. R., Sharkey, J., Pitts, S. B. J., and Rodriguez, D. A. (2013). Validity of secondary retail food outlet data: A systematic review. *American Journal of Preventive Medicine*, 45(4):462–473.

Frank, L. D., Schmid, T. L., Sallis, J. F., Chapman, J., and Saelens, B. E. (2005). Linking objectively measured physical activity with objectively measured urban form: Findings from smartraq. *American Journal of Preventive Medicine*, 28(2):117–125.

Galea, S., Ahern, J., Rudenstine, S., Wallace, Z., and Vlahov, D. (2005). Urban built environment and depression: A multilevel analysis. *Journal of Epidemiology & Community Health*, 59(10):822–827.

Gao, S., Yang, K., Shi, H., Wang, K., and Bai, J. (2022). Review on panoramic imaging and its applications in scene understanding. *IEEE Transactions on Instrumentation and Measurement*, 71:1–34.

Gebru, T., Krause, J., Wang, Y., Chen, D., Deng, J., Aiden, E. L., and Fei-Fei, L. (2017). Using deep learning and Google street view to estimate the demographic makeup of neighborhoods across the United States. *Proceedings of the National Academy of Sciences*, 114(50):13108–13113.

Gehl, J. (2013). *Cities for People*. Island Press.

Gong, F.-Y., Zeng, Z.-C., Ng, E., and Norford, L. K. (2019). Spatiotemporal patterns of streetlevel solar radiation estimated using Google street view in a high-density urban environment. *Building and Environment*, 148:547–566.

Gong, F.-Y., Zeng, Z.-C., Zhang, F., Li, X., Ng, E., and Norford, L. K. (2018). Mapping sky, tree, and building view factors of street canyons in a high-density urban environment. *Building and Environment*, 134:155–167.

Goodchild, M. F. (2007). Citizens as sensors: The world of volunteered geography. *GeoJournal*, 69(4):211–221.

Goodchild, M. F. and Li, L. (2012). Assuring the quality of volunteered geographic information. *Spatial Statistics*, 1:110–120.

Goodfellow, I., Pouget-Abadie, J., Mirza, M., Xu, B., Warde-Farley, D., Ozair, S., Courville, A., and Bengio, Y. (2014). Generative adversarial nets. In Advances in Neural Information Processing Systems, pages 2672–2680.

Grabaskas, N. and Wang, Z. (2022). Intrinsic rewards for reinforcement learning within complex 2d environments. In Intelligent Systems and Applications: Proceedings of the 2021 Intelligent Systems Conference (IntelliSys) Volume 1, pages 425–437. Springer.

Haddawy, P., Wettayakorn, P., Nonthaleerak, B., Su Yin, M., Wiratsudakul, A., Schoning, J., Laosiritaworn, Y., Balla, K., Euaungkanakul, S., Quengdaeng, P., et al. (2019). Large scale detailed mapping of dengue vector breeding sites using street view images. *PLoS Neglected Tropical Diseases*, 13(7):e0007555.

Hak, T., Janou skovˇ a, S., and Moldan, B. (2016). Sustainable development goals: A need for relevant indicators. *Ecological Indicators*, 60:565–573.

Handy, S. L., Boarnet, M. G., Ewing, R., and Killingsworth, R. E. (2002). How the built environment affects physical activity: Views from urban planning. *American Journal of Preventive Medicine*, 23(2):64–73.

Harley, J. B. (1990). Cartography, ethics and social theory. *Cartographica: The International Journal for Geographic Information and Geovisualization*, 27(2):1–23.

He, K., Gkioxari, G., Dollar, P., and Girshick, R. (2017). Mask R-CNN. In IEEE International Conference on Computer Vision, pages 2980–2988. IEEE.

He, Z., Jiang, S., Zhang, J., and Wu, G. (2022). Automatic damage detection using anchor-free method and unmanned surface vessel. *Automation in Construction*, 133:104017.

Hidalgo, M. C. and Hernandez, B. (2001). Place attachment: Conceptual and empirical questions. *Journal of Environmental Psychology*, 21(3):273–281.

Hoehner, C. M., Ramirez, L. K. B., Elliott, M. B., Handy, S. L., and Brownson, R. C. (2005). Perceived and objective environmental measures and physical activity among urban adults. *American Journal of Preventive Medicine*, 28(2):105–116.

Hou, Y. and Biljecki, F. (2022). A comprehensive framework for evaluating the quality of street view imagery. *International Journal of Applied Earth Observation and Geoinformation*, 115:103094.

Hu, C.-B., Zhang, F., Gong, F.-Y., Ratti, C., and Li, X. (2020). Classification and mapping of urban canyon geometry using Google street view images and deep multitask learning. *Building and Environment*, 167:106424.

Hwang, J. and Sampson, R. J. (2014). Divergent pathways of gentrification: Racial inequality and the social order of renewal in Chicago neighborhoods. *American Sociological Review*, 79(4):726–751.

Ilic, L., Sawada, M., and Zarzelli, A. (2019). Deep mapping gentrification in a large Canadian city using deep learning and Google street view. *PloS One*, 14(3):e0212814.

Jacobs, J. (2016). *The death and life of great American cities*. Vintage.

Jat, M. K., Garg, P. K., and Khare, D. (2008). Monitoring and modelling of urban sprawl using remote sensing and GIS techniques. *International Journal of Applied Earth Observation and Geoinformation*, 10(1):26–43.

Ji, W., Ma, J., Twibell, R. W., and Underhill, K. (2006). Characterizing urban sprawl using multi-stage remote sensing images and landscape metrics. *Computers, Environment and Urban Systems*, 30(6):861–879.

Jiang, B., Claramunt, C., and Klarqvist, B. (2000). Integration of space syntax into gis for modelling urban spaces. *International Journal of Applied Earth Observation and Geoinformation*, 2(3-4):161–171.

Jin, X., Long, Y., Sun, W., Lu, Y., Yang, X., and Tang, J. (2017). Evaluating cities' vitality and identifying ghost cities in China with emerging geographical data. *Cities*, 63:98–109.

Johnson, J. (2019). Jumping into the world of virtual and augmented reality. *Knowledge Quest*, 47(4):22–27.

Johnson, N. S. and Gabler, H. C. (2015). Injury outcome in crashes with guardrail end terminals. *Traffic Injury Prevention*, 16(sup2):S103–S108.

Juhasz, L. and Hochmair, H. H. (2016). User contribution patterns and completeness evaluation of Mapillary, a crowdsourced street level photo service. *Transactions in GIS*, 20(6):925–947.

Kamal, I. M. and Bae, H. (2022). Super-encoder with cooperative autoencoder networks. *Pattern Recognition*, 126:108562.

Kamran, S. and Haas, O. (2007). A multilevel traffic incidents detection approach: Identifying traffic patterns and vehicle behaviours using real-time GPS data. In 2007 IEEE Intelligent Vehicles Symposium, pages 912–917.

Kang, J., Korner, M., Wang, Y., Taubenb¨ ock, H., and Zhu, X. X. (2018). Building instance classification using street view images. *ISPRS Journal of Photogrammetry and Remote Sensing*, 145:44–59.

Kang, Y., Abraham, J., Ceccato, V., Duarte, F., Gao, S., Ljungqvist, L., Zhang, F., Nasman, P., and Ratti, C. (2023). Assessing differences in safety perceptions using GeoAI and survey across neighbourhoods in Stockholm, Sweden. *Landscape and Urban Planning*, 236:104768.

Kang, Y., Zhang, F., Gao, S., Lin, H., and Liu, Y. (2020). A review of urban physical environment sensing using street view imagery in public health studies. *Annals of GIS*, 26(3):261–275.

Kang, Y., Zhang, F., Peng, W., Gao, S., Rao, J., Duarte, F., and Ratti, C. (2021). Understanding house price appreciation using multi-source big geo-data and machine learning. *Land Use Policy*, 111:104919.

Ke, T.-W., Hwang, J.-J., Guo, Y., Wang, X., and Yu, S. X. (2022). Unsupervised hierarchical semantic segmentation with multiview cosegmentation and clustering transformers. In Proceedings of the IEEE/CVF Conference on Computer Vision and Pattern Recognition, pages 2571–2581.

Khorshidi, S., Carter, J., Mohler, G., and Tita, G. (2021). Explaining crime diversity with Google street view. *Journal of Quantitative Criminology*, 37:361–391.

Khosla, A., An An, B., Lim, J. J., and Torralba, A. (2014). Looking beyond the visible scene. In Proceedings of the IEEE Conference on Computer Vision and Pattern Recognition, pages 3710–3717.

Ki, D. and Lee, S. (2021). Analyzing the effects of green view index of neighborhood streets on walking time using Google street view and deep learning. *Landscape and Urban Planning*, 205:103920.

Kostakos, P., Alavesa, P., Oppenlaender, J., and Hosio, S. (2019). Vr ethnography: a pilot study on the use of virtual reality 'go-along' interviews in Google street view. In Proceedings of the 18th International Conference on Mobile and Ubiquitous Multimedia, pages 1–5.

Krylov, V. A., Kenny, E., and Dahyot, R. (2018). Automatic discovery and geotagging of objects from street view imagery. *Remote Sensing*, 10(5):661.

Lander, C., Wiehr, F., Herbig, N., Kruger, A., and L¨ ochtefeld, M. (2017). Inferring landmarks¨ for pedestrian navigation from mobile eye-tracking data and Google street view. In Proceedings of the 2017 CHI Conference Extended Abstracts on Human Factors in Computing Systems, pages 2721–2729.

Lauko, I. G., Honts, A., Beihoff, J., and Rupprecht, S. (2020). Local color and morphological image feature based vegetation identification and its application to human environment street view vegetation mapping, or how green is our county? *Geo-spatial Information Science*, 23(3):222–236.

Law, S., Paige, B., and Russell, C. (2019). Take a look around: using street view and satellite images to estimate house prices. *ACM Transactions on Intelligent Systems and Technology*, 10(5):1–19.

Lehtola, V. V., Koeva, M., Elberink, S. O., Raposo, P., Virtanen, J.-P., Vahdatikhaki, F., and Borsci, S. (2022). Digital twin of a city: Review of technology serving city needs. *International Journal of Applied Earth Observation and Geoinformation*, page 102915.

Less, E. L., McKee, P., Toomey, T., Nelson, T., Erickson, D., Xiong, S., and Jones-Webb, R. (2015). Matching study areas using Google street view: A new application for an emerging technology. *Evaluation and Program Planning*, 53:72–79.

Li, D., Deal, B., Zhou, X., Slavenas, M., and Sullivan, W. C. (2018a). Moving beyond the neighborhood: Daily exposure to nature and adolescents' mood. *Landscape and Urban Planning*, 173:33–43.

Li, M. and Yao, W. (2020). 3D map system for tree monitoring in Hong Kong using Google street view imagery and deep learning. *ISPRS Annals of the Photogrammetry, Remote Sensing and Spatial Information Sciences*, 3:765–772.

Li, X., Cai, B. Y., Qiu, W., Zhao, J., and Ratti, C. (2019). A novel method for predicting and mapping the occurrence of sun glare using Google street view. *Transportation Research Part C: Emerging Technologies*, 106:132–144.

Li, X., Duarte, F., and Ratti, C. (2021). Analyzing the obstruction effects of obstacles on light pollution caused by street lighting system in Cambridge, Massachusetts. *Environment and Planning B: Urban Analytics and City Science*, 48(2):216–230.

Li, X. and Ratti, C. (2019). Using Google street view for street-level urban form analysis, a case study in Cambridge, Massachusetts. *The Mathematics of Urban Morphology*, 5:457–470.

Li, X., Ratti, C., and Seiferling, I. (2018b). Quantifying the shade provision of street trees in urban landscape: A case study in Boston, USA, using Google street view. *Landscape and Urban Planning*, 169:81–91.

Li, X., Zhang, C., and Li, W. (2015a). Does the visibility of greenery increase perceived safety in urban areas? evidence from the place pulse 1.0 dataset. *ISPRS International Journal of Geo-Information*, 4(3):1166–1183.

Li, X., Zhang, C., Li, W., Ricard, R., Meng, Q., and Zhang, W. (2015b). Assessing street-level urban greenery using Google street view and a modified green view index. *Urban Forestry & Urban Greening*, 14(3):675–685.

Li, Z. and Ma, J. (2022). Discussing street tree planning based on pedestrian volume using machine learning and computer vision. *Building and Environment*, 219: 109178.

Liang, B. and Weng, Q. (2010). Assessing urban environmental quality change of Indianapolis, United States, by the remote sensing and GIS integration. *IEEE Journal of Selected Topics in Applied Earth Observations and Remote Sensing*, 4(1):43–55.

Liang, C., Serge, A., Zhang, X., Wang, H., and Wang, W. (2023). Assessment of street forest characteristics in four African cities using Google street view measurement: Potentials and implications. *Environmental Research*, 221:115261.

Lin, J., Zheng, Z., Zhong, Z., Luo, Z., Li, S., Yang, Y., and Sebe, N. (2022). Joint representation learning and keypoint detection for cross-view geo-localization. *IEEE Transactions on Image Processing*, 31:3780–3792.

Liu, W., Anguelov, D., Erhan, D., Szegedy, C., Reed, S., Fu, C.-Y., and Berg, A. C. (2016). SSD: Single shot multibox detector. In European Conference on Computer Vision, pages 21–37.

Liu, Y., Zhang, X., Ding, J., Xi, Y., and Li, Y. (2023). Knowledge-infused contrastive learning for urban imagery-based socioeconomic prediction. In Proceedings of the ACM Web Conference 2023, pages 4150–4160.

Lynch, K. (1964). *The Image of the City*. MIT Press.

Masson, V., Lemonsu, A., Hidalgo, J., and Voogt, J. (2020). Urban climates and climate change. *Annual Review of Environment and Resources*, 45:411–444.

Mayer, K., Haas, L., Huang, T., Bernabe-Moreno, J., Rajagopal, R., and Fischer, M. (2023). Estimating building energy efficiency from street view imagery, aerial imagery, and land surface temperature data. *Applied Energy*, 333:120542.

Meyer, M. A. and Hendricks, M. D. (2018). Using photography to assess housing damage and rebuilding progress for disaster recovery planning. *Journal of the American Planning Association*, 84(2):127–144.

Micusik, B. and Kosecka, J. (2009). Piecewise planar city 3D modeling from street view panoramic sequences. In Proceedings of the IEEE Conference on Computer Vision and Pattern Recognition.

Miller, R. B. and Small, C. (2003). Cities from space: potential applications of remote sensing in urban environmental research and policy. *Environmental Science & Policy*, 6(2):129–137.

Mohamed, A., Lee, H.-y., Borgholt, L., Havtorn, J. D., Edin, J., Igel, C., Kirchhoff, K., Li, S.-W., Livescu, K., Maaløe, L., et al. (2022). Self-supervised speech representation learning: A review. *IEEE Journal of Selected Topics in Signal Processing*, 16, 1179–1210.

Mooney, S. J., DiMaggio, C. J., Lovasi, G. S., Neckerman, K. M., Bader, M. D., Teitler, J. O., Sheehan, D. M., Jack, D. W., and Rundle, A. G. (2016). Use of Google street view to assess environmental contributions to pedestrian injury. *American Journal of Public Health*, 106(3):462–469.

Nagata, S., Nakaya, T., Hanibuchi, T., Amagasa, S., Kikuchi, H., and Inoue, S. (2020). Objective scoring of streetscape walkability related to leisure walking: Statistical modeling approach with semantic segmentation of Google street view images. *Health & Place*, 66:102428.

Naik, N., Kominers, S. D., Raskar, R., Glaeser, E. L., and Hidalgo, C. A. (2017). Computer vision uncovers predictors of physical urban change. *Proceedings of the National Academy of Sciences*, 114(29):7571–7576.

Naik, N., Philipoom, J., Raskar, R., and Hidalgo, C. (2014). Streetscore-predicting the perceived safety of one million streetscapes In Proceedings of the IEEE Conference on Computer Vision and Pattern Recognition Workshops. 779–785.

Najmi, A., Gevaert, C. M., Kohli, D., Kuffer, M., and Pratomo, J. (2022). Integrating remote sensing and street view imagery for mapping slums. *ISPRS International Journal of GeoInformation*, 11(12):631.

Nguyen, Q. C., Huang, Y., Kumar, A., Duan, H., Keralis, J. M., Dwivedi, P., Meng, H.-W., Brunisholz, K. D., Jay, J., Javanmardi, M., et al. (2020). Using 164 million Google street view images to derive built environment predictors of covid-19 cases. *International Journal of Environmental Research and Public Health*, 17(17):6359.

Ning, H., Ye, X., Chen, Z., Liu, T., and Cao, T. (2022). Sidewalk extraction using aerial and street view images. *Environment and Planning B: Urban Analytics and City Science*, 49(1):7–22.

Noyman, A. and Larson, K. (2020). A deep image of the city: Generative urban-design visualization. *Challenge*, 7:30.

Oliveira, A. A. A., Wang, Z., and Hirata, R. (2022). Locating urban trees near electric wires using Google street view photos: A new dataset and a semi-supervised learning approach in the wild. In Proceedings of the IEEE/CVF Conference on Computer Vision and Pattern Recognition, pages 4286–4294.

O'Regan, A. C., Byrne, R., Hellebust, S., and Nyhan, M. M. (2022). Associations between Google street view-derived urban greenspace metrics and air pollution measured using a distributed sensor network. *Sustainable Cities and Society*, 87:104221.

Pang, H. E. and Biljecki, F. (2022). 3D building reconstruction from single street view images using deep learning. *International Journal of Applied Earth Observation and Geoinformation*, 112:102859.

Perdue, W. C., Stone, L. A., and Gostin, L. O. (2003). The built environment and its relationship to the public's health: The legal framework. *American Journal of Public Health*, 93(9):1390–1394.

Richter, S. R., AlHaija, H. A., and Koltun, V. (2022). *Enhancing photorealism enhancement*, 45(2):1700–1715.

Rundle, A. G., Bader, M. D., Richards, C. A., Neckerman, K. M., and Teitler, J. O. (2011). Using Google street view to audit neighborhood environments. *American Journal of Preventive Medicine*, 40(1):94–100.

Rzotkiewicz, A., Pearson, A. L., Dougherty, B. V., Shortridge, A., and Wilson, N. (2018). Systematic review of the use of Google street view in health research: Major themes, strengths, weaknesses and possibilities for future research. *Health & place*, 52:240–246.

Sampson, R. J. and Raudenbush, S. W. (1999). Systematic social observation of public spaces: A new look at disorder in urban neighborhoods. *American Journal of Sociology*, 105(3):603–651.

Seekins, T., Rennie, B., and Hammond, J. (2014). Exploring the feasibility of using "google street view" to assess the accessibility of community environments: developing definitions and observational protocol for image recognition and classification. In Environmental Contexts and Disability, volume 8, pages 123–139.

Shi, L., Chu, E., Anguelovski, I., Aylett, A., Debats, J., Goh, K., Schenk, T., Seto, K. C., Dodman, D., Roberts, D., et al. (2016). Roadmap towards justice in urban climate adaptation research. *Nature Climate Change*, 6(2):131–137.

Soga, M., Evans, M. J., Tsuchiya, K., and Fukano, Y. (2021). A room with a green view: The importance of nearby nature for mental health during the covid-19 pandemic. *Ecological Applications*, 31(2):e2248.

Suel, E., Polak, J. W., Bennett, J. E., and Ezzati, M. (2019). Measuring social, environmental and health inequalities using deep learning and street imagery. *Scientific Reports*, 9(1):1–10.

Sun, M., Zhang, F., and Duarte, F. (2021). Automatic building age prediction from street view images. In 2021 7th IEEE International Conference on Network Intelligence and Digital Content (IC-NIDC), pages 102–106. IEEE.

Sun, M., Zhang, F., Duarte, F., and Ratti, C. (2022). Understanding architecture age and style through deep learning. *Cities*, 128:103787.

Szczesniak, J. T., Ang, Y. Q., Letellier-Duchesne, S., and Reinhart, C. F. (2022). A method for using street view imagery to auto-extract window-to-wall ratios and its relevance for urbanlevel daylighting and energy simulations. *Building and Environment*, 207: 108108.

Takizawa, A. and Kinugawa, H. (2020). Deep learning model to reconstruct 3d cityscapes by generating depth maps from omnidirectional images and its application to visual preference prediction. *Design Science*, 6:e28.

Thackway, W., Ng, M., Lee, C.-L., and Pettit, C. (2023). Implementing a deep-learning model using Google street view to combine social and physical indicators of gentrification. *Computers, Environment and Urban Systems*, 102:101970.

Van der Spek, S., Van Schaick, J., De Bois, P., and De Haan, R. (2009). Sensing human activity: GPS tracking. *Sensors*, 9(4):3033–3055.

van Nes, A., Yamu, C., van Nes, A., and Yamu, C. (2021). Space syntax applied in urban practice. *Introduction to Space Syntax in Urban Studies*, pages 213–237.

Verma, D., Jana, A., and Ramamritham, K. (2020). Predicting human perception of the urban environment in a spatiotemporal urban setting using locally acquired street view images and audio clips. *Building and Environment*, 186:107340.

Vo, A. V., Bertolotto, M., Ofterdinger, U., and Laefer, D. F. (2023). In search of basement indicators from street view imagery data: An investigation of data sources and analysis strategies. *KI-Künstliche Intelligenz*, pages 1–13.

Voumard, J., Abellan, A., Nicolet, P., Penna, I., Chanut, M.-A., Derron, M.-H., and Jaboyedo ff, M. (2017). Using street view imagery for 3-d survey of rock slope failures. *Natural Hazards and Earth System Sciences*, 17(12):2093–2107.

Wang, B., Xu, T., Gao, H., Ta, N., Chai, Y., and Wu, J. (2021a). Can daily mobility alleviate green inequality from living and working environments? *Landscape and Urban Planning*, 214:104179.

Wang, J. and Biljecki, F. (2022). Unsupervised machine learning in urban studies: A Systematic Review of Applications. *Cities*, 129:103925.

Wang, J., Sun, K., Cheng, T., Jiang, B., Deng, C., Zhao, Y., Liu, D., Mu, Y., Tan, M., Wang, X., et al. (2020). Deep high-resolution representation learning for visual recognition. *IEEE Transactions on Pattern Analysis and Machine Intelligence*, 43:3349–3364.

Wang, L., Han, X., He, J., and Jung, T. (2022a). Measuring residents' perceptions of city streets to inform better street planning through deep learning and space syntax. *ISPRS Journal of Photogrammetry and Remote Sensing*, 190:215–230.

Wang, R., Feng, Z., Pearce, J., Yao, Y., Li, X., and Liu, Y. (2021b). The distribution of greenspace quantity and quality and their association with neighbourhood socio-economic conditions in Guangzhou, China: A new approach using deep learning method and street view images. *Sustainable Cities and Society*, 66:102664.

Wang, R., Lu, Y., Zhang, J., Liu, P., Yao, Y., and Liu, Y. (2019). The relationship between visual enclosure for neighbourhood street walkability and elders' mental health in China: Using street view images. *Journal of Transport & Health*, 13:90–102.

Wang, Y., Zeng, Z., Li, Q., and Deng, Y. (2022b). A complete reinforcement-learning-based framework for urban-safety perception. *ISPRS International Journal of Geo-Information*, 11(9):465.

Whyte, W. H. et al. (1980). *The Social Life of Small Urban Spaces*. Conservation Foundation Washington, DC.

Wilson, J. Q. and Kelling, G. L. (2015). Broken windows. Routledge.

Wilson, N. and Thomson, G. (2015). Suboptimal smokefree signage at some hospitals: field observations and the use of google street view. *The New Zealand Medical Journal*, 128(1415):56.

Wilson, W. H. (1989). *The City Beautiful Movement*. Johns Hopkins University Press Baltimore.

Xia, Y., Yabuki, N., and Fukuda, T. (2021). Sky view factor estimation from street view images based on semantic segmentation. *Urban Climate*, 40:100999.

Xiang, L., Cai, M., Ren, C., and Ng, E. (2021). Modeling pedestrian emotion in high-density cities using visual exposure and machine learning: Tracking real-time physiology and psychology in Hong Kong. *Building and Environment*, 205:108273.

Xiao, J. and Quan, L. (2009). Multiple view semantic segmentation for street view images. In 2009 IEEE 12th International Conference on Computer Vision, pages 686–693. IEEE.

Xie, Y., Sha, Z., and Yu, M. (2008). Remote sensing imagery in vegetation mapping: a review. *Journal of Plant Ecology*, 1(1):9–23.

Xing, Z., Yang, S., Zan, X., Dong, X., Yao, Y., Liu, Z., and Zhang, X. (2023). Flood vulnerability assessment of urban buildings based on integrating high-resolution remote sensing and street view images. *Sustainable Cities and Society*, 92:104467.

Yan, Y., Feng, C.-C., Huang, W., Fan, H., Wang, Y.-C., and Zipf, A. (2020). Volunteered geographic information research in the first decade: A narrative review of selected journal articles in GIScience. *International Journal of Geographical Information Science*, 34(9):1765–1791.

Yang, J., Zhao, L., Mcbride, J., and Gong, P. (2009). Can you see green? assessing the visibility of urban forests in cities. *Landscape and Urban Planning*, 91(2):97–104.

Yang, X., Song, Z., King, I., and Xu, Z. (2022). A survey on deep semi-supervised learning. *IEEE Transactions on Knowledge and Data Engineering*, 35:8934–8954.

Yang, Y., Lu, Y., Yang, L., Gou, Z., and Zhang, X. (2020). Urban greenery, active school transport, and body weight among Hong Kong children. *Travel Behaviour and Society*, 20:104–113.

Yao, Y., Zhang, J., Qian, C., Wang, Y., Ren, S., Yuan, Z., and Guan, Q. (2021). Delineating urban job-housing patterns at a parcel scale with street view imagery. *International Journal of Geographical Information Science*, 35(10):1927–1950.

Ye, C., Zhang, F., Mu, L., Gao, Y., and Liu, Y. (2020). Urban function recognition by integrating social media and street-level imagery. *Environment and Planning B: Urban Analytics and City Science*.

Ye, C., Zhang, F., Mu, L., Gao, Y., and Liu, Y. (2021). Urban function recognition by integrating social media and street-level imagery. *Environment and Planning B: Urban Analytics and City Science*, 48(6):1430–1444.

Ye, Y., Zeng, W., Shen, Q., Zhang, X., and Lu, Y. (2019). The visual quality of streets: A human-centred continuous measurement based on machine learning algorithms and street view images. *Environment and Planning B: Urban Analytics and City Science*, 46(8):1439– 1457.

Yin, L., Cheng, Q., Wang, Z., and Shao, Z. (2015). 'Big data' for pedestrian volume: Exploring the use of Google street view images for pedestrian counts. *Applied Geography*, 63:337–345.

Yin, L. and Wang, Z. (2016). Measuring visual enclosure for street walkability: Using machine learning algorithms and Google street view imagery. *Applied Geography*, 76:147–153.

Yu, X., Her, Y., Huo, W., Chen, G., and Qi, W. (2022). Spatio-temporal monitoring of urban street-side vegetation greenery using Baidu street view images. *Urban Forestry & Urban Greening*, 73:127617.

Yuan, Y., Wang, R., Niu, T., and Liu, Y. (2023). Using street view images and a geographical detector to understand how street-level built environment is associated with urban poverty: A case study in Guangzhou. *Applied Geography*, 156:102980.

Zhang, A., Song, L., and Zhang, F. (2022). Perception of pleasure in the urban running environment with street view images and running routes. *Journal of Geographical Sciences*, 32(12):2624–2640.

Zhang, F., Fan, Z., Kang, Y., Hu, Y., and Ratti, C. (2021). "Perception bias": Deciphering a mismatch between urban crime and perception of safety. *Landscape and Urban Planning*, 207:104003.

Zhang, F., Mirandaa, A. S., Duarte, F., Vale, L., Hack, G., Liu, Y., Batty, M., and Ratti, C. (2023a). Urban visual intelligence: Studying cities with AI and street-level imagery. *Annals of the American Association of Geographers*, 114(2):1–18.

Zhang, F., Wegner, J. D., Yang, B., and Liu, Y. (2023b). Street-level imagery analytics and applications. *ISPRS Journal of Photogrammetry and Remote Sensing*, 199:195–196.

Zhang, F., Wu, L., Zhu, D., and Liu, Y. (2019). Social sensing from street-level imagery: A case study in learning spatio-temporal urban mobility patterns. *ISPRS Journal of Photogrammetry and Remote Sensing*, 153:48–58.

Zhang, F., Zhou, B., Liu, L., Liu, Y., Fung, H. H., Lin, H., and Ratti, C. (2018). Measuring human perceptions of a large-scale urban region using machine learning. *Landscape and Urban Planning*, 180:148–160.

Zhang, F., Zu, J., Hu, M., Zhu, D., Kang, Y., Gao, S., Zhang, Y., and Huang, Z. (2020a). Uncovering inconspicuous places using social media check-ins and street view images. *Computers, Environment and Urban Systems*, 81:101478.

Zhang, G., Yi, J., Yuan, J., Li, Y., and Jin, D. (2023c). Das: Efficient street view image sampling for urban prediction. *ACM Transactions on Intelligent Systems and Technology*, 14:1–20.

Zhang, L., Pei, T., Wang, X., Wu, M., Song, C., Guo, S., and Chen, Y. (2020b). Quantifying the urban visual perception of Chinese traditional-style building with street view images. *Applied Sciences*, 10(17):5963.

Zhao, F., Zhang, C., Dong, N., You, Z., and Wu, Z. (2022). A uniform framework for anomaly detection in deep neural networks. *Neural Processing Letters*, 54(4): 3467–3488.

Zhao, H., Shi, J., Qi, X., Wang, X., and Jia, J. (2017). Pyramid scene parsing network. In Proceedings of IEEE Conference on Computer Vision and Pattern Recognition, pages 2881–2890.

Zheng, Y., Zhang, L., Xie, X., and Ma, W.-Y. (2009). Mining interesting locations and travel sequences from GPS trajectories. In Proceedings of the 18th International Conference on World Wide Web, pages 791–800.

Zhou, B., Lapedriza, A., Khosla, A., Oliva, A., and Torralba, A. (2017). Places: A 10 million image database for scene recognition. *IEEE Transactions on Pattern Analysis and Machine Intelligence*, 40:1452–1464.

Zhou, H., He, S., Cai, Y., Wang, M., and Su, S. (2019). Social inequalities in neighborhood visual walkability: Using street view imagery and deep learning technologies to facilitate healthy city planning. *Sustainable Cities and Society*, 50:101605.

Zhu, D., Zhang, F., Wang, S., Wang, Y., Cheng, X., Huang, Z., and Liu, Y. (2020a). Understanding place characteristics in geographic contexts through graph convolutional neural networks. *Annals of the American Association of Geographers*, 110(2):408–420.

Zhu, J., Shen, Y., Zhao, D., and Zhou, B. (2020b). In-domain gan inversion for real image editing. In European Conference on Computer Vision, pages 592–608.

Zou, S. and Wang, L. (2022). Mapping individual abandoned houses across cities by integrating VHR remote sensing and street view imagery. *International Journal of Applied Earth Observation and Geoinformation*, 113:103018.

2 Geospatial Big Data for Urban Sustainability Science

Zidong Yu and Xintao Liu

URBAN SUSTAINABILITY SCIENCE

Humans are experiencing unprecedentedly rapid growth regarding the urbanization process, which can be ascribed to a large volume of the population migrating to and permanently concentrating in cities or urban areas. Besides the potential and opportunities of the urbanization process, dramatic changes in cities have also exacerbated sustainable issues, such as traffic congestion, exploding population, socioeconomic inequality, and a series of housing problems. With the rising concerns on urban sustainability, the idea of planning a sustainable city has been a main topic in not only the academic discipline of urban studies but also recent discussions from policymakers and industry communities (Chiesura, 2004; Jenks & Jones, 2009). The pressing demand from emerging and expanding cities has promoted a novel bifurcation of sustainability science that is to apply sustainability science to urban areas by focusing on tackling challenges derived from complex urban environments. Urban sustainability can be considered as a multidisciplinary field to understand and tackle increasing challenges in cities and urban areas, in achieving sustainable development that maintains a symbiosis among social, economic, and environmental aspects in cities. Current literature has contributed to this new sub-field of sustainability science by introducing interdisciplinary methodologies and research perspectives, with the aim of better knowing how human behaviors and urban environments jointly shape our cities (Klopp & Petretta, 2017; Huang et al., 2015).

Geography has long been recognized as a closely associated discipline contributing to urban sustainability science, of which its derivative insights allow scholars to unveil the sophisticated spatial patterns and relationships manifested by urban sustainability issues (Baerwald 2010; Bettencourt & Kaur, 2011; Mooney et al., 2013). Specifically speaking, topics from urban sustainability are inherently interdisciplinary and related to social-ecological interactions, implying that studying various human behaviors and socioeconomic phenomena can be vital. In this particular sense, utilizing geographic expertise is able to reveal spatial knowledge by investigating the distribution patterns of diverse socioeconomic activities and their human–environment relationships (Turner & Kaplan, 2019; Yu et al., 2021). Compared to universal solutions to sustainability issues, the local knowledge

DOI: 10.1201/9781003244561-3

provided via geographic insights can generate more beneficial, flexible, and robust information in real-world practices given different urban contexts. While qualitative research provides valuable insights, its intrinsic restrictions such as limited generalizability and potential participant bias can be challenged in studying some specific topics, yielding a need for complementary quantitative approaches. Geospatial quantitative methodologies can deliver more rigorous and effective paradigms to understand sustainability knowledge in cities, by reporting large-scale and detailed analytical results from collective human activities.

GEOSPATIAL BIG DATA

Big data has been proposed to refer to tackling and analytics of massive datasets, producing significant impacts on multiple research communities, in particular, social science largely related to human socioeconomic activities. Given the concept of big data, current literature has contended that big data exhibits inherently qualitative differences compared to conventional small datasets. In this connection, big data can be characterized with 5Vs such as volume, velocity, variety, veracity, and value, for example, volume represents the large size of the dataset and velocity denotes that data can be created or accumulated in almost real time (Batty, 2017; Kitchin & McArdle, 2016). On account of various geospatial technologies including global positioning system (GPS) and geographic information systems (GIS), geographic information has been the main source of extant big data (Jiang & Thill, 2015). In this regard, a massive volume of Volunteered Geographic Information (VGI) has been collected via crowdsourcing modes involving a large group of dispersed participants, denoting a novel phenomenon in current urban systems. VGI can be seen as a representative type of user-generated content with detailed geographic information such as coordinates and location tags, supplementing the conventional spatial data mainly collected by officials or organizations. This type of geospatial big data is largely distinct from the traditional datasets previously used in geographic research, of which traditional datasets, for instance, census and survey data, are mostly manually collected and represented at an aggregated scale.

The popularity of information and communications technology (ICT) and the Internet of Things (IoT) has guided our society toward a digital era. In this context, geospatial technologies have provided effective solutions that enable the researcher to obtain huge volumes of location-based information related to human behaviors, with the aim of understanding city and urban systems. Different kinds of geospatial big data, generated from the proliferation of personal devices and ubiquitous sensing networks, have gained growing interest from geospatial literature, enlarging the evidence-based understanding of real-world human activities (Heipke, 2010; Li et al., 2016). Notably, a wealth of information on subjects of the socioeconomics and built environment within an urban area can be indicated by user-generated content specified with spatial coordinates, for instance, GPS trajectory, transportation transaction records, social media, and points of interest (POIs). These datasets can be automatically created by daily human activities (e.g., travel, calling, or using social media) and featured with fine spatial and temporal resolutions.

By combining advent geospatial analytics and visualization, scholars can better recognize the spatiotemporal patterns of human activities and their relationships with physical environmental factors. For example, previous studies have demonstrated the potential of using phone location data as well as vehicle GPS records to quantitatively specify the relationship between human mobility and urban emissions hotspots (Gately et al., 2017; Yu et al., 2020). These findings are worthwhile for helping local stakeholders from different fields such as planning, transportation, and environmental management, to learn the impacts of city growth on communities and the urban environment, therefore allocating resources more efficiently and reasonably. While geospatial big data has been extensively applied in current urban geography studies focusing on human mobility, socioeconomic activities, and health issues, comprehensive accounts for urban sustainability planning are still scanty.

APPLICATION OF USING GEOSPATIAL BIG DATA

Having recognized the necessity of contemporary geospatial big data for capturing complex urban contexts and dynamics, a principal research question is therefore proposed: How can geospatial big data help us to better understand the growing complexities of cities, with the purpose of generating novel insights that can enhance decision-making in the pursuit of urban sustainability. This chapter attempts to present an in-depth investigation by reviewing the use of geospatial big data in the context of urban sustainability science. Recent advancement in geospatial technology has demonstrated the possibility to collect and analyze vast amounts of information from urban environments, from the dynamics of people and goods to the spatial distribution of resources and amenities. Such valuable datasets can be employed to gain alternative insights into the complex dynamics of urban sustainability, including the role of agglomeration economies, the patterns of urban freight, and the vitalities and functions of different urban areas. Drawing on case studies from current literature, this chapter will illustrate three types of applications that rely on geospatial big data, shedding light on how such knowledge can help address the challenges of urban sustainability.

Three specific applications of employing geospatial big data are introduced to demonstrate the potential in the study of urban sustainability. By reporting the concept of agglomeration economies in cities, an empirical case study on how geospatial data can be used to assess their impacts on socioeconomic issues is first reported. Then, this chapter explores the feasible capability of geospatial big data that helps to improve transportation planning. Last but not least, the use of geospatial analytics to investigate urban functions is discussed, such as residential, commercial, and industrial zones.

AGGLOMERATION ECONOMIES

Agglomeration economies refer to firms and industries that are located near each other in a cluster or urban agglomeration. These economies can have important implications for urban sustainability science, as they can influence the distribution of economic activity, environmental impacts, and social outcomes in urban areas. To capture agglomeration economies, POIs could be one of the potential geospatial big datasets to include valuable information that is probably to be applied to various research topics in urban studies (Gao et al., 2017). Previous literature has taken advantage to better understand human activities within cities, by revealing nuanced urban functions (Yu et al., 2022a).

Yu et al. (2022c) propose a case study using POIs to depict the geographic boundaries of agglomeration economies and characterize their industrial functional patterns. The Guangdong-Hong Kong-Macao Greater Bay Area (GBA) is considered the study area that is featured strong economic performance. Kernel density estimation is initially utilized to evaluate the spatial extent of concentrated industrial activities and identify agglomeration economies. Subsequently, they investigate the local industry sectors by employing the TF-IDF (Term Frequency-Inverse Document Frequency) algorithm.

Speaking of the manufacturing activities, peripheral areas of GBA exhibit a high level of specialization, which is evident as presented in Figure 2.1. The word clouds reveal that these regions specialize in particular industrial chain systems and production processes, with a strong emphasis on specialized manufacturing preferences. Peripheral agglomerations 1 and 5, for example, are closely associated with low-tech industries of bathroom accessories and footwear products, respectively. On the contrary, agglomerations 6– 8 from central areas reveal diverse manufacturing sectors, with keywords related to various industrial functions. These patterns indicate the existence of diversification economies, by providing straightforward evidence.

Functional attributes of industrial agglomerations that engage in specialized manufacturing activities are illustrated in Figure 2.2. Remarkably, it can be seen that manufacturing sectors are predominantly related to low-tech industries such as

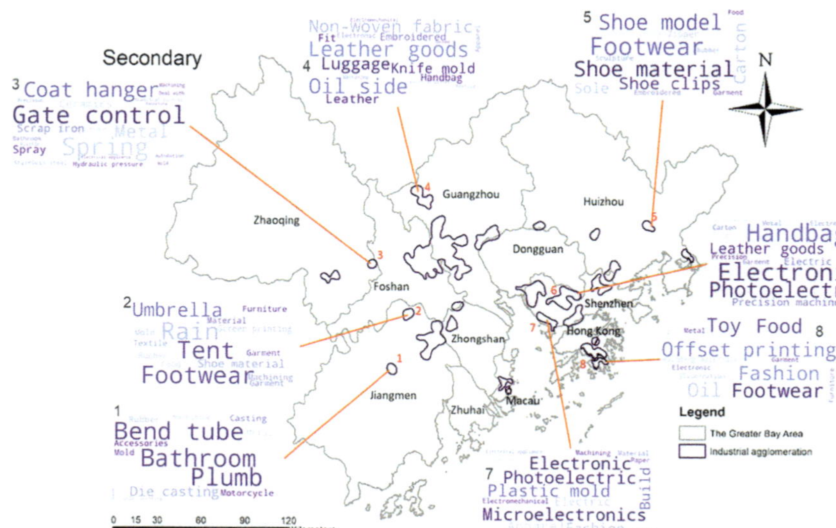

FIGURE 2.1 Spatial and functional characteristics reported by agglomeration economies in terms of manufacturing sector.

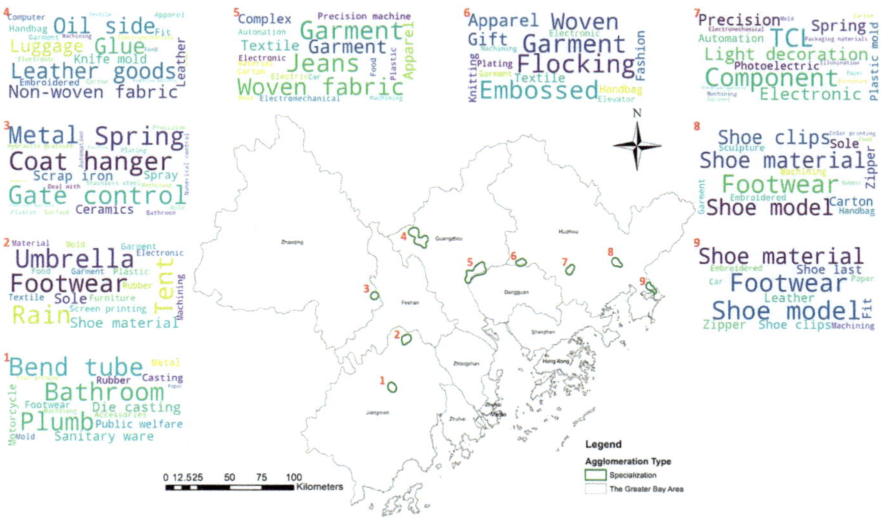

FIGURE 2.2 Functional characteristics reported by localization (i.e., specialization) agglomerations in terms of manufacturing sector.

bathroom hardware, leather, and footwear. The semantic analysis reveals that keywords within an identified agglomeration economy closely relate to specific manufacturing processes. For instance, agglomeration 4 in northern Guangzhou specializes in leather production, and a majority of its keywords are related to leather byproducts like handbags as well as leather edge painting. It is noteworthy that some agglomerations have a specialization in overlapping manufacturing sectors, such as footwear production reported in agglomerations 2, 8, and 9.

This proposed case study not only demonstrates the spatial diversity of human activities but also highlights the distinct functional preferences that are exhibited across agglomerations within cities. The analytical findings can provide valuable insights for future research that seeks to develop effective sustainable planning and development strategies.

Urban Freight

Urban freight logistics is a critical component of the city's daily operations, exerting significant efforts on the urban sustainability of transportation, socioeconomic, and environmental aspects (Melo & Baptista, 2017). In recent decades, cities have witnessed a dramatic surge in urban freight demand. Hence, it is essential to identify sustainable mobility solutions that can effectively diminish the adverse effects of freight transportation services (Nuzzolo et al., 2016). To achieve sustainable transportation, intra-urban shared freight systems have been widely acknowledged as having the capability to reduce the number of operating vehicles and improve their efficiency. Nevertheless, speaking of urban shared freight networks, scant research focus has been placed on their spatial regularities and how socioeconomic factors can relate to the network properties exhibited by a shared freight network.

Guo et al. (2023) present a research framework focusing on the structural properties derived from a shared freight network, via an intra-urban quantitative perspective based on a complex network. A data-driven case analysis is performed utilizing the vast amount of GPS trajectories generated by freight vehicles within Hong Kong.

Spatial interaction regularities of freight origins and destinations are firstly visualized via the urban freight networks. The spatial interactions on weekdays and weekends shown in Figures 2.3 and 2.4 illustrate that numerous freight trips occur between particular pairs of TAZs, for instance, TAZ 326 (Central Kwai Chung) and TAZ 329 (Tsuen Wan) as well as TAZ 295 (Kwun Tong) and TAZ 280 (Kowloon Bay). The observation of high-frequency goods movement between the above-mentioned TAZs suggests a potential perspective for optimizing de facto shared freight systems. For example, by merging orders with the same vehicle given the overlapping order time and origins–destinations, the operation number of vehicles can be significantly reduced. This can help lower the operation costs for both carriers and users, more importantly, reducing the on-road emissions by freight vehicles. As can be seen, the knowledge of spatial interactions in urban freight activities can produce practical implications for urban and transportation planners to improve current freight services that meet sustainability requirements.

Besides, the associations between the network properties and various socioeconomic conditions are also investigated, by leveraging ordinary least squares (OLS) regression and geographically weighted regression (GWR) models. The analytical results suggest that the structural characteristics of shared freight networks are strongly correlated with some socioeconomic factors selected in this

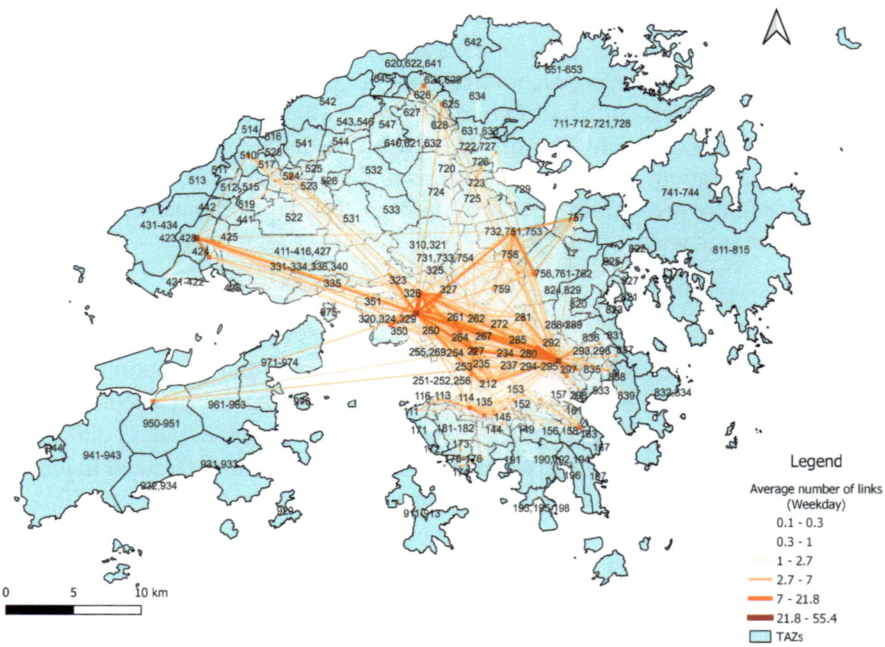

FIGURE 2.3 The spatial layout of shared freight network on weekdays.

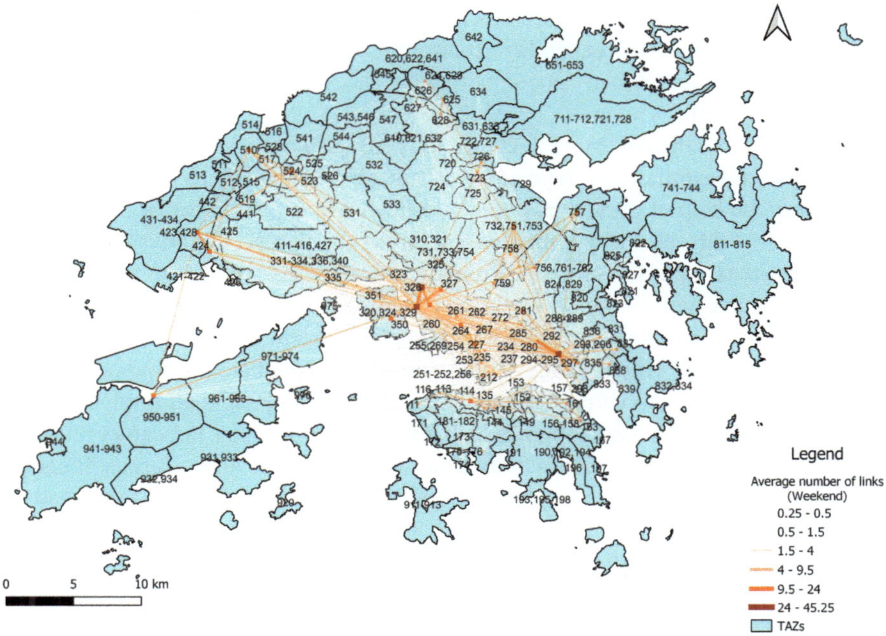

FIGURE 2.4 The spatial layout of shared freight network on weekends.

study. For illustration, socioeconomic factors such as marital status (i.e., divorced, separated, and widowed population) are relatively related to the degree of centrality, implying that this kind of user may prefer to consider shared freight services. This may be explained by the lack of family support received from other family members. It is recommended that the planning process for future shared freight systems should take into account not only the economic advantages but also the social benefits to local communities and minor population groups, to improve social equality. That is to say, by implementing shared freight systems, relevant stakeholders and policymakers should emphasize freight demand from different population groups, including vulnerable populations.

Urban Functions and Transit-Oriented Development (TOD)

Understanding the complex function within cities is another essential topic for urban sustainability science, as it provides us with in-depth knowledge of how cities work and how they are spatially associated with the economy, society, and environment. The striking growth of population, urbanization, and overwhelming private vehicles have led to severe traffic congestion, triggering environmental concerns for cities to sustainable development. To address these challenges, scholars and urban planners have introduced various strategies for reducing the use of private vehicles in cities (Redman et al., 2013). Among them, transit-oriented development (TOD) has emerged as an effective urban planning strategy in recent decades and has been widely recognized as a potential solution (Cervero, 2004).

Research in the existing literature has extensively demonstrated that metro station catchment areas have the capability to attract and contain a significant number of human activities and urban functions, acting as essential and dynamic areas for everyday urban life.

The study conducted by Yu et al. (2022b) proposes an empirical research framework to quantitatively represent catchment areas of metro stations in Hong Kong, with the help of geospatial big data collected in the dense urban environment. They utilize POIs as the comprehensive proxy to perceive human activities in catchment areas. Individuals visit different types of POIs for various reasons, such as work, leisure, and dining.

POIs data are employed to delineate the spatial extent of TOD catchment areas for each station in Hong Kong. Topic modeling technique of Latent Dirichlet Allocation (LDA) from natural language processing (NLP) is then implemented to identify functional thematic topics, which are seen as important indicators reflecting the diverse thematic activities within identified catchment areas. Urban functions generated from LDA can specify high-frequency urban functions that are spatially co-located together, thus highlighting their thematic characteristics. Meanwhile, in order to figure out the urban functions significantly associated with metro stations, this study proposes a standard deviation metric to compute the quantitative representativeness (i.e., uniqueness) of topics based on the catchment-to-topic matrix. Four distinctive thematic functions are mainly observed within TOD areas, as reported in Figure 2.5. Four thematic urban functions are revealed that are

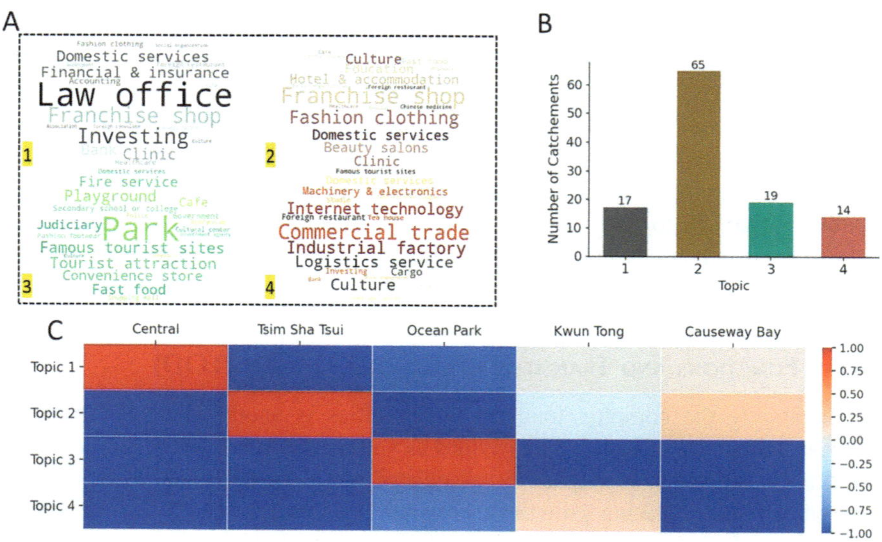

FIGURE 2.5 (a) Visualization via word cloud for urban functions identified from LDA topic modeling. (b) Number of station catchment areas for the four topics. (c) Quantitative intensity of thematic topics in selected metro stations (columns) with representative cognitions, using standard deviation metric. Red topics are the ones the metro station contains much more than average level, indicating a particular interest.

connected to commercial, residential, tourism, and industrial activities. For example, commercial functions are prevalent among the identified thematic topics that could be seen as the dominant urban functions. Furthermore, the distribution of thematic topics also reveals intriguing insights. A large number of metro stations have urban functions related to residential activities, indicating the vital role shown by TOD in the daily lives of local residents.

This study employs hierarchical clustering to aggregate different catchment areas according to their thematic topic weights. Metro stations classified into one group have similar topic compositions, indicating similar types of urban functions. The clustering heat map reported in Figure 2.6 clearly visualizes and explains the differences in topic composition. A large portion of catchment areas has one thematic urban function (i.e., residential function). Moreover, an interesting result indicates a spatial clustering of metro stations featured by comparable functional

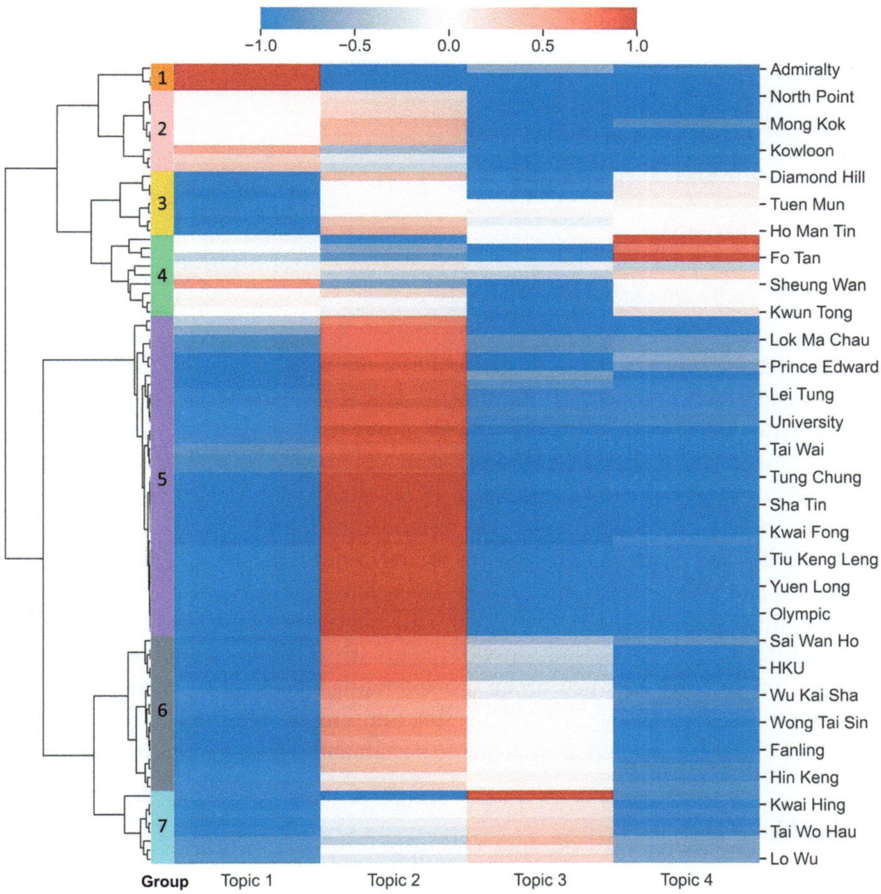

FIGURE 2.6 Clustering results of urban thematic topics associated with metro station catchment areas. Colored numbers shown in the first column represent the classification result.

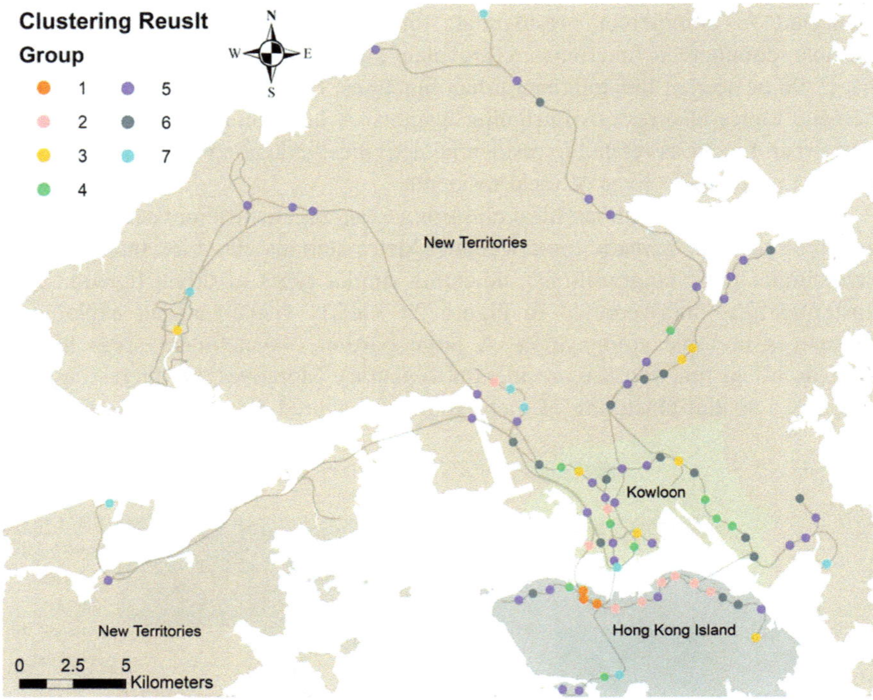

FIGURE 2.7 The spatial distribution of metro stations projected from the clustering results.

patterns, as shown in Figure 2.7. Particularly, the residential-based metro stations reveal insights into the spatial disparity regarding housing issues in the region of New Territories. The empirical findings from this study support the implementation of the new town program in New Territories, which intends to alleviate the overpopulation of downtown areas (He et al., 2020). Also, the spatial clusters of industrial-based catchment areas can provide direct evidence of where urban sustainability and environmental efforts are most needed. Targeted interventions to reduce air and noise pollution should be accordingly developed for such areas.

Analytical results proposed by this case study would be critical for urban sustainability science because it provides alternative insights for understanding the complex interactions between urban transportation systems and sustainability. By improving our knowledge of urban function, more reasonable and effective strategies could be developed to promote sustainable urban transport solutions.

CONCLUSIONS

Urban sustainability science has close relationships with cities that are recognized as complex systems shaped by a wide range of environmental and socioeconomic factors. Utilizing geographic expertise can be a promising perspective to reveal spatial knowledge related to diverse socioeconomic activities and their human–environment relationships. Meanwhile, as geospatial big data is being extensively applied in urban geography studies, limited attention has been paid to urban sustainability studies.

This chapter, therefore, aims to present an in-depth investigation by reviewing the use of geospatial big data in the context of urban sustainability science. Having referred to extant case studies, three practical themes using geospatial big data to understand urban sustainable issues are illustrated, including agglomeration economies, human mobility, and urban functions. The first proposed case study demonstrates the potential of using geospatial analytics and big data to understand agglomeration economies, along with several theoretical and practical implications. With detailed insights into industrial clustering activities, agglomeration economies, and their interdependencies featured by specialization, competitiveness, and potential synergies can be unveiled. Such patterns can allow relevant stakeholders to improve policy and planning, designing targeted interventions, for instance, infrastructure development, zoning regulations for high-emitting industries, and industrial investment incentives, to promote economic growth and productivity together with sustainability considerations. The second case study is to investigate the structural properties derived from an intra-urban shared freight network, via a vast amount of GPS trajectories generated by freight vehicles in cities. By depicting the shared freight network, urban planners and policymakers can identify the busy areas, optimizing the route and transportation modes for good movements. This kind of knowledge enables more reasonable vehicle allocation strategies to mitigate congestion and reduce fuel consumption, which can thereby lower greenhouse gas emissions. The last case study is to present an empirical research framework to quantitatively represent catchment areas of metro stations in Hong Kong, via geospatial big data collected in the dense urban environment. The analytical results confirm a strong spatial heterogeneity across TOD catchment areas surrounding metro stations, of which four thematic functions are reported and related to commercial, residential, tourism, and industrial activities. Accessibility to essential services within TOD areas can reduce the reliance on private vehicles, mitigating the environmental impact related to car-dependent communities, such as traffic congestion and air pollution. Ultimately, the evidence presented above has demonstrated that the massive volume and multi-dimensional sensing capability shown by geospatial big data allow us to show unique superiorities to better perceive the human dynamics and socioeconomic issues in cities.

FUTURE CHALLENGES

While geospatial big data can provide promising opportunities in studying urban sustainability, several limitations and challenges should be figured out. One challenge is the quality of the data that can be ascribed to the incomplete or inaccurate. Current social media markets are influenced and segmented by a series of factors such as age, gender, socioeconomic status, and geography. Such disparities indicate that using geospatial big data such as VGI can still lead to biased or misleading analytical results. Another drawback is the data formats and processing methods. Its size and complexity require specialized skills and computing resources, for storage, processing, analysis, and visualization. More importantly, the use of geospatial data also raises privacy concerns, of which users' individual information such as movement and activities can be easily collected and disclosed. In this sense, geoprivacy-preserving methods and tighter data collection policies should be introduced for preserving personal privacy.

REFERENCES

Baerwald, T. J. (2010). Prospects for geography as an interdisciplinary discipline. *Annals of the Association of American Geographers*, 100(3), 493–501.

Batty, M. (2017). Data about cities: Redefining big, recasting small. In Kitchin, R., Lariault, T. P., & McArdle, G. (Eds.), *Data and the City* (1st ed., pp. 31–43). Routledge. https://doi.org/10.4324/9781315407388

Bettencourt, L. M., & Kaur, J. (2011). Evolution and structure of sustainability science. *Proceedings of the National Academy of Sciences*, 108(49), 19540–19545.

Chiesura, A. (2004). The role of urban parks for the sustainable city. *Landscape and urban planning*, 68(1), 129–138.

Cervero, R., Murphy, S., Ferrell, C., Goguts, N., Tsai, Y., Arrington, G.B. et al. (2004). Transit-orienteddevelopment in the United States: Experiences, challenges, and prospects (TCRP 102). Washington, DC: Transportation Research Board.

Gao, S., Janowicz, K., & Couclelis, H. (2017). Extracting urban functional regions from points of interest and human activities on location-based social networks. *Transactions in GIS*, 21(3), 446–467.

Gately, C. K., Hutyra, L. R., Peterson, S., & Wing, I. S. (2017). Urban emissions hotspots: Quantifying vehicle congestion and air pollution using mobile phone GPS data. *Environmental Pollution*, 229, 496–504.

Guo, Z., Zhao, P., Senousi, A. M., Liu, X., & Mansourian, A. (2023). Exploring the structural characteristics of intra-urban shared freight network and their associations with socioeconomic status. *Travel Behaviour and Society*, 32, 100576.

He, S. Y., Wu, D., Chen, H., Hou, Y., & Ng, M. K. (2020). New towns and the local agglomeration economy. *Habitat International*, 98, 102153.

Heipke, C. (2010). Crowdsourcing geospatial data. *ISPRS Journal of Photogrammetry and Remote Sensing*, 65(6), 550–557.

Huang, L., Wu, J., & Yan, L. (2015). Defining and measuring urban sustainability: A review of indicators. *Landscape Ecology*, 30, 1175–1193.

Jenks, M., & Jones, C. (Eds.). (2009). Dimensions of the sustainable city. Vol. 2, Springer, Science & Business Media.

Jiang, B., & Thill, J. C. (2015). Volunteered Geographic Information: Towards the establishment of a new paradigm. *Computers, Environment and Urban Systems*, 53, 1–3.

Kitchin, R., & McArdle, G. (2016). What makes Big Data, Big Data? Exploring the ontological characteristics of 26 datasets. *Big Data & Society*, 3(1), 1–10.

Klopp, J. M., & Petretta, D. L. (2017). The urban sustainable development goal: Indicators, complexity and the politics of measuring cities. *Cities*, 63, 92–97.

Li, S., Dragicevic, S., Castro, F. A., Sester, M., Winter, S., Coltekin, A., … & Cheng, T. (2016). Geospatial big data handling theory and methods: A review and research challenges. *ISPRS Journal of Photogrammetry and Remote Sensing*, 115, 119–133.

Melo, S., & Baptista, P. (2017). Evaluating the impacts of using cargo cycles on urban logistics: Integrating traffic, environmental and operational boundaries. *European Transport Research Review*, 9(2), 1–10.

Mooney, H. A., Duraiappah, A., & Larigauderie, A. (2013). Evolution of natural and social science interactions in global change research programs. *Proceedings of the National Academy of Sciences*, 110 (supplement_1), 3665–3672.

Nuzzolo, A., Comi, A., Ibeas, A., & Moura, J. L. (2016). Urban freight transport and city logistics policies: Indications from Rome, Barcelona, and Santander. *International Journal of Sustainable Transportation*, 10(6), 552–566.

Redman, L., Friman, M., Gärling, T., & Hartig, T. (2013). Quality attributes of public transport that attract car users: A research review. *Transport Policy*, 25, 119–127.

Turner, V. K., & Kaplan, D. H. (2019). Geographic perspectives on urban sustainability: Past, current, and future research trajectories. *Urban Geography*, 40(3), 267–278.

Yu, X., Ivey, C., Huang, Z., Gurram, S., Sivaraman, V., Shen, H., … & Zheng, J. (2020). Quantifying the impact of daily mobility on errors in air pollution exposure estimation using mobile phone location data. *Environment International*, 141, 105772.

Yu, Z., & Liu, X. (2021). Urban agglomeration economies and their relationships to built environment and socio-demographic characteristics in Hong Kong. *Habitat International*, 117, 102417.

Yu, Z., Xiao, Z., & Liu, X. (2022a). A data-driven perspective for sensing urban functional images: Place-based evidence in Hong Kong. *Habitat International*, 130, 102707.

Yu, Z., Zhu, X., & Liu, X. (2022b). Characterizing metro stations via urban function: Thematic evidence from transit-oriented development (TOD) in Hong Kong. *Journal of Transport Geography*, 99, 103299.

Yu, Z., Zu, J., Xu, Y., Chen, Y., & Liu, X. (2022c). Spatial and functional organizations of industrial agglomerations in China's Greater Bay Area. *Environment and Planning B: Urban Analytics and City Science*, 49(7), 1995–2010.

3 Urban Flooding Monitoring and Management in Geospatial Perspective

Data, Techniques, and Platforms

Zhiwei Li, Cheolhee Yoo, and Qihao Weng

URBAN SUSTAINABILITY SCIENCE

As one of the most prevalent natural disasters worldwide, floods have been posing severe threats to human life, infrastructure, and ecosystems, and also leading to hindrances to socially sustainable development (Jongman et al., 2015; Francesch-Huidobro et al., 2017; Wagenaar et al., 2020). With the influences of ongoing global climate change, floods caused by extreme precipitation have become more frequent and intensive, which in turn increases flood risks, particularly for densely populated urban areas (Hallegatte et al., 2013; UNDRR, 2015; Alfieri et al., 2017; IPCC, 2021). Recent studies reveal that 23% of the world's population is directly exposed to 1-in-100-year floods, and estimates suggest that the proportion of people affected by floods has increased by 24% globally since the turn of the century (Tellman et al., 2021; Rentschler et al., 2022). Urban regions with dense populations and infrastructure are especially susceptible to flood disasters. In many urban areas worldwide, the concurrent issue of rapid urbanization exacerbating flood risks makes this problem more serious. The reduced water permeability due to increased impervious surfaces and often poorly planned urban drainage systems leads to enhanced vulnerability of urban areas to flooding (Douglas et al., 2008; Sampson et al., 2015).

Floods in urban environments can result in significant physical damage to human lives, properties, and infrastructure, as well as economic losses, social disruption, and health problems (Ashley et al., 2005; Hallegatte et al., 2013; Foudi et al., 2017). The current situation of increasingly frequent and intensive floods reveals an urgent need for effective monitoring and management strategies to mitigate the adverse effects of urban flooding and ensure life and property

 DOI: 10.1201/9781003244561-4

security. Effective urban flood monitoring and management have become critical in disaster response, facilitating a better understanding of the risks and helping people prepare and adapt more effectively.

In the field of flood monitoring and management, advancements in geospatial techniques have revolutionized the approach. Specifically, flood monitoring provides vital information that can assist in disaster response, as real-time data acquisition capabilities enable a timely response to flood events (Schumann et al., 2007). Real-time flood information, offered by remote sensing technologies, can facilitate timely evacuations, emergency response planning, and post-disaster relief actions (Fakhruddin et al., 2015). Similarly, comprehensive flood management plans can enhance urban resilience, reduce the vulnerability of populations, and ensure a more efficient allocation of resources for disaster risk reduction (Di Baldassarre et al., 2014). These geospatial techniques leverage data from various sources to offer a holistic view of urban flood situations, providing invaluable tools for predicting, monitoring, and managing flooding events in urban areas. Furthermore, geospatial technologies enable the development of powerful flood systems and platforms, which have been designed to integrate data from multiple sources to support urban flood monitoring and management.

Nevertheless, flood monitoring and management are still challenging tasks due to their dynamic processes and the variety of complex environments, especially in urban areas. For example, the effectiveness of geospatial techniques in flood monitoring and management is significantly influenced by data quality and availability. Additionally, gaps exist in the technical capabilities of geospatial techniques and their practical applications (Paprotny et al., 2018). These challenges necessitate continued research and development of geospatial techniques in the field, particularly in the incorporation of multi-source and multi-sensor data, as well as the integration of geospatial techniques with emerging technologies such as artificial intelligence and digital twin. The subsequent content of this chapter aims to provide a brief review of recent achievements in urban flood monitoring and management from a geospatial perspective, focusing primarily on three aspects as shown in Figure 3.1, i.e., data, techniques, and platforms. The goal is to identify the current problems and future prospects in the field.

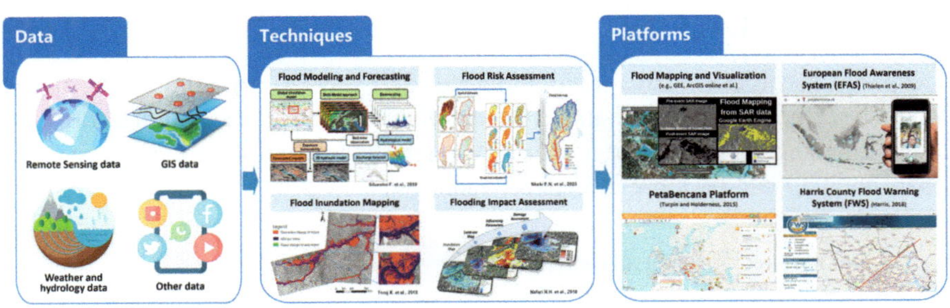

FIGURE 3.1 Framework of flood monitoring and management: data, techniques, and platforms.

GEOSPATIAL DATA FOR URBAN FLOODING MONITORING

Geospatial data sources for flood monitoring can be broadly divided into four categories, namely, remote sensing data, Geographic Information Systems (GIS) data, weather and hydrological data, and other data. Each data source plays an important role in urban flooding monitoring and management. 1) Remote sensing data acquired through satellite and aerial platforms, such as MODIS, Landsat, Sentinel-1/2, and unmanned aerial vehicles (UAVs), plays a vital role in flood monitoring and is commonly used to acquire essential information such as surface water extent, topography, and land cover and use. They offer images over a wide range of areas, facilitating the detection of flooded regions and tracking their changes over time. 2) GIS data provides necessary spatial and attribute information for flood monitoring, such as population, urban infrastructure, road networks, and digital elevation models (DEMs). Particularly, DEMs are essential for the modeling of flood dynamics and flood risk assessment (Li et al., 2022). 3) Weather and hydrology data, including rainfall, river discharge, groundwater levels, and soil moisture content, provide essential information for understanding the hydrological processes leading to flooding and are vital for forecasting future flood events. Weather and hydrological data are usually available as real-time data feeds or historical archives. 4) The other data, including social media data, can also serve as auxiliary information sources that promote the monitoring of floods (Li et al., 2021; Sadiq et al., 2022). For example, through the analysis of real-time social media data, relevant information about flood events, including time, location, and affected areas, can be obtained, which is important for flood monitoring, impact assessment, and emergency response.

Integrating multi-source geospatial data is necessary for comprehensive flood monitoring and assessment, in which data from different sources is combined to provide more complete and diverse information on the flood situation. Techniques such as image fusion, data assimilation, and machine learning algorithms are frequently used for integrating remote sensing, GIS, weather and hydrology, and other data. The integration of these various data sources and the subsequent analysis have created a new paradigm in flood monitoring, enriching the understanding of flood dynamics and enhancing flood prediction and response capabilities.

GEOSPATIAL TECHNIQUES FOR URBAN FLOODING MANAGEMENT

Geospatial techniques have revolutionized the way we monitor and manage flooding scenarios. In this chapter, we mainly focus on the techniques for urban flooding management in three aspects. Through the utilization of these techniques, urban flooding can be more effectively managed.

Flood Modeling and Forecasting

The use of hydrological models in predicting flood occurrences has gained prominence over the years (Silvestro et al., 2019). These models use geospatial data to estimate how rainfall, evapotranspiration, and land use can impact flood dynamics. As one of the commonly used hydrological models, the physically based model involves the numerical representation of hydrological processes and needs high-resolution geospatial data (Bates et al., 2010). While computationally intensive, the physically based model provides a more accurate prediction of the spatial and temporal distribution of floodwaters. Besides, rainfall-runoff models can also be utilized for predicting urban flooding. They function by simulating the transformation of rainfall into runoff and its conveyance to the outlet of a watershed. The integration of rainfall-runoff models with geospatial data enhances the accuracy of flood forecasting. For example, the model has been integrated into the Soil and Water Assessment Tool (SWAT) (Arnold et al., 1998), which uses geospatial data like land use, soil characteristics, and topography to simulate the hydrological cycle.

Flood Mapping and Risk Assessment

Satellite remote sensing provides timely and up-to-date data for flood mapping over large areas. As the two main types of remote sensing data sources applied in flood mapping, optical and synthetic aperture radar (SAR) images are applied in different aspects for their respective characteristics (Tong et al., 2018). On the one hand, in general, SAR data have been widely used in flood mapping because they are independent of weather conditions and imaging periods. However, flood monitoring with SAR data in urban areas is relatively constrained due to the limitations of the SAR imaging mechanism (Notti et al., 2018; Liang & Liu, 2020; McCormack et al., 2022). On the other hand, optical satellite images can provide rich spectral information for surface information extraction and flood detection; however, they are easily affected by clouds, leading to the missing information of the ground surface in images and reducing the frequency of valid observations (Notti et al., 2018; Goffi et al., 2020; Zeng et al., 2020). Considering the revisit cycles of satellites and the duration of floods with their respective intervals, the frequency of satellite observation is crucial for flood monitoring (Tulbure et al., 2022), which therefore necessitates improvements in flood monitoring frequency by combining optical satellite data and SAR data obtained from multiple sensors.

Besides, flood risk assessments under the framework of Crichton's Risk Triangle, which is defined as a function of three key components: hazard, exposure, and vulnerability (Crichton, 1999), have been studied at global and regional scales (Merz et al., 2004; Haynes et al., 2008; Darabi et al., 2019). The generated flood risk maps can be integrated into spatial decision support systems that guide urban planning. These systems can provide insights into the suitability of areas for infrastructure development, the best locations for flood defenses, and areas that require evacuation plans (Zhang et al., 2019; He et al., 2021). Particularly as one of the indicators of flood risk, flood vulnerability can be mapped by considering factors such as population density, infrastructure, and land use. Vulnerability maps can show areas with higher exposure to flood hazards and populations that may be more impacted by floods. Such vulnerability information can guide response and evacuation efforts and assist in infrastructure development and land-use planning. In this aspect, multi-criteria decision analysis (MCDA) is a common technique used to evaluate flood risk by integrating various flood-related factors (Chowdhury et al., 2009; Pham et al., 2021; Nkeki et al., 2023).

Flooding Impact Assessment, Recovery, and Adaptation

After a flood event, geospatial techniques can be employed to evaluate its impact and monitor the recovery process. A better understanding of the impacts of severe floods is essential for policymaking and site-specific adaptation and response plans, especially in urban areas. Flood damage to infrastructure, such as buildings, can be effectively monitored from remote sensing images (Chen & Yu, 2019; Nex et al., 2019; Adriano et al., 2021). High-resolution satellite images can be used to identify flood-damaged areas, quantify the extent of the flood, and assess the recovery process. For example, a study by Adriano et al. (2021) suggested the building damage mapping method can use the fusion of high-resolution optical imagery and high-to-moderate-resolution SAR data obtained before and after a disaster. Based on a deep convolutional neural network, the damaged buildings can be identified and compared with the pre-disaster situation. The assessment of flood impact can inform future flood risk management and mitigation strategies (Joyce et al., 2009; Nafari, 2018). These techniques could be applied not only for flood disaster damage monitoring but also for analyzing the flood recovery progress by continuously monitoring with time-series satellite images. In addition, geospatial techniques can also aid in flood response, such as evacuation and rescue. Flood extent maps can be used to identify safe routes for evacuation and to locate stranded individuals. Besides, evacuation routes can be optimized with geospatial techniques, which can also be used to help distribute resources effectively during the rescue process. Furthermore, geospatial techniques can support adaptation strategies for urban flooding, which can be applied to long-term planning, such as the development of flood defense infrastructure (e.g., levees, dams, and flood walls), land use planning, and the design of efficient drainage systems.

GEOSPATIAL PLATFORMS FOR URBAN FLOODING MANAGEMENT

Geospatial platforms play an increasingly crucial role in urban flood monitoring and management by providing a sophisticated interface for data collection, analysis, visualization, and dissemination. In this chapter, these platforms are broadly introduced, including web-based mapping and visualization platforms, decision support platforms for flood risk assessment and management, and real-time flood monitoring and warning platforms.

Web-based mapping and visualization platforms are designed to process large amounts of geospatial data, providing visual representations of complex and dynamic flood situations. These platforms allow the tracking of flooding events and serve as a valuable tool for both disaster response teams and the public. They can provide flooding information, including water levels, flood hazard predictions, vulnerable areas, and evacuation routes, improving the overall efficiency of disaster response and adaptation (Sheffield et al., 2018). Google Earth Engine (GEE) is an example of a web-based platform that has been widely used for flood monitoring

(DeVries et al., 2020). It provides access to a comprehensive catalog of satellite imagery and geospatial datasets, allowing users to perform online image processing and spatial data analysis and create dynamic visualizations of flood events (Gorelick et al., 2017). Similar platforms include Esri ArcGIS Online, QGIS Cloud, and Mapbox, each with its own unique features and advantages, also showing their potential for flood monitoring.

In addition, geospatial platforms play an integral role in decision support systems designed for flood risk assessment and management. These systems are designed to provide relevant and timely information to facilitate informed decision-making by integrating geospatial data with hydrological models and risk assessment methodologies (Tingsanchali, 2012). For example, the Hydrologic Engineering Center's River Analysis System (HEC-RAS), developed by the U.S. Army Corps of Engineers, is an integrated system for floodplain management that includes a GIS component for geospatial data handling and analysis (USACE, 2010). The DSS-Flood, developed by the National Institute of Water and Atmospheric Research (NIWA) in New Zealand, combines hydrodynamic modeling, geospatial data, and a user-friendly interface to support decision-making during flood events (Fewtrell et al., 2011).

Real-time flood monitoring platforms are crucial for timely and effective disaster responses. Real-time systems incorporate data from remote sensing, GIS, and hydrometeorological networks to provide up-to-date information on the flood situation. Sensor networks, the Internet of Things (IoT), telemetry systems, and crowd-sourced data platforms also contribute to real-time monitoring. The information gathered can be disseminated to the public and relevant agencies as early warnings to prepare for incoming floods. The European Flood Awareness System (EFAS) is an example of a system that uses real-time data for flood monitoring and early warning (Thielen et al., 2009). Besides, several cities worldwide have successfully leveraged geospatial platforms for urban flood management. In Jakarta, Indonesia, the PetaBencana platform, which means "disaster map" in Indonesian, collects real-time reports from social media and combines them with remote sensing data to create dynamic, crowd-sourced flood maps (Turpin and Holderness, 2015). These maps are accessible to both the public and decision-makers, enhancing the city's resilience to flooding. Another application example is in the city of Houston, Texas, USA, where a real-time flood alert system has been developed using the Harris County Flood Warning System (FWS) (Harris, 2018). The FWS employs geospatial platforms to collect, analyze, and disseminate flood data, thus providing timely flood warnings to residents and enabling proactive measures.

Generally, geospatial platforms are invaluable tools and offer numerous benefits for urban flood management. They facilitate real-time monitoring, foster informed decision-making, and promote public participation. However, their effective implementation requires overcoming challenges such as data accuracy, system complexity, and issues related to privacy and data security. Furthermore, the utility and effectiveness of these platforms often depend on the availability of reliable, high-quality geospatial data and the technical capacity of their users. Thus, addressing these issues would harness the full potential of geospatial platforms for urban flood management.

PROBLEMS AND PROSPECTS

Despite the advancements in geospatial techniques and platforms for urban flooding monitoring and management, challenges in this field remain. Meanwhile, ongoing technological breakthroughs suggest promising prospects for the future. This section presents an overview of these challenges, emerging trends, and future prospects. The goal is to explore the potential for continued progress in this field.

DATA AVAILABILITY AND QUALITY ISSUES

Data is the foundation of geospatial techniques. However, challenges related to its availability and quality persist. Accessibility to high-resolution, up-to-date data, especially in less-developed regions, remains limited. Similarly, spatial and temporal variations in data quality and the integration of multiple heterogeneous data sources can pose significant challenges (Feng et al., 2016; Zeng et al., 2020). Furthermore, the trade-offs between spatial and temporal resolutions of multi-source satellite images necessitate the incorporation of multi-sensor images and multi-source data fusion for more detailed and timely monitoring of flood situations (DeVries et al., 2020; Huang and Jin, 2020; Tulbure et al., 2022). Another issue pertains to the large volume of data produced by massive remote sensing satellites. Despite the abundance of data, managing, processing, and extracting valuable information in real time remains a challenge due to computational constraints. In this aspect, the ongoing advancement in remote sensing and GIS technologies presents a promising future for urban flood management. Besides, more powerful satellites, such as those part of the Copernicus Program, Sentinel-1/2 provide higher resolution imagery, and rapid data acquisition capabilities, thereby improving the accuracy and timeliness of flood monitoring (Giuliani et al., 2017). Similarly, advances in LiDAR (Light Detection and Ranging) technology are providing unprecedented levels of detail in topographic data, significantly enhancing the precision of flood modeling and risk assessment (Glenn et al., 2016).

TECHNICAL LIMITATIONS AND GAPS

Limitations and gaps exist in geospatial techniques and their practical applications (Paprotny et al., 2018). On the one hand, current modeling techniques for flood prediction and risk assessment are inherently complex, necessitating a high degree of technical expertise. Furthermore, these models often rely on a multitude of assumptions and approximations, resulting in inevitable uncertainties (Beven and Hall, 2014). For example, hydrological models have difficulty accurately representing urban areas due to complex urban morphology and the effects of infrastructural developments on water flow. Also, these models face challenges in integrating dynamic urban landscapes, such as changing land use patterns, population densities, and infrastructure development. This limits their prediction accuracy and their ability to adapt to the evolving nature of urban environments. On the other hand, most critically, most of the previous research results on flooding have not been turned into practical systems and tools for users. The gap exists between flood monitoring technologies and their applications, and the development of intelligent real-time flood monitoring platforms is crucial for governments and individuals to benefit urban flood management, especially in flood-prone areas worldwide.

Geospatial Artificial Intelligence-Enabled Flood Monitoring and Management

The integration of geospatial techniques with artificial intelligence is an emerging trend with considerable potential for flood monitoring. As a subset of artificial intelligence, deep learning algorithms can help manage large volumes of remote sensing data, improve the accuracy of data interpretation, and enhance real-time monitoring capabilities (Zhang et al., 2016; Yuan et al., 2020). Furthermore, machine learning algorithms can help refine hydrological models, reducing their uncertainty and improving their predictive performance. These technologies can also aid in identifying vulnerable areas and populations, thereby informing targeted disaster response and risk reduction strategies. In addition, the combination of physical models (e.g., hydrological models) and machine learning has shown its potential in the fields, which will enhance the advantages of both kinds of methods and thus promote their efficiency in urban flooding monitoring and management.

Specifically, as for flood hazard prediction, studies on the application of machine learning to flood prediction are at an early stage of development (Mosavi et al., 2018), and improving the quantity and quality of data as well as the hybridization or integration of algorithmic models are deemed to be promising directions to enhance the effectiveness of machine learning methods for flood prediction. In terms of flood mapping, studies using machine learning methods have been proposed in the literature, including supervised learning (Moya et al., 2020), semi-supervised learning (Li et al., 2019a), and unsupervised learning (Li et al., 2019b). Due to the limitations of urban flood datasets, these studies are often trained and tested using the data acquired during the same flood events, hampering testing robustness and the generalization of the developed methods with respect to different urban typologies and severities of flood events. Besides, as urban flood risk is a dynamic process that is mainly affected by climate change and urban land use changes, access to real-time meteorological data and high-frequency land use data is essential for the dynamic assessment and monitoring of flood risks (Zhang et al., 2019). However, the difficulty of accessing real-time hydro-meteorological data and the efficiency of flood forecasting limit the wider applications of most previous flood research results. Therefore, there is an inevitable trend to fuse real-time social media data with machine learning methods in flood risk assessment (Wagenaar et al., 2020). Generally, the leverage of geospatial artificial intelligence will foster improvements in the accuracy and efficiency of geospatial techniques applied for flood monitoring and management.

Integration of Geospatial Platforms with Other Urban Management Systems

To create a more robust and comprehensive approach to urban flood management, it is essential to integrate geospatial platforms with other urban management systems, such as urban planning and infrastructure management systems. This integration can help to align flood management efforts with broader urban development goals, improve the use of resources, and foster a more resilient urban environment (Dottori et al., 2018). For example, in the Netherlands, the 3Di Water Management System integrates geospatial platforms with urban water management and spatial planning systems, enabling more holistic and sustainable flood management solutions. Besides, the digital twin has gained popularity recently due to its strong abilities in 3D visualization and scenario simulation. With the combination of GIS and remote sensing techniques, urban areas can be simulated as a digital twin in which terrain, buildings, infrastructure, and mobility information are integrated (White et al., 2021). Based on the integrated platform with this state-of-the-art flow, tools can be created to assess the impacts of floods by building a digital twin for areas with high flood risk and by realizing flood situations.

Generally, while challenges persist in the implementation of geospatial techniques for urban flood management, advances in remote sensing and GIS technologies, and artificial intelligence, coupled with urban management system integration, are reshaping urban flood management and paving the way for more resilient and flood-adaptive urban environments.

CONCLUSIONS

Due to climate change, recurring urban flooding worldwide increases concerns for urban areas, severely threatening the lives, property, and sustainability of urban environments. Sufficient data, advanced techniques, and effective platforms for urban flood monitoring and management are critical for mitigating the impacts of flood disasters. The importance of geospatial data for urban flooding monitoring is explicit. Various data sources, such as remote sensing imagery, GIS data, weather and hydrological data, and other data, contribute to a comprehensive understanding of flooding dynamics. However, it should be noted that not only data collection but also the integration and analysis of multi-source geospatial data provide detailed and real-time monitoring of flood situations. These techniques, which range from multi-source data fusion to spatial analysis methods, are geared toward reducing the impacts of flooding. Moreover, geospatial techniques are core to urban flooding monitoring and management, which further assist in flood modeling and forecasting, mapping of flooding areas, and identification of high-risk populations, thereby aiding in strategic infrastructure planning and flood mitigation efforts. Furthermore, geospatial techniques also provide a route for post-flood impact assessment, aiding recovery operations, and shaping future adaptation strategies. Geospatial platforms are the integration point for geospatial techniques, which range from web-based mapping and visualization tools and decision support systems to real-time monitoring and warning platforms, all of which contribute to dealing with urban flooding. The successful implementation of geospatial platforms in various cities shows the potential and value they add to flood monitoring and management.

However, there are still challenges in urban flooding monitoring and management. Issues related to data availability and quality, and technical limitations in modeling and analysis often impede the application of geospatial techniques in urban flooding monitoring and management. Encouragingly, the field is advancing with novel remote sensing and GIS technologies, and the integration of geospatial techniques with artificial intelligence and other urban management systems. There is an ongoing need for continued research and development in this field to overcome the current limitations and explore new opportunities. The refinement and advancement of geospatial data, techniques, and platforms would further enhance our ability to deal with future urban floods.

REFERENCES

Adriano, B., Yokoya, N., Xia, J., Miura, H., Liu, W., Matsuoka, M., Koshimura, S., 2021. Learning from multimodal and multitemporal earth observation data for building damage mapping. *ISPRS J. Photogramm. Remote Sens.* 175, 132–143. 10.1016/j.isprsjprs.2021.02.016

Alfieri, L., Bisselink, B., Dottori, F., Naumann, G., de Roo, A., Salamon, P., Wyser, K., Feyen, L., 2017. Global projections of river flood risk in a warmer world. *Earth's Future.* 5, 171–182. 10.1002/2016EF000485

Arnold, J.G., Srinivasan, R., Muttiah, R.S., Williams, J.R., 1998. Large area hydrologic modeling and assessment part I: Model development. *J. Am. Water Resour. Assoc.* 34, 73–89. 10.1111/j.1752-1688.1998.tb05961.x

Ashley, R.M., Balmfort, D.J., Saul, A.J., Blanskby, J.D., 2005. Flooding in the future – Predicting climate change, risks and responses in urban areas. *Water Sci. Technol.* 52, 265–273. 10.2166/wst.2005.0142

Bates, P.D., Horritt, M.S., Fewtrell, T.J., 2010. A simple inertial formulation of the shallow water equations for efficient two-dimensional flood inundation modelling. *J. Hydrol.* 387, 33–45. 10.1016/j.jhydrol.2010.03.027

Beven, K., Hall, J., 2014. *Applied Uncertainty Analysis for Flood Risk Management.* Imperial College Press. 10.1142/P588

Chen, F., Yu, B., 2019. Earthquake-induced building damage mapping based on multi-task deep learning framework. *IEEE Access* 7, 181396–181404. 10.1109/ACCESS.2019.2958983

Chowdhury, R., Flentje, P., Bhattacharya, G., 2009. *Geotechnical Slope Analysis.* CRC Press. 10.1201/9780203864203

Crichton, D., 1999. The Risk Triangle. Natural disaster management: a presentation to commemorate the International Decade for Natural Disaster Reduction (IDNDR), 1990-2000. *Inglet. J Tudor Rose* 76.

Darabi, H., Choubin, B., Rahmati, O., Torabi Haghighi, A., Pradhan, B., Kløve, B., 2019. Urban flood risk mapping using the GARP and QUEST models: A comparative study of machine learning techniques. *J. Hydrol.* 569, 142–154. 10.1016/j.jhydrol.2018.12.002

DeVries, B., Huang, C., Armston, J., Huang, W., Jones, J.W., Lang, M.W., 2020. Rapid and robust monitoring of flood events using Sentinel-1 and Landsat data on the Google Earth Engine. *Remote Sens. Environ.* 240, 111664. 10.1016/j.rse.2020.111664

Di Baldassarre, G., Kemerink, J.S., Kooy, M., Brandimarte, L., 2014. Floods and societies: the spatial distribution of water-related disaster risk and its dynamics. *WIREs Water* 1, 133–139. 10.1002/wat2.1015

Dottori, F., Szewczyk, W., Ciscar, J.-C., Zhao, F., Alfieri, L., Hirabayashi, Y., Bianchi, A., Mongelli, I., Frieler, K., Betts, R.A., Feyen, L., 2018. Increased human and economic losses from river flooding with anthropogenic warming. *Nat. Clim. Change* 8, 781–786. 10.1038/s41558-018-0257-z

Douglas, I., Alam, K., Maghenda, M., Mcdonnell, Y., Mclean, L., Campbell, J., 2008. Unjust waters: Climate change, flooding and the urban poor in Africa. *Environ. Urban.* 20, 187–205. 10.1177/0956247808089156

Fakhruddin, S.H.M., Kawasaki, A., Babel, M.S., 2015. Community responses to flood early warning system: Case study in Kaijuri Union, Bangladesh. *Int. J. Disaster Risk Reduct.* 14, 323–331. 10.1016/j.ijdrr.2015.08.004

Feng, M., Sexton, J.O., Huang, C., Anand, A., Channan, S., Song, X.P., Song, D.X., Kim, D.H., Noojipady, P., Townshend, J.R., 2016. Earth science data records of global forest cover and change: Assessment of accuracy in 1990, 2000, and 2005 epochs. *Remote Sens. Environ.* 184, 73–85. 10.1016/j.rse.2016.06.012

Fewtrell, T.J., Duncan, A., Sampson, C.C., Neal, J.C., Bates, P.D., 2011. Benchmarking urban flood models of varying complexity and scale using high resolution terrestrial LiDAR data. *Phys. Chem. Earth* 36, 281–291. 10.1016/j.pce.2010.12.011

Foudi, S., Osés-Eraso, N., Galarraga, I., 2017. The effect of flooding on mental health: Lessons learned for building resilience. *Water Resour. Res.* 53, 5831–5844. 10.1002/2017WR020435

Francesch-Huidobro, M., Dabrowski, M., Tai, Y., Chan, F., Stead, D., 2017. Governance challenges of flood-prone delta cities: Integrating flood risk management and climate change in spatial planning. *Prog. Plann.* 114, 1–27. 10.1016/j.progress.2015.11.001

Giuliani, G., Chatenoux, B., De Bono, A., Rodila, D., Richard, J.P., Allenbach, K., Dao, H., Peduzzi, P., 2017. Building an Earth Observations Data Cube: lessons learned from the Swiss Data Cube (SDC) on generating Analysis Ready Data (ARD). *Big Earth Data* 1, 100–117. 10.1080/20964471.2017.1398903

Glenn, N.F., Neuenschwander, A., Vierling, L.A., Spaete, L., Li, A., Shinneman, D.J., Pilliod, D.S., Arkle, R.S., McIlroy, S.K., 2016. Landsat 8 and ICESat-2: Performance and potential synergies for quantifying dryland ecosystem vegetation cover and biomass. *Remote Sens. Environ.* 185, 233–242. 10.1016/j.rse.2016.02.039

Goffi, A., Stroppiana, D., Brivio, P.A., Bordogna, G., Boschetti, M., 2020. Towards an automated approach to map flooded areas from Sentinel-2 MSI data and soft integration of water spectral features. *Int. J. Appl. Earth Obs. Geoinf.* 84, 101951. 10.1016/j.jag.2019.101951

Gorelick, N., Hancher, M., Dixon, M., Ilyushchenko, S., Thau, D., Moore, R., 2017. Google Earth Engine: Planetary-scale geospatial analysis for everyone. *Remote Sens. Environ.* 202, 18–27. 10.1016/j.rse.2017.06.031

Hallegatte, S., Green, C., Nicholls, R.J., Corfee-Morlot, J., 2013. Future flood losses in major coastal cities. *Nat. Clim. Change* 3, 802–806. 10.1038/nclimate1979

Harris County Flood Control District, 2018. The Harris County Flood Warning System. https://www.harriscountyfws.org/ [WWW Document].

Haynes, H., Haynes, R., Pender, G., 2008. Integrating socio-economic analysis into decision-support methodology for flood risk management at the development scale (Scotland). *Water Environ. J.* 22, 117–124. 10.1111/j.1747-6593.2007.00086.x

He, M., Chen, C., Zheng, F., Chen, Q., Zhang, J., Yan, H., Lin, Y., 2021. An efficient dynamic route optimization for urban flooding evacuation based on cellular automata. *Comput. Environ. Urban Syst.* 87, 101622. 10.1016/j.compenvurbsys.2021.101622

Huang, M., Jin, S., 2020. Rapid flood mapping and evaluation with a supervised classifier and change detection in Shouguang using Sentinel-1 SAR and Sentinel-2 optical data. *Remote Sens.* 12, 2073. 10.3390/rs12132073

IPCC, 2021. *Climate Change 2021: The Physical Science Basis*. Cambridge University Press.

Jongman, B., Winsemius, H.C., Aerts, J.C.J.H., Coughlan De Perez, E., Van Aalst, M.K., Kron, W., Ward, P.J., 2015. Declining vulnerability to river floods and the global benefits of adaptation. *Proc. Natl. Acad. Sci. USA.* 112, E2271–E2280. 10.1073/pnas.1414439112

Joyce, K.E., Belliss, S.E., Samsonov, S. V., McNeill, S.J., Glassey, P.J., 2009. A review of the status of satellite remote sensing and image processing techniques for mapping natural hazards and disasters. *Prog. Phys. Geogr.* 33, 183–207. 10.1177/03091333 09339563

Li, J., Liu, Z., Lei, X., Wang, L., 2021. Distributed fusion of heterogeneous remote sensing and social media data: A review and new developments. *Proc. IEEE* 109, 1350–1363. 10.1109/JPROC.2021.3079176

Li, S., Sun, D., Goldberg, M.D., Kalluri, S., Sjoberg, B., Lindsey, D., Hoffman, J.P., DeWeese, M., Connelly, B., Mckee, P., Lander, K., 2022. A downscaling model for derivation of 3-D flood products from VIIRS imagery and SRTM/DEM. *ISPRS J. Photogramm. Remote Sens.* 192, 279–298. 10.1016/j.isprsjprs.2022.08.025

Li, Y., Martinis, S., Wieland, M., 2019a. Urban flood mapping with an active self-learning convolutional neural network based on TerraSAR-X intensity and interferometric coherence. *ISPRS J. Photogramm. Remote Sens.* 152, 178–191. 10.1016/j.isprsjprs. 2019.04.014

Li, Y., Martinis, S., Wieland, M., Schlaffer, S., Natsuaki, R., 2019b. Urban flood mapping using SAR intensity and interferometric coherence via Bayesian network fusion. *Remote Sens.* 11, 2231. 10.3390/rs11192231

Liang, J., Liu, D., 2020. A local thresholding approach to flood water delineation using Sentinel-1 SAR imagery. *ISPRS J. Photogramm. Remote Sens.* 159, 53–62. 10.1016/ j.isprsjprs.2019.10.017

McCormack, T., Campanyà, J., Naughton, O., 2022. A methodology for mapping annual flood extent using multi-temporal Sentinel-1 imagery. *Remote Sens. Environ.* 282, 113273. 10.1016/j.rse.2022.113273

Merz, B., Kreibich, H., Thieken, A., Schmidtke, R., 2004. Estimation uncertainty of direct monetary flood damage to buildings. *Nat. Hazards Earth Syst. Sci.* 4, 153–163. 10.5194/nhess-4-153-2004

Mosavi, A., Ozturk, P., Chau, K., 2018. Flood Prediction Using Machine Learning Models: Literature Review. *Water* 10, 1536. 10.3390/w10111536

Moya, L., Mas, E., Koshimura, S., 2020. Learning from the 2018 Western Japan heavy rains to detect floods during the 2019 Hagibis Typhoon. *Remote Sens.* 12, 2244. 10.3390/ rs12142244

Nafari, R.H., Mendis, P., 2018. Flood Damage Assessment in Urban Areas. [Doctoral dissertation, University of Melbourne]. https://www.bnhcrc.com.au/publications/ biblio/bnh-5019

Nex, F., Duarte, D., Tonolo, F.G., Kerle, N., 2019. Structural building damage detection with deep learning: assessment of a state-of-the-art CNN in operational conditions. *Remote Sens.* 11, 2765. 10.3390/rs11232765

Nkeki, F.N., Bello, E.I., Agbaje, I.G., 2023. Is the existing methods sustainable? A hybrid approach to flood risk mapping. *MethodsX* 11, 102348. 10.1016/j.mex.2023.102348

Notti, D., Giordan, D., Caló, F., Pepe, A., Zucca, F., Galve, J.P., 2018. Potential and limitations of open satellite data for flood mapping. *Remote Sens.* 10, 1673. 10.3390/ rs10111673

Paprotny, D., Sebastian, A., Morales-Nápoles, O., Jonkman, S.N., 2018. Trends in flood losses in Europe over the past 150 years. *Nat. Commun.* 9. 10.1038/s41467-018-04253-1

Pham, B.T., Luu, C., Dao, D. Van, Phong, T. Van, Nguyen, H.D., Le, H. Van, von Meding, J., Prakash, I., 2021. Flood risk assessment using deep learning integrated with multi-criteria decision analysis. *Knowl.-Based Syst.* 219. 10.1016/j.knosys. 2021.106899

Rentschler, J., Salhab, M., Jafino, B.A., 2022. Flood exposure and poverty in 188 countries. *Nat. Commun.* 13. 10.1038/s41467-022-30727-4

Sadiq, R., Akhtar, Z., Imran, M., Ofli, F., 2022. Integrating remote sensing and social sensing for flood mapping. *Remote Sens. Appl. Soc. Environ.* 25, 100697. 10.1016/j.rsase.2022.100697

Sampson, C.C., Smith, A.M., Bates, P.D., Neal, J.C., Alfieri, L., Freer, J.E., 2015. A high-resolution global flood hazard model. *Water Resour. Res.* 51, 7358–7381. 10.1002/2015WR016954

Schumann, G., Hostache, R., Puech, C., Hoffmann, L., Matgen, P., Pappenberger, F., Pfister, L., 2007. High-resolution 3-D flood information from radar imagery for flood hazard management. *IEEE Trans. Geosci. Remote Sens.* 45, 1715–1725. 10.1109/TGRS.2006.888103

Sheffield, J., Wood, E.F., Pan, M., Beck, H., Coccia, G., Serrat-Capdevila, A., Verbist, K., 2018. Satellite remote sensing for water resources management: potential for supporting sustainable development in data-poor regions. *Water Resour. Res.* 54, 9724–9758. 10.1029/2017WR022437

Silvestro, F., Rossi, L., Campo, L., Parodi, A., Fiori, E., Rudari, R., Ferraris, L., 2019. Impact-based flash-flood forecasting system: Sensitivity to high resolution numerical weather prediction systems and soil moisture. *J. Hydrol.* 572, 388–402. 10.1016/j.jhydrol.2019.02.055

Tellman, B., Sullivan, J.A., Kuhn, C., Kettner, A.J., Doyle, C.S., Brakenridge, G.R., Erickson, T.A., Slayback, D.A., 2021. Satellite imaging reveals increased proportion of population exposed to floods. *Nature* 596, 80–86. 10.1038/s41586-021-03695-w

Thielen, J., Bartholmes, J., Ramos, M.H., De Roo, A., 2009. The European flood alert system – Part 1: Concept and development. *Hydrol. Earth Syst. Sci.* 13, 125–140. 10.5194/hess-13-125-2009

Tingsanchali, T., 2012. Urban flood disaster management. *Procedia Eng.* 32, 25–37. 10.1016/j.proeng.2012.01.1233

Tong, X., Luo, X., Liu, Shuguang, Xie, H., Chao, W., Liu, Shuang, Liu, Shijie, Makhinov, A.N., Makhinova, A.F., Jiang, Y., 2018. An approach for flood monitoring by the combined use of Landsat 8 optical imagery and COSMO-SkyMed radar imagery. *ISPRS J. Photogramm. Remote Sens.* 136, 144–153. 10.1016/j.isprsjprs.2017.11.006

Tulbure, M.G., Broich, M., Perin, V., Gaines, M., Ju, J., Stehman, S. V., Pavelsky, T., Masek, J.G., Yin, S., Mai, J., Betbeder-Matibet, L., 2022. Can we detect more ephemeral floods with higher density harmonized Landsat Sentinel 2 data compared to Landsat 8 alone? *ISPRS J. Photogramm. Remote Sens.* 185, 232–246. 10.1016/j.isprsjprs.2022.01.021

Turpin, E., Holderness, T., 2015. From social media to geosocial intelligence: Crowdsourcing civic co-management for flood response in Jakarta, Indonesia. *Soc. Media Gov. Serv.* 115–133. 10.1007/978-3-319-27237-5_6

UNDRR, 2015. Global assessment report on disaster risk reduction 2015. *Int. Strat. Disaster Reduct.* 1–311. United Nations. https://doi.org/10.18356/919076d9-en. https://www.un-ilibrary.org/content/books/9789210572699

USACE, 2010. HEC-RAS River Analysis System, User's Manual, Version 4.1.

Wagenaar, D., Curran, A., Balbi, M., Bhardwaj, A., Soden, R., Hartato, E., Mestav Sarica, G., Ruangpan, L., Molinario, G., Lallemant, D., 2020. Invited perspectives: How machine learning will change flood risk and impact assessment. *Nat. Hazards Earth Syst. Sci.* 20, 1149–1161. 10.5194/nhess-20-1149-2020

White, G., Zink, A., Codecá, L., Clarke, S., 2021. A digital twin smart city for citizen feedback. *Cities* 110, 103064. 10.1016/j.cities.2020.103064

Yuan, Q., Shen, H., Li, T., Li, Z., Li, S., Jiang, Y., Xu, H., Tan, W., Yang, Q., Wang, J., Gao, J., Zhang, L., 2020. Deep learning in environmental remote sensing: Achievements and challenges. *Remote Sens. Environ.* 241, 111716. 10.1016/j.rse.2020.111716

Zeng, Z., Gan, Y., Kettner, A.J., Yang, Q., Zeng, C., Brakenridge, G.R., Hong, Y., 2020. Towards high resolution flood monitoring: An integrated methodology using passive microwave brightness temperatures and Sentinel synthetic aperture radar imagery. *J. Hydrol.* 582, 124377. 10.1016/j.jhydrol.2019.124377

Zhang, H., Cheng, L.I., Cheng, J., Zhifeng, W.U., Yanyan, W.U., 2019. A review of urban flood risk assessment based on the framework of hazard-exposure-vulnerability. *Prog. Geogr.* 38, 175–190. 10.18306/dlkxjz.2019.02.003

Zhang, L., Zhang, L., Du, B., 2016. Deep learning for remote sensing data: A technical tutorial on the state of the art. *IEEE Geosci. Remote Sens. Mag.* 4, 22–40. 10.1109/MGRS.2016.2540798

Part II

Geospatial Techniques for Renewable Cities

4 Assessing Sustainability in China's Resource-Based Cities

Leveraging Remote Sensing Techniques for Evaluating Environment-Related SDG Progress

Huijuan Xiao and Song Xue

INTRODUCTION

The Sustainable Development Goals (SDGs) proposed by the United Nations provide a framework for promoting sustainable development (United Nations, 2015). Compared with national analysis, exploring city-level SDGs holds significant importance since this localized context is essential for developing targeted strategies and interventions that effectively address the unique needs of a particular city (Lafortune et al., 2019; Schmidt-Traub et al., 2017; Wiedmann & Allen, 2021). However, achieving these goals can be particularly challenging in resource-based cities (Xiao et al., 2022). Many of these cities in China are currently facing complex environmental challenges that threaten their sustainability and make it difficult to achieve the SDGs (Dong et al., 2007; Li et al., 2013; Xiao et al., 2022).

Resource-based cities in China, which rely on the exploitation of natural resources, have played a significant role in the country's economic development as key regional centers for several decades fuel (The Chinese Government, 2013). These cities have made substantial contributions to the national economy. In 2013, the State Council of China introduced the National Sustainable Development Plan for Resource-based Cities, which identified 262 cities out of 334 prefecture-level administrative units as resource-based cities. Since 1949, these resource-based cities have collectively produced an impressive volume of natural resources, including over 52.9 billion tons of raw coal, 5.5 billion tons of crude oil, 5.8 billion tons of iron ore, and 2 billion cubic meters of timber. This tremendous output has played a crucial role in establishing a comprehensive industrial system and driving economic

DOI: 10.1201/9781003244561-6

development in China. However, it is important to note that resource-based cities also face challenges as they consume significant amounts of energy and generate substantial CO_2 emissions and air pollutants emissions (Ma et al., 2018; Tan et al., 2020). To overcome these challenges, it is crucial to evaluate the sustainability of these cities to know the environmental situations and promote the achievement of SDGs.

Many studies have used statistical yearbooks, surveys, and censuses as data sources to evaluate the sustainability of cities in China (Liu et al., 2021; Xiao et al., 2022, 2023). Statistical yearbooks are annual publications produced by governments or organizations, providing comprehensive statistical data on various aspects of a country, region, or industry, such as China City Statistical Yearbooks and The China Urban Construction Statistical Yearbook. National or regional surveys and censuses are conducted periodically to collect demographic, social, and economic data, such as China's national population census in 2010 and China national population sample survey in 2005 and 2015.

In some cases, data for certain cities may be unavailable in statistical yearbooks, surveys, and censuses, particularly for remote or underdeveloped regions where data collection through surveys or other conventional methods can be challenging. For instance, China has a total of 293 prefecture-level cities as of 2016; however, some studies only cover 285 Chinese cities (Xiao et al., 2022, 2023) and 206 prefecture-level cities (Cheng et al., 2022) due to data unavailability. Remote sensing data can help address some of the limitations of data obtained from statistical yearbooks, surveys, and censuses by offering more comprehensive spatial coverage. This allows researchers to gather data from areas where traditional data sources might be insufficient or lacking.

Remote sensing data not only address these coverage gaps but also provide a valuable tool for monitoring and analyzing deforestation, air pollutants, climate change, and environmental conditions. This becomes particularly important when assessing the progress of resource-based cities in China towards achieving environment-related SDG indicators.

This chapter will explore the use of geospatial techniques in evaluating environment-related SDG indicators for resource-based cities in China. Through this analysis, it is possible to gain a deeper understanding of the environmental challenges faced by these cities and identify potential solutions to promote sustainable urban development. The application of geospatial techniques can provide valuable insights into deforestation, air pollutants, climate change, and other environmental issues that directly impact the sustainability of resource-based cities.

METHODS

City Categorization

This chapter presents an analysis of a total of 128 resource-based cities in China, which includes 125 prefecture-level administrative divisions (comprising prefecture-level cities, prefectures, autonomous prefectures, and leagues) and 3 provincial-level municipalities (Qianjiang, Dongfang, and Hami). The classification of these resource-based cities is based on the Sustainable Development Plan for Resource-based Cities in China (2013–2020), issued by the State Council of China. This plan provides a strategic framework for the sustainable development of cities that are primarily driven by resource extraction and processing industries, such as mining, forestry, and energy production.

The list of resource-based cities under investigation is provided in Table 4.1. These cities represent a diverse range of geographic locations, resource types, and development stages, offering valuable insights into the challenges and opportunities associated with the sustainable development of resource-based cities in China. The year 2018 is selected in the analysis of these resource-based cities.

Data Collection

Indicators and data sources of environmental-related SDG indicators are listed in Table 4.2. The collection of data for SDG indicators, such as SO_2 emission concentration and PM2.5 concentration, is essential for monitoring progress towards achieving environmental targets. Two widely recognized sources for such data are the Emissions Database for Global Atmospheric Research (EDGAR) and the Socioeconomic Data and Applications Center (SEDAC) (Table 4.2). For SDG 13.2 (CO_2 emission per capita), the emission data was collected from the ODIAC-Open-source Data Inventory for Anthropogenic CO_2 and population data was obtained from WorldPop Global Project Population Data (Table 4.2).

EDGAR, developed by the European Commission's Joint Research Centre (JRC) and the Netherlands Environmental Assessment Agency (PBL), is a comprehensive global database providing a consistent set of emission data for various air pollutants and greenhouse gases, including SO_2 and CO_2 emissions. This database covers emissions from multiple sectors, such as energy, industrial processes, and agriculture, which has been widely used in many studies about air pollutants (Friedlingstein et al., 2019; Janssens-Maenhout et al., 2019; Smith et al., 2011). SEDAC, operated by the Center for International Earth Science Information Network (CIESIN) at Columbia University, is another valuable resource for collecting data related to SDG indicators. SEDAC offers a wide range of datasets, including the Global Annual PM2.5 Grids, which provide estimates of global particulate matter concentrations (PM2.5) at a high spatial resolution. These datasets can be used to evaluate the air quality and health-related SDG indicators (Apte et al., 2018; Shaddick et al., 2018; Van Donkelaar et al., 2015).

The SDG indicator 15.1.1 focuses on tracking the change in forest share, which is a critical aspect of assessing progress towards the sustainable management of forests. To collect data on this indicator, this chapter used the MCD12Q1.061 MODIS Land Cover Type Yearly Global dataset (Table 4.2), provided by the NASA Land Processes Distributed Active Archive Center (LP DAAC) that has been widely used by many studies in forest field (Pritchard et al., 2018; Song et al., 2018; Zeng et al., 2018). This dataset is derived from the Moderate Resolution Imaging Spectroradiometer

TABLE 4.1
Scores of Environment-Related SDG Indicators for Resource-Based Cities in China

No	City	Environment-Related SDG scores	No	City	Environment-Related SDG scores
1	Tangshan	90.74	37	Yichun-HLJ	96.08
2	Handan	90.82	38	Qitaihe	93.73
3	Xingtai	92.76	39	Mudanjiang	96.31
4	Zhangjiakou	95.51	40	Heihe	96.64
5	Chengde	97.10	41	Daxing'anling	96.83
6	Datong	90.50	42	Xuzhou	88.21
7	Yangquan	91.90	43	Suqian	93.87
8	Changzhi	91.81	44	Huzhou	94.14
9	Jincheng	92.36	45	Huainan	86.88
10	Shuozhou	91.36	46	Maanshan	90.09
11	Jinzhong	92.57	47	Huaibei	90.90
12	Yuncheng	92.81	48	Tongling	87.84
13	Xinzhou	94.26	49	Chuzhou	94.94
14	Linfen	94.03	50	Suzhou-AH	93.91
15	Lvliang	94.72	51	Bozhou	94.64
16	Baotou	92.78	52	Chizhou	94.96
17	Wuhai	80.76	53	Xuancheng	95.91
18	Chifeng	96.60	54	Sanming	97.10
19	Erdos	87.86	55	Nanping	97.65
20	Hulunbuir	92.79	56	Longyan	96.39
21	Anshan	71.33	57	Jingdezhen	94.96
22	Fushun	94.42	58	Pingxiang	95.09
23	Benxi	95.88	59	Xinyu	93.64
24	Fuxin	95.52	60	Ganzhou	97.52
25	Panjin	92.81	61	Yichun-JX	95.87
26	Huludao	94.61	62	Zibo	89.54
27	Jilin	94.76	63	Zaozhuang	90.55
28	Liaoyuan	94.92	64	Dongying	91.38
29	Tonghua	96.15	65	Jining	89.90
30	Baishan	95.59	66	Tai'an	90.97
31	Songyuan	95.83	67	Linyi	93.39
32	Yanbian Korean Autonomous Prefecture	97.03	68	Luoyang	92.47
33	Jixi	96.01	69	Pingdingshan	90.83
34	Hegang	94.05	70	Hebi	90.17
35	Shuangyashan	94.08	71	Jiaozuo	90.11
36	Daqing	92.07	72	Puyang	91.58

TABLE 4.1 *(Continued)*

Scores of Environment-Related SDG Indicators for Resource-Based Cities in China

No	City	Environment-Related SDG scores	No	City	Environment-Related SDG scores
73	Sanmenxia	93.33	101	Qianxinan Buyei and Miao Autonomous Prefecture	96.77
74	Nanyang	94.74	102	Qiannan Buyei and Miao Autonomous Prefecture	97.18
75	Huangshi	64.64	103	Qujing	95.80
76	Ezhou	73.05	104	Baoshan	98.16
77	Qianjiang	94.11	105	Zhaotong	97.11
78	Hengyang	95.55	106	Lijiang	98.05
79	Shaoyang	96.14	107	Pu'er	98.24
80	Chenzhou	96.04	108	Lincang	98.19
81	Loudi	94.47	109	Chuxiong Yi Autonomous Prefecture	97.75
82	Shaoguan	95.95	110	Tongchuan	93.46
83	Yunfu	93.52	111	Baoji	95.59
84	Baise	96.93	112	Xianyang	94.11
85	Hezhou	96.22	113	Weinan	92.49
86	Hechi	97.83	114	Yan'an	93.79
87	Dongfang	93.74	115	Yulin-SX	91.41
88	Zigong	94.96	116	Jinchang	92.13
89	Panzhihua	96.11	117	Baiyin	92.98
90	Luzhou	95.34	118	Wuwei	97.42
91	Guangyuan	96.98	119	Zhangye	96.85
92	Nanchong	96.43	120	Pingliang	94.73
93	Guang'an	94.63	121	Qingyang	97.22
94	Dazhou	96.82	122	Longnan	98.13
95	Ya'an	97.23	123	Haixi Mongol and Tibetan Autonomous Prefecture	96.09
96	Ngawa Tibetan and Qiang Autonomous Prefecture	98.60	124	Shizuishan	88.45
97	Liangshan Yi Autonomous Prefecture	98.53	125	Karamay	91.19
98	Liupanshui	93.40	126	Hami	87.85
99	Anshun	96.46	127	Bayingolin Mongol Autonomous Prefecture	92.65
100	Bijie	96.31	128	Altay Prefecture	95.79
				Average	93.61

TABLE 4.2
Indicator List and Data Sources of Environmental-Related SDG Indicators

No	SDG	SDG Indicator	Unit	Data Source
1	SDG 3.9	SO_2 emissions	kg substance/m^2/s	EDGAR - Emissions Database for Global Atmospheric Research
2	SDG 11.6	PM2.5	kg substance/m^2/s	Socioeconomic data and applications center (SEDAC)
4	SDG 13.2	CO_2 emission per capita	kg/person	ODIAC-Open-source Data Inventory for Anthropogenic CO_2; WorldPop Global Project Population Data: Estimated Residential Population per 100x100m Grid Square
5	SDG 15.1	Change in forest share from 2001 to 2018	kg substance/m^2/s	MCD12Q1.061 MODIS Land Cover Type Yearly Global

(MODIS) sensor aboard NASA's Terra and Aqua satellites and offers global land cover classifications at a 500-meter spatial resolution. These datasets contain multiple land cover classification schemes, including the International Geosphere-Biosphere Programme (IGBP) classification system, which provides information on various forest types such as evergreen, deciduous, and mixed forests. By comparing the forest area data for the years 2001 and 2018, this chapter calculated the change in forest share over this period and assessed progress towards the SDG target of promoting sustainable forest management.

Multi-criteria Decision Making

In accordance with Rezaei (2015), this chapter applies the best-worst multi-criteria decision-making method to determine the weight of each SDG indicator. As a contrast to several other multi-criteria decision-making approaches, the best-worst multi-criteria decision-making method is a comparison-based MCDM technique that contrasts the best and worst criteria with other criteria. Due to its ability to produce dependable outcomes with minimal data requirements (Rezaei, 2016), the best-worst multi-criteria decision-making method has gained popularity across numerous studies involving multi-criteria decision-making in areas such as supply chain management, the airline industry, and research and development performance (Badri Ahmadi et al., 2017; Gupta, 2018; Salimi & Rezaei, 2018). The best-worst multi-criteria decision-making method can be outlined in the following steps (Rezaei, 2016; Rezaei et al., 2015):

1. Choose a set of decision criteria, represented as $\{C_1, C_2, C_3, \ldots, C_n\}$.
2. Identify the best and worst criteria among all options, as determined by the decision-makers.

3. Establish the preference of the best criteria over other alternatives, which is conveyed as a score between one and nine. A score of one signifies equal preference between the best criteria and an alternative, while a score of nine implies a strong preference of the best criterion over an alternative. These preferences can be represented as a Best-to-Others vector, $A_B = (a_{B1}, \ a_{B2}, \ a_{B3}, \ldots, \ a_{Bn})$. a_{Bj} denotes the preference of the best criterion B over alternative j.
4. Ascertain the preference of all alternatives over the worst criteria, also expressed as a score ranging from one to nine. These preferences can be depicted as Others-to-Worst, $A_w = (a_{1w}, \ a_{2w}, \ a_{3w}, \ldots, \ a_{nW})^T$. a_{jW} signifies the preference of alternative j over the worst criterion W.
5. Compute the optimal weights, $(w_1^*, \ w_2^*, \ w_3^*, \ldots, \ w_n^*)$. The weight solution must adhere to specific conditions, namely $w_{B/W_j} = a_{Bj}$ and $w_{j/W_W} = a_{jw}$ for all alternatives. Consequently, the maximum of $\{|w_B - a_{Bj}w_j|, \ |w_j - a_{jw}w_w|\}$ must be minimized. The objective function and constraints can be expressed as follows:

$$\begin{aligned} & \min max_j \{|w_B - a_{Bj}w_j|, \ |w_j - a_{jw}w_w|\} \\ & \text{s.t. } \textstyle\sum_j w_j = 1 \\ & w_j \geq 0, \ \ for \ \ all \ \ j \end{aligned} \tag{4.1}$$

Since model (4.1) is not a linear problem, it can be reformulated as follows:

$$\begin{aligned} & \min \xi^L \\ & \text{s.t. } |w_B - a_{Bj}w_j| \leq \xi^L, \ \ for \ \ all \ \ j \\ & |w_j - a_{jw}w_w| \leq \xi^L, \ \ for \ \ all \ \ j \\ & \textstyle\sum_j w_j = 1 \\ & w_j \geq 0, \ \ for \ \ all \ \ j \end{aligned} \tag{4.2}$$

By solving model (4.2), the weight $(w_1^*, \ w_2^*, \ w_3^*, \ldots, \ w_n^*)$ and ξ^{L*} can be obtained. ξ^{L*} serves as a measure of the solution's robustness. As ξ^{L*} approaches to zero, the derived weight becomes more reliable.

Next, the SDG indicators are standardized using an improved min-max method, which prevents the occurrence of zero values (Zhu & Wei, 2015). For benefit-type SDG indicators, the standardized indicator can be calculated as follows:

$$C_j^S = 0.9 \times 100 \times \frac{x_{i,j} - \min\limits_j(x_{i,j})}{\max\limits_j(x_{i,j}) - \min\limits_j(x_{i,j})} + 10 \tag{4.3}$$

C_j^S represents the value of indicator j after standardization. For cost-type indicators, the standardized indicator can be determined using this formula (Zhu & Wei, 2015):

$$C_j^S = 0.9 \times 100 \times \frac{\max_j(x_{i,j}) - x_{i,j}}{\max_j(x_{i,j}) - \min_j(x_{i,j})} + 10 \tag{4.4}$$

Utilizing the standardized indicators obtained above, the sustainability of environment-related SDG indicators can be assessed as follows:

$$S = \sum_{j=1}^{J} w_j^* C_j^S \tag{4.5}$$

RESULTS

Weight of SDG Indicators

In this chapter, the Best Worst Method (BWM) is employed to assess the relative importance of five SDG indicators. This methodology enables a robust and efficient evaluation of the weights associated with each indicator, allowing for a better understanding of their respective contributions to the overall sustainable development assessment. The weights of the five SDG indicators can be derived from Table 4.3. The results reveal that the weights for SDG 3.9, SDG 11.6, SDG 12.4, SDG 13.2, and SDG 15.1 were 0.0472, 0.0849, 0.1698, 0.2547, and 0.4433, respectively. These values indicate the relative significance of each indicator in the

TABLE 4.3
Best-to-Others and Others-to-Worst Vectors, the Weights, and the Consistency Test Using Best-Worst Method

Criteria Number = 5	Criterion 1	Criterion 2	Criterion 3	Criterion 4	Criterion 5
Names of Criteria	SDG 3.9	SDG 11.6	SDG 12.4	SDG 13.2	SDG 15.1
Select the Best	SDG 15.1				
Select the Worst	SDG 3.9				
Best to Others	SDG 3.9	SDG 11.6	SDG 12.4	SDG 13.2	SDG 15.1
SDG 15.1	8	6	3	2	1
Others to the Worst	SDG 3.9				
SDG 3.9	1				
SDG 11.6	3				
SDG 12.4	5				
SDG 13.2	6				
SDG 15.1	8				
Weights	SDG 3.9	SDG 11.6	SDG 12.4	SDG 13.2	SDG 15.1
	0.0472	0.0849	0.1698	0.2547	0.4433
Ksi*	0.0660				

context of sustainable development, with higher weights corresponding to greater importance. Furthermore, the consistency ratio (CR) of the evaluation was calculated to be 0.0660, which was below 0.1 (Table 4.3). This implies that the weighting results are consistent and reliable, providing a solid foundation for further analysis and decision-making.

In summary, the application of the Best Worst Method in this chapter allows for a comprehensive and reliable assessment of the weights of different SDG indicators. This information can be instrumental in guiding policy decisions and resource allocation to effectively address the diverse aspects of sustainable development.

Score of Each Environment-Related SDG

Figure 4.1 presents the scores for each environment-related SDG indicator of the top 10 best resource-based performers in 2018. These top performers include Ngawa Tibetan and Qiang Autonomous Prefecture, Liangshan Yi Autonomous Prefecture, Pu'er, Lincang, Baoshan, Longnan, Lijiang, Hechi, Chuxiong Yi Autonomous Prefecture, and Nanping (Figure 4.1). Their overall scores in 2018

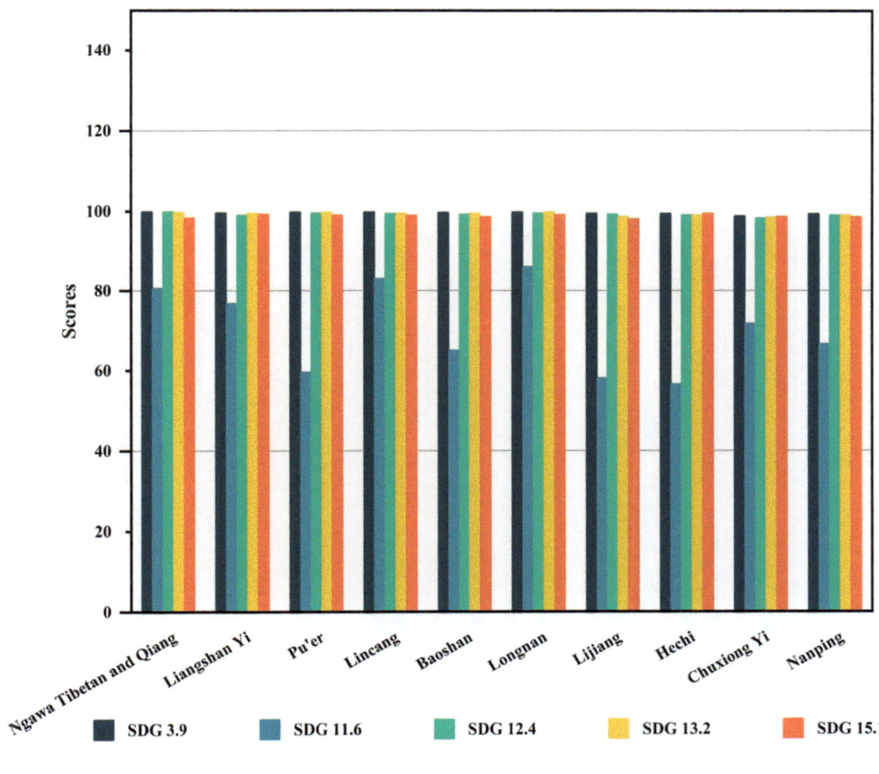

FIGURE 4.1 Scores of each environment-related SDG indicator of top 10 best performer in 2018.

range from 97.65 to 98.60, showcasing a high level of achievement in terms of sustainable development (Figure 4.1). A closer look at Ngawa Tibetan and Qiang Autonomous Prefecture, the best performer, reveals that its scores for SDG 3.9, SDG 11.6, SDG 12.4, SDG 13.2, and SDG 15.1 are 99.78, 80.50, 99.80, 99.54, and 98.18, respectively (Figure 4.1). These impressive scores exemplify the significant strides made by the region in addressing environmental challenges and promoting sustainable development.

On the other end of the spectrum, the bottom 10 resource-based performers in 2018 include Shizuishan, Xuzhou, Erdos, Hami, Tongling, Huainan, Wuhai, Ezhou, Anshan, and Huangshi (Figure 4.2). Their overall scores range from a low of 64.64 to a high of 88.45 (Figure 4.2). Huangshi, the worst performer among these, has an overall score of 64.64, with its scores for SDG 3.9, SDG 11.6, SDG 12.4, SDG 13.2, and SDG 15.1 being 25.04, 71.54, 26.71, 39.69, and 96.72, respectively (Figure 4.2).

These contrasting results highlight the disparities in environmental performance among various resource-based cities. The top performers serve as models for other cities to emulate, while the underperforming cities require targeted interventions and support to improve their environmental outcomes. By examining these scores,

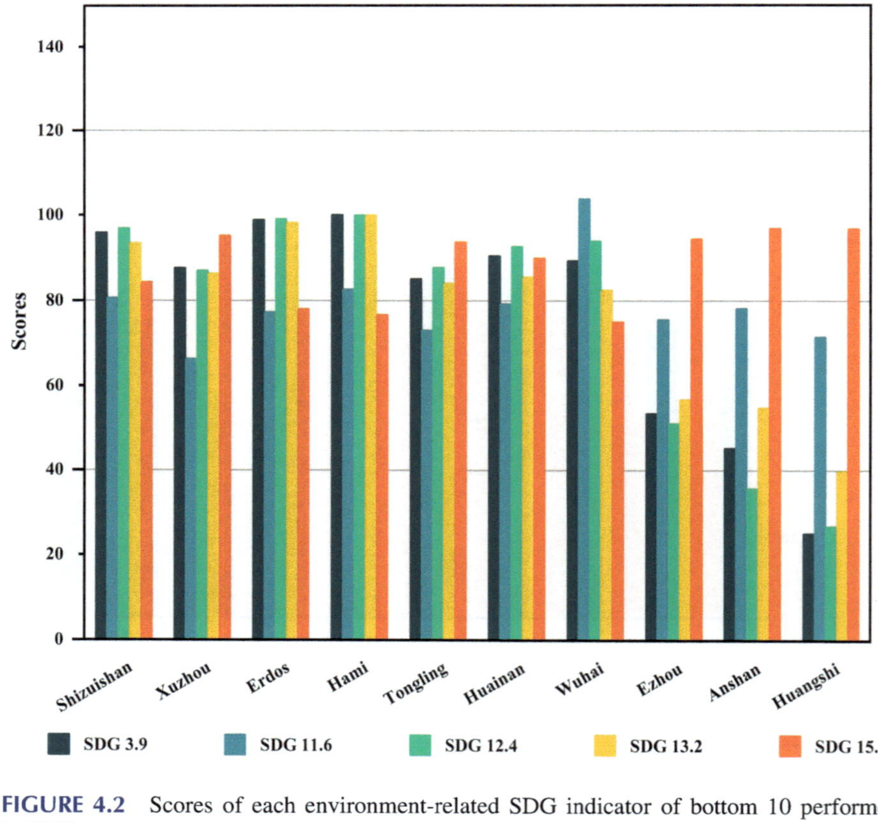

FIGURE 4.2 Scores of each environment-related SDG indicator of bottom 10 performer in 2018.

policymakers and stakeholders can identify areas of strength and weakness, informing the development of tailored strategies to address the unique challenges faced by each city in achieving sustainable development.

Overall Score of Environment-Related SDGs

Figure 4.3 displays the overall scores of environment-related SDG indicators for 128 resource-based cities, which are distinguished by gridlines on the map in Figure 4.3. The results in Figure 4.3 reveal that the average overall score for environment-related SDG indicators in resource-based cities is 93.61, which is lower than the average score of 94.19 for non-resource–based cities. This indicates that resource-based cities, on average, have lower performance in addressing environmental challenges and achieving sustainable development compared to non-resource–based cities.

Upon examining the scores for individual SDG indicators, it is evident that resource-based cities have lower scores for SDG 11.6 and SDG 15.1, with values of 73.18 and 94.88, respectively (Table 4.4). These scores are lower than those of non-resource-based cities, which stand at 75.71 and 96.49, respectively (Table 4.4). This suggests that resource-based cities generally face greater challenges in addressing air pollution and land degradation. Conversely, resource-based cities outperform non-resource–based cities in SDG 3.9, SDG 12.4, and SDG 13.2, with scores of 95.81, 95.39, and 95.31, respectively (Table 4.4). These scores are higher than those

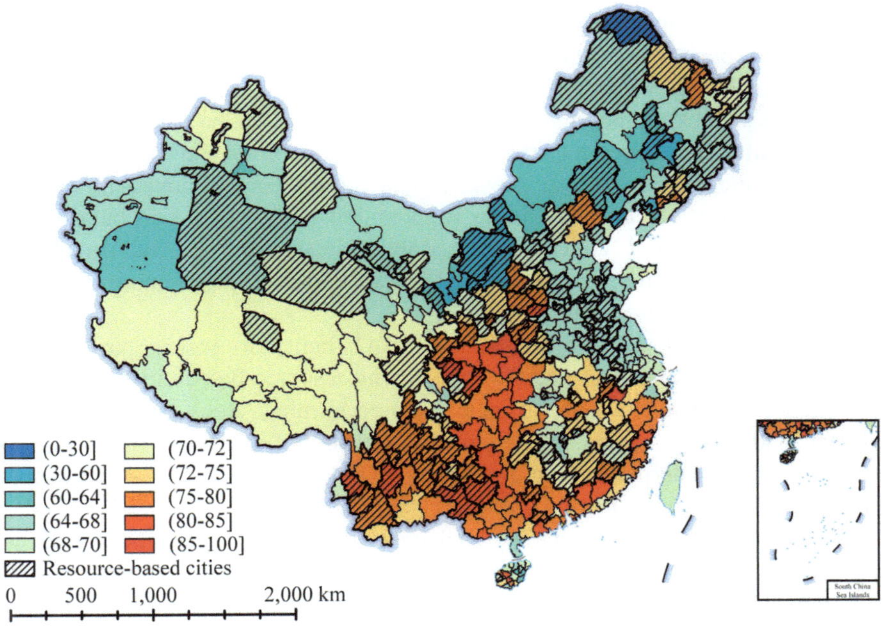

FIGURE 4.3 The overall scores of environment-related SDG indicators of resource-based and non-resource–based cities of China in 2018.

TABLE 4.4
Comparison between Resource-Based and Non-Resource–Based Cities Regarding Environment-Related SDG Indicators in 2018

	Overall Scores	SDG 3.9	SDG 11.6	SDG 12.4	SDG 13.2	SDG 15.1
Resource-based cities	93.61	95.81	73.18	95.39	95.31	94.88
Non-resource–based	94.19	95.05	75.71	94.90	93.92	96.49

of non-resource–based cities, which have scores of 95.05, 94.90, and 93.92, respectively (Table 4.4).

In summary, the comparison of environment-related SDG indicator scores between resource-based and non-resource–based cities reveals a mixed performance. While resource-based cities lag behind in some areas, they excel in others. Policymakers and stakeholders must recognize these differences to develop targeted strategies and interventions that address the specific challenges and opportunities faced by resource-based cities in their pursuit of sustainable development.

CONCLUSION

This chapter has explored the challenges faced by resource-based cities in China for achieving the SDGs and the potential of geospatial techniques in evaluating environment-related SDG indicators for these cities. Resource-based cities have played a significant role in China's economic development, providing valuable natural resources that have driven the country's growth. However, the exploitation of these resources has also led to complex environmental challenges that threaten their sustainability and make it difficult to achieve the SDGs. Remote sensing data have been employed to evaluate the sustainability of cities in China. Remote sensing data, in particular, offers comprehensive spatial coverage that can address gaps in traditional data sources, enabling a more accurate evaluation of the environmental challenges faced by resource-based cities.

The analysis of environment-related SDG indicator scores has revealed disparities in environmental performance among resource-based cities, with some cities excelling while others struggle. Top-performing cities, such as Ngawa Tibetan and Qiang Autonomous Prefecture, serve as models for other cities to emulate, while underperforming cities require targeted interventions and support to improve their environmental outcomes.

A comparison between resource-based and non-resource–based cities has shown a mixed performance, with resource-based cities lagging behind in some environment-related SDGs but outperforming in others. Policymakers and stakeholders must recognize these differences to develop targeted strategies and interventions that address the specific challenges and opportunities faced by resource-based cities in their pursuit of sustainable development.

Finally, the potential of remote sensing techniques in evaluating other SDGs in the future has been discussed. As the availability and quality of remote sensing data continue to improve, these techniques can be applied to monitor progress towards various SDGs, such as those related to climate change, agriculture, and water resources management. By leveraging the power of remote sensing, researchers and policymakers can make more informed decisions and develop effective strategies to address the complex challenges of sustainable development in resource-based cities.

There are some limitations of the application of the best-worst method. This method can be influenced by individual biases and subjectivity, given its dependence on expert opinion to assign importance to different criteria. The method becomes more complicated and labor-intensive with a rise in the number of criteria, due to the surge in pairwise comparisons needed.

REFERENCES

Apte, J. S., Brauer, M., Cohen, A. J., Ezzati, M., & Pope III, C. A. (2018). Ambient PM2. 5 reduces global and regional life expectancy. *Environmental Science & Technology Letters*, 5(9), 546–551.

Badri Ahmadi, H., Kusi-Sarpong, S., & Rezaei, J. (2017). Assessing the social sustainability of supply chains using Best Worst Method. *Resources, Conservation and Recycling*, 126, 99–106.

Cheng, D., Xue, Q., Hubacek, K., Fan, J., Shan, Y., Zhou, Y., Coffman, D., Managi, S., & Zhang, X. (2022). Inclusive wealth index measuring sustainable development potentials for Chinese cities. *Global Environmental Change*, 72, 102417.

Dong, S., Li, Z., Li, B., & Xue, M. (2007). Problems and Strategies of Industrial Transformation of China's Resource-based Cities. China Population. *Resources and Environment*, 17(5), 12–17.

Friedlingstein, P., Jones, M. W., O'Sullivan, M., Andrew, R. M., Hauck, J., Peters, G. P., Peters, W., Pongratz, J., Sitch, S., Le Quéré, C., Bakker, D. C. E., Canadell, J. G., Ciais, P., Jackson, R. B., Anthoni, P., Barbero, L., Bastos, A., Bastrikov, V., Becker, M., … Zaehle, S. (2019). Global Carbon Budget 2019. *Earth System Science Data*, 11(4), 1783–1838.

Gupta, H. (2018). Evaluating service quality of airline industry using hybrid best worst method and VIKOR. *Journal of Air Transport Management*, 68, 35–47.

Janssens-Maenhout, G., Crippa, M., Guizzardi, D., Muntean, M., Schaaf, E., Dentener, F., Bergamaschi, P., Pagliari, V., Olivier, J. G. J., Peters, J. A. H. W., van Aardenne, J. A., Monni, S., Doering, U., Petrescu, A. M. R., Solazzo, E., & Oreggioni, G. D. (2019). EDGAR v4.3.2 Global Atlas of the three major greenhouse gas emissions for the period 1970–2012. *Earth System Science Data*, 11(3), 959–1002.

Lafortune, G., Zoeteman, K., Fuller, G., Mulder, R., Dagevos, J., & Schmidt-Traub, G. (2019). 2019 European cities SDG index and dashboards report. *Sustainable Development Solutions Network (SDSN).*

Li, H., Long, R., & Chen, H. (2013). Economic transition policies in Chinese resource-based cities: An overview of government efforts. *Energy Policy*, 55, 251–260.

Liu, Y., Du, J., Wang, Y., Cui, X., Dong, J., Hao, Y., Xue, K., Duan, H., Xia, A., Hu, Y., Dong, Z., Wu, B., Zhao, X., & Fu, B. (2021). Evenness is important in assessing progress towards sustainable development goals. *National Science Review*, 8(8), nwaa238.

Ma, Y., Yan, J., Sha, J., He, G., Song, C., Fan, S., & Ke, W. (2018). Dynamic simulation of the atmospheric environment improved by a focus on clean energy utilization of resource-based cities in China. *Journal of Cleaner Production*, 192, 396–410.

Pritchard, R., Ryan, C. M., Grundy, I., & van Der Horst, D. (2018). Human appropriation of net primary productivity and rural livelihoods: Findings from six villages in Zimbabwe. *Ecological Economics*, 146, 115–124.

Rezaei, J. (2015). Best-worst multi-criteria decision-making method. *Omega*, 53, 49–57.

Rezaei, J. (2016). Best-worst multi-criteria decision-making method: Some properties and a linear model. *Omega*, 64, 126–130.

Rezaei, J., Wang, J., & Tavasszy, L. (2015). Linking supplier development to supplier segmentation using Best Worst Method. *Expert Systems with Applications*, 42(23), 9152–9164.

Salimi, N., & Rezaei, J. (2018). Evaluating firms' R&D performance using best worst method. *Evaluation and Program Planning*, 66, 147–155.

Schmidt-Traub, G., Kroll, C., Teksoz, K., Durand-Delacre, D., & Sachs, J. D. (2017). National baselines for the Sustainable Development Goals assessed in the SDG Index and Dashboards. *Nature Geoscience*, 10(8), 547–555.

Shaddick, G., Thomas, M. L., Green, A., Brauer, M., Donkelaar, A., Burnett, R., Chang, H. H., Cohen, A., Dingenen, R. V., & Dora, C. (2018). Data integration model for air quality: A hierarchical approach to the global estimation of exposures to ambient air pollution. *Journal of the Royal Statistical Society Series C: Applied Statistics*, 67(1), 231–253.

Smith, S. J., van Aardenne, J., Klimont, Z., Andres, R. J., Volke, A., & Delgado Arias, S. (2011). Anthropogenic sulfur dioxide emissions: 1850–2005. *Atmospheric Chemistry and Physics*, 11(3), 1101–1116.

Song, X.-P., Hansen, M. C., Stehman, S. V., Potapov, P. V., Tyukavina, A., Vermote, E. F., & Townshend, J. R. (2018). Global land change from 1982 to 2016. *Nature*, 560(7720), Article 7720.

Tan, M., Zhao, H., Li, G., & Qu, J. (2020). Assessment of potentially toxic pollutants and urban livability in a typical resource-based city, China. *Environmental Science and Pollution Research*, 27(15), 18640–18649.

The Chinese Government. (2013). The Sustainable Development Plan for Resources-Based Cities in China (2013–2020) [File Number: State Council, 2013/45]. http://www.gov.cn/zwgk/2013-12/03/content_2540070.htm

United Nations. (2015). Transforming our world: The 2030 Agenda for sustainable development.

Van Donkelaar, A., Martin, R. V., Brauer, M., & Boys, B. L. (2015). Use of satellite observations for long-term exposure assessment of global concentrations of fine particulate matter. *Environmental Health Perspectives*, 123(2), 135–143.

Wiedmann, T., & Allen, C. (2021). City footprints and SDGs provide untapped potential for assessing city sustainability. *Nature Communications*, 12(1), 1–10.

Xiao, H., Bao, S., Ren, J., & Xu, Z. (2023). Transboundary impacts on SDG progress across Chinese cities: A spatial econometric analysis. *Sustainable Cities and Society*, 92, 104496.

Xiao, H., Xu, Z., Ren, J., Zhou, Y., Lin, R., Bao, S., Zhang, L., Lu, S., Lee, C. K. M., & Liu, J. (2022). Navigating Chinese cities to achieve sustainable development goals by 2030. *The Innovation*, 3(5), 100288.

Zeng, Z., Estes, L., Ziegler, A. D., Chen, A., Searchinger, T., Hua, F., Guan, K., Jintrawet, A., & Wood, E. F. (2018). Highland cropland expansion and forest loss in Southeast Asia in the twenty-first century. *Nature Geoscience*, 11(8), 556–562.

Zhu, X., & Wei, G. (2015). Discussion on excellent standard of dimensionless method in entropy method. *Statistics & Decision*, 2, 12–15.

5 Urban Green Space Maps Based on GeoAI

Yang Chen

INTRODUCTION: BACKGROUND AND DRIVING FORCES

Urban green spaces (UGS), such as rivers, green lands, and wetlands, play a key role in the water cycle (Abbott et al., 2019), carbon cycle (Yue et al., 2020), and climate change (Li et al., 2020). In addition, green spaces in cities provide ecosystems services (Tavares et al., 2019) and health benefits for residents (Dai et al., 2018; Selway et al., 2020; Astell-Burt et al., 2020). Mapping of UGS is an important task for environmental conservation and management, urban planning, and urban ecological monitoring.

The traditional UGS mapping employs field measurements (Voltersen et al., 2014), but this method is time-consuming. Furthermore, UGS may change over time rapidly and therefore it is necessary to update UGS maps frequently (Kuang and Dou, 2020). With the development of moderate-high and very high-resolution satellite sensors (such as, Sentinel-2, Ziyuan-3, and WorldView-4), a great number of remote sensing images can be acquired, which provides rapid and accurate information over urban areas (Xu and Somers, 2021).

The UGS maps are frequently generated by using image classification methods. The image classification methods may include 1) unsupervised classification method (Lee et al., 1999; Liang et al., 2018), 2) supervised classification method (Wang, 1990), and 3) deep learning method (Chen et al., 2014). The unsupervised classification methods such as K-means and ISODATA are used for urban land-cover mapping, but the accuracy of UGS mapping is lower due to insufficient use of spectral or spatial information (Zhang and Kerekes, 2011). Supervised classification methods such as support vector machine (SVM), random forest, and neural networks (NN) are widely used to extract UGS (i.e., water, vegetation, and forest) (Kranjčić et al., 2019; Di et al., 2019; Ma et al., 2014). The SVM method is used to extract UGS from Sentinel-2A multispectral imagery (Kranjčić et al., 2019). Di et al. proposed the random forest method based on the object-based image analysis (OBIA) technique to map UGS (Di et al., 2019). The neural networks method based on OBIA technique is used for urban mapping from unmanned aerial vehicle (UAV) imagery (Ma et al., 2014; Teodoro and Araujo, 2016). However, the above classification methods use low-level features, such as spectral, textural, geometrical, and contextual features, to extract UGS that may lead to serious extract errors. In addition, building shadow, artificial grassland, and farmland can all confuse with urban green spaces.

DOI: 10.1201/9781003244561-7

In recent years, deep learning methods are widely used for recognizing objects, image classification, semantic segmentation, and so on (Reichstein et al., 2019; Chen et al., 2021). The convolutional neural network (CNN) is an advanced deep learning algorithm, which can automatically extract high-level features from input data by some convolution and pooling operations (LeCun et al., 2015). The CNN was widely used for urban land-use and green space mapping (Zhang et al., 2018; Fu and Qu., 2018; Liu et al., 2019; Xu et al., 2020). Zhang et al (2018) developed an object-based convolutional neural network (OCNN) for urban land-use mapping. It is found that the performance of the OCNN method relies on segmented objects. The full convolutional network (FCN) is used for high-resolution remote sensing image classification (Fu and Qu., 2018). They found the accuracy of FCN method is higher than 85%. Liu et al (2019) used that DeepLab v3plus semantic segmentation model for UGS mapping. Huang et al (2018) developed a semi-transfer deep convolutional neural network (STDCNN) method for urban land-use mapping from high spatial resolution remote sensing data. Xu et al (2020) proposed a method for combining deep learning network-based classification with vegetation phenological features for the UGS mapping from Gaofen-2 images. The overall accuracy of deep learning network yielded higher than 93%. The result of this study suggests deep learning methods have achieved excellent classification in remote sensing data. Although the above deep learning methods can get more high accuracy classification, their direct application for UGS mapping can lead to the blurring of UGS boundaries due to the stride convolution and pooling operations. In addition, good quality training samples for UGS mapping can be time-consuming and costly to obtain in spite of the need.

Previous studies further suggest that the prior information of class events can be beneficial in remote sensing image classification to improve mapping accuracy (Foody,1995; Zhao et al., 2019; Fonte et al., 2020; Du et al., 2020). Crowdsourced geospatial datasets can provide prior information of useful for urban land-cover (LULC) mapping (Johnson and Iizuka, 2016). The OpenStreetMap (OSM) platform was one of the most widely used sources of crowdsourced geographic datasets, which can provide freely available the prior information of remote sensing image mapping for the whole globe. The OSM data consists of polygon data and point of interest (POI) data. The OSM polygon data provide prior knowledge on urban surface objects, while the POI data provide geographic information (such as, position, place, category, and distribution) for urban land-use mapping accuracy (Yao et al., 2017). In addition, the OSM data and remote sensing images had been integrated to improve the accuracy of urban land mapping because OSM data can provide auxiliary and even crucial information for satellite image classification (Wan et al., 2017; Luo et al., 2019). Although many research works have been conducted to utilized the OSM data in geographic information mining and spatial data infrastructure (Elwood et al., 2012), a few have drawn attention to the potential use of OSM information for UGS mapping with remote sensing imagery to improve the accuracy of urban mapping (Arsanjani et al., 2013; Liu et al., 2020; Luo et al., 2019).

Another common limitation is that while existing deep learning methods can be used for UGS mapping in small areas (small than city scale), few deep learning methods can be extended for mapping at the city scale. Such as, a deep belief network (DBN) is used to generate UGS mapping with an area of about16 km2 (Zou et al., 2017). Hartling et al (2019) used a dense convolutional network (DenseNet) model for UGS mapping with an area of about 523 ha (the western part of the city of St. Louis, USA). The accuracy of DenseNet method yielded higher than 82.6%.

To overcome these limitations, this study designs an automatic urban green space mapping GeoAI method that integrates remote sensing image with crowd-sourced OpenStreetMap (OSM) data, aiming to improve the accuracy of the UGS mapping. The remainder of this chapter is organized as follows: second section describes the proposed GeoAI method workflow and the implementation details; third section introduces the study area and datasets; fourth section analysis discusses the importance of the multi-scale feature extraction module, boundary enhancement module, and OSM data for UGS mapping.

GEOAI METHOD

Figure 5.1 describes the proposed methodology of this research, which consists of three components: 1) multi-scale feature extraction module, 2) multi-modal information fuse module, and 3) boundary enhancement module. The multi-scale feature extraction module is proposed to extract the multi-scale semantic information of UGS from remote sensing images and OSM data. Then, the multi-modal information fuse module is used to integrate prior information (OSM) and

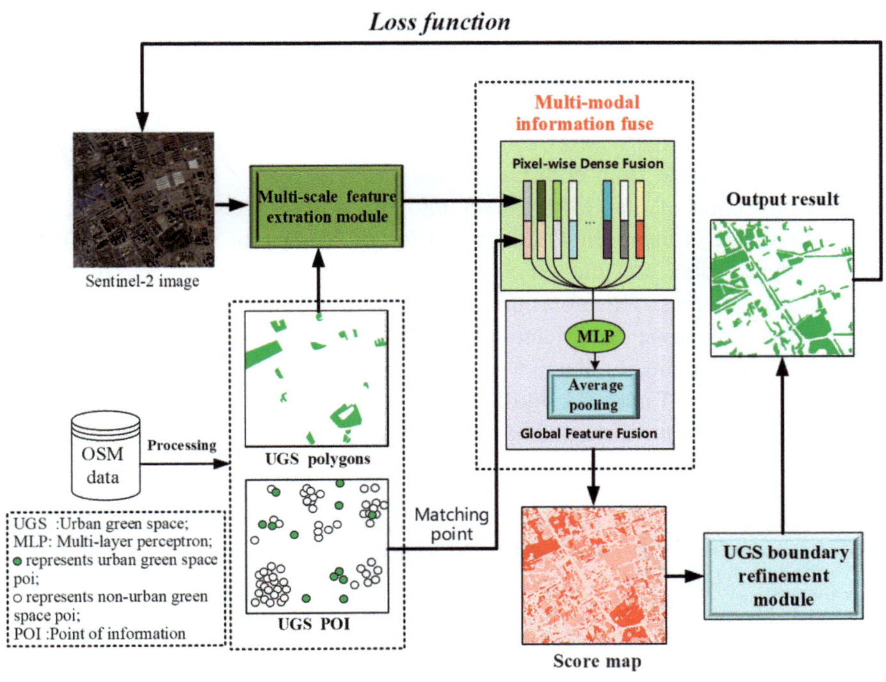

FIGURE 5.1 The workflow of GeoAI method.

spectral-spatial information, and to enhance UGS features. We use the boundary enhancement module to clear up the blurring of UGS boundaries caused by stride convolution and pooling operations.

Data Preprocessing

The workflow OSM preprocessing are described here. The OSM data was created by volunteers, including polygon data (urban wetland, residential green space, and natural reserve, etc.) and point of interest (POI) data (theme park, supermarket, hotel, etc.) (Haklay and Weber., 2008; Zhao et al., 2019). In this study, UGS refer to open-space areas reserved for parks and other "green spaces", including plant life, water and other kinds of natural environment (Herzele and Wiedemann, 2003).

Because the OSM polygon data and UGS polygon data have different classification nomenclatures, a polygon classification rule was developed to associate the datasets. In addition, we used the "Landuse", "natural", and "water" polygon data from the OSM polygon data as testing samples. the POI data provide rich geographic information, such as position, place, category, distribution, etc. However, the OSM POI classes did not correspond with UGS point classes, so a point of interest classification rule was established. The OSM classes were converted into the UGS classes, which consists of the following steps:

i. The OSM polygon topological inconsistency processing: Since the UGS classes can contain multiple OSM classes, such as urban wetland, residential green space, and natural reserve belong to urban green spaces; therefore, the spatial topology rule was established in ArcGIS to remove the inconsistency of the OSM polygon topological. In case of a single layer, the topological check was conducted to remove any intersected polygons; in the case of two layers, the overlapping parts of two layers were extracted from both layers (Liu et al., 2020).
ii. Conversion of the OSM classes into the UGS classes: There existed a challenge in using the OSM dataset for UGS mapping because the OSM polygon data and the UGS polygon data have different classification nomenclatures. Therefore, the OSM classes were converted into the UGS classes by some classification rules. In each class, the subclasses (OSM classes) that did not match with were removed. The classification rules are described in Tables 5.1 and 5.2.
iii. UGS classes were created based on OSM, UGS polygons and UGS POIs.

TABLE 5.1
UGS Classes Derived from the OSM "Landuse"

UGS Classes	Corresponding OSM "Landuse"
UGS polygons	water, reservoir, pond, basin, grass meadow, urban forest, wood, green belt, urban wetland, residential green space, and natural reserve.
NonUGS polygons	residential, building, roads, industrial, farm, farmyard, retail, and bare land, etc.

TABLE 5.2
UGS Point Classes Derived from the OSM "POI"

UGS Point Classes	Corresponding OSM "POI"
UGS points	pond, rivers, theme park, wetland park, outdoor swimming pool, botanical garden, and forest park.
NonUGS points	bicycle parking, apartment, bank, bar, bus stop, crossing, restaurant, supermarket, and hotel, etc.

Multi-scale Feature Extraction Module

Multiscale semantic information, such as the wetland area around the lakes and roadside green belts are located on both sides of a road, is widely used in remote sensing image classification (Peng et al., 2019). This multi-scale semantic information is key for UGS mapping tasks. However, it is not easy to extract the multi-scale semantic information of UGS when traditional machine learning methods were employed. Considering the work developed by multi-scale convolutional architecture (Eigen and Fergus., 2015), we built a multi-scale feature extraction module to extract multi-scale semantic information from remote sensing images and OSM data.

The multi-scale feature extraction module consists of convolutional layer and pyramid mix-pooling layer. The overall structure of the module is shown in Figure 5.2.

Previous research shows that a smaller size convolutional kernel is better for extracting local semantic information, while a larger size convolutional kernel can extract global semantic information (Peng et al., 2017; Ji et al., 2018). Therefore, convolution kernels of different sizes were used to extract multi-scale semantic information in the multi-scale feature extraction module. Each convolution layer is linked by a rectified linear unit (ReLU). The convolutional kernel is defined by Equation (5.1).

$$F^k = f\left(\sum_{n=1}^{N} (\mathrm{W}^{kn} \otimes F^n) + \mathrm{b}^{(k)}\right) \tag{5.1}$$

where $f(\cdot)$ is rectified linear unit (ReLU) function, N is the number of this module feature maps, and k represents the convolutional kernel sizes of 1, 3, and 5. F^n and W^{kn} are the learnable parameters in the multi-scale feature extraction module. $\mathrm{b}^{(k)}$ is the convolutional bias vector.

Some research suggest that the pooling operation can avoid over-fitting and reduce the function of feature dimensionality (Studer et al., 2020; Zheng et al., 2019; Sun et al., 2017). The multi-scale feature extraction module uses convolution kernels of different sizes lead to generate feature mapping of different scales. However, traditional pooling operation requires a fixed-size input, usually includes max pooling, min pooling, and the average pooling, but traditional pooling methods are no longer practical for the feature mapping of different scales.

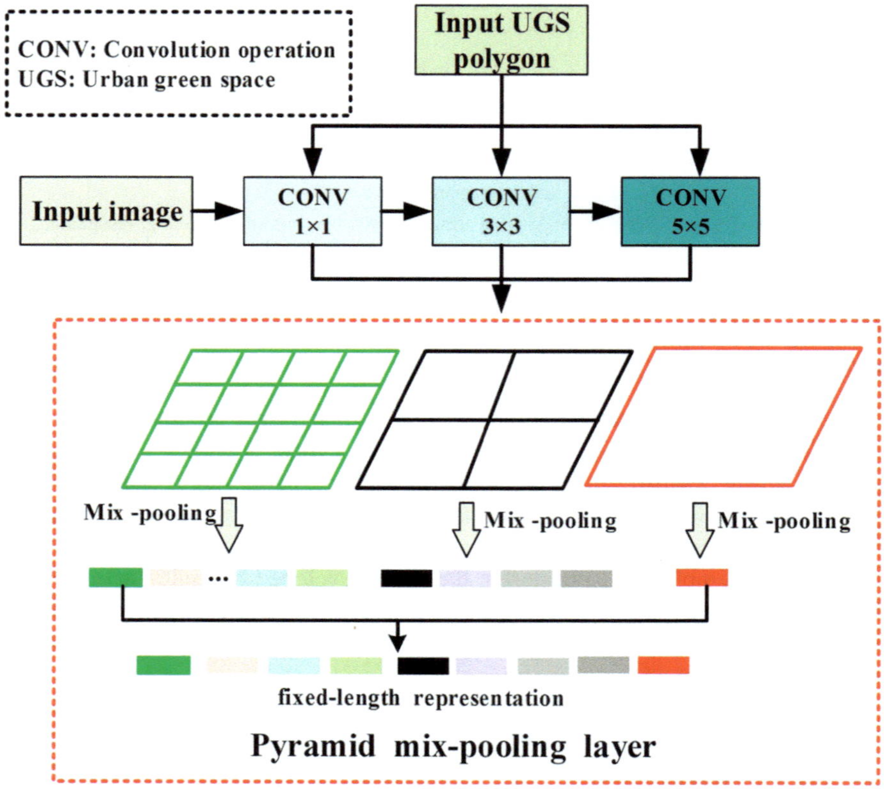

FIGURE 5.2 The architecture of the multi-scale feature extraction module.

The spatial pyramid pooling (SPP) layer cannot require a fixed-size input, which can generate fixed-size vectors as the output (He et al., 2015). In each spatial bin, max pooling is used to get the responses of each filter. However, some existing research showed that max pooling could lead to the loss of a many details (Chen et al., 2018a). To solve the above limit, we utilized a spatial pyramid pooling (SPP) layer by replacing max pooling with novel mix-pooling (He et al., 2015).

Some typical pooling strategies include min pooling, average pooling, and max pooling (Zheng et al., 2019; He et al., 2015; Scherer at al., 2010). If pooling operation is done with the average pooling or min pooling, the high-intensity features are weakened (Zheng et al., 2019; He et al., 2015). If pooling operation is done with max pooling, the low-intensity features are weakened (Scherer at al., 2010). Inspired by a hybrid pooling idea (Tong and Tanaka, 2019), this study presents a novel mix-pooling layer based on the max pooling, min pooling, and the average pooling to reduce the loss of image features during the process of pooling. The novel mix-pooling can be defined by Equation (5.2).

$$Pooling_{mix} = \alpha Pooling_{max} + \beta Pooling_{min} + (1 - \alpha - \beta) Pooling_{ave} \quad (5.2)$$

where Pooling max is the max pooling layer, Pooling min is the min pooling layer, Poolingave is the average pooling layer. α and β are the pooling factors.

The function of α and β are to dynamically optimize the traditional pooling based on different feature map areas. Inspired by formulating the pooling distribution idea (Liu et al., 2016), α and β are mathematically expressed by Equation (5.3), and Equation (5.4), respectively.

$$\alpha = \frac{K_{max}}{K}, \quad \alpha \in (0, 1) \tag{5.3}$$

$$\beta = \frac{k_{min}}{K}, \quad \beta \in (0, 1) \tag{5.4}$$

K is the total number of the element in the entire feature map, k max is the number of the greater than the average of all elements, and kmin is the number of the less than the average of all elements.

Multi-modal Information Fuse Module

The main technical challenge in the multi-modal information fuse field is the correct integration of multi-scale semantic information from remote sensing data and the OSM point data. Therefore, we processed the data separately to generate spectral-spatial and prior poi features. The PointNet has two key modules: 1) the max pooling layer as a symmetric function to aggregate information from all the points; and 2) a local and global information combination structure (Qi et al., 2017).

After obtaining semantic information from both remote sensing data and OSM data, we fused the semantic information. The information fusion module can generally be classified into two types: (1) the concatenation fusion method; (2) the elementwise fusion method (Wang et al., 2005). The elementwise fusion method does not contain any parameter and can fuse a different number of the feature vector (Zhang, 2019). However, the elementwise fusion method cannot dense fusion semantic information from dense feature vector. To solve the above challenges, we designed a multi-modal information fuse module with the multi-layer perceptron (MLP) and average pooling. The core idea of the multi-modal information fuse module is per-pixel fusion. PointNet is used to produce a dense per POI feature by mapping each of the marked OSM points to a $\mathbf{I}_{in2}$-dimensional feature space. The spectral-spatial feature maps by spatial pyramid mix-pooling layer are reshaped to produce $\mathbf{I}_{in1}$-dimensional vector from remote sensing data and OSM polygons. First, the addition strategy was used to fuse local feature. Its formula can be expressed by Equation (5.5).

$$\mathbf{I} = \mathbf{I}_{in1} + \mathbf{I}_{in2} \tag{5.5}$$

Then, MLP networks was used to fuse global feature. The multi-modal information fuse module can be mathematically expressed by Equation (5.6).

$$G(\mathbf{I}_{out}) = f(\mathrm{MLP}(avg(\mathbf{I}))) \tag{5.6}$$

where $f(\cdot)$ is rectified linear unit (ReLU) function, avg represents the average pooling operation, $G(\mathbf{I}_{out})$ is the final fusion result. In this study, the multi-layer perceptron (MLP) is composed of two convolution kernels of size 1 × 1.

Boundary Enhancement Module

The proposed method use convolution and pooling operations, leading to the blurring of UGS boundaries. Conditional random field (CRF) was widely adopted as a post-processing step to refine object boundaries (Chen et al., 2017), which is a suboptimal choice for the global schema inference process. A residual refinement module was developed to solve the global schema inference process for boundary refinement (Islam et al., 2017; Qin et al., 2019). However, the above boundary refinement methods are hard to obtain both region and global boundaries in score maps. Therefore, we propose a novel boundary enhancement module with three dilated convolution layers to refine the score map boundaries on different scales. As shown in Figure 5.3(a), it contains 2-dilated kernel with sizes of 5 × 5, 3 × 3, 1 × 1. In the proposed boundary enhancement module, 2-dilated kernel (see Figure 5.3(b)) of different sizes were used to refine the score map boundaries at different scales. In this study, the boundary enhancement module can be expressed by Equation (5.7).

$$G_2 = G_1 + R(G_1) \tag{5.7}$$

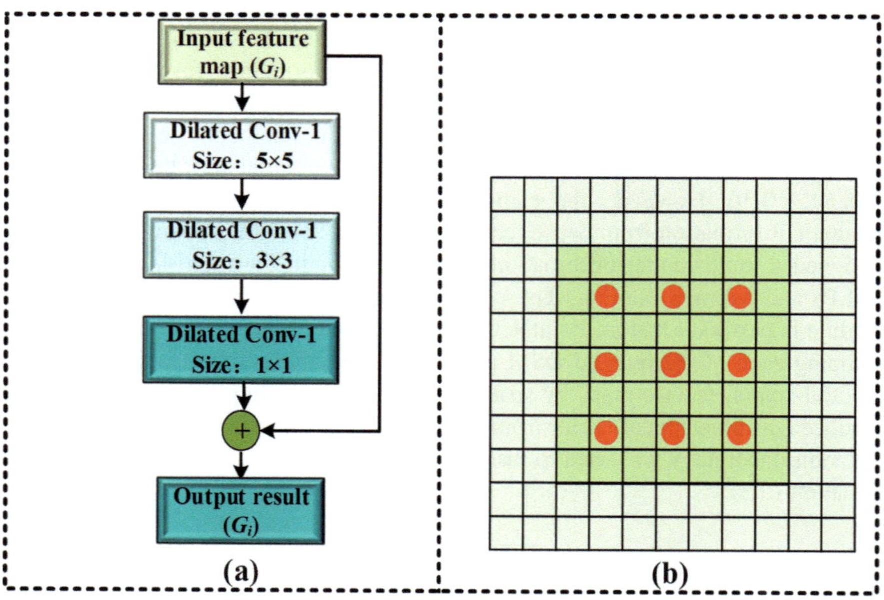

FIGURE 5.3 The boundary enhancement module and 2-dilated kernel: (a) The structure of boundary enhancement module; (b) The explanation for 2-dilated kernel.

where the G1 is a coarse score map, G2 is the refined score map, and $R(G_1)$ is the residual branch.

STUDY AREA AND DATASETS

STUDY AREA

Wuhan city, China, was selected as the study area (114.16° N and 30.56° E) (Figure 5.4(a)). It covers an area of 8,494.41 km^2 and had 11,212,000 permanent residents in 2019 (Xiao et al., 2020). Owing to economic development and the rapid growth of urban populations (In 2020, Wuhan urban population of 1418 million), the built-up lands of the Wuhan city are continuously expanding (Wang et al., 2020). The climate in Wuhan is subtropical monsoon, with four distinct seasons and abundant rainfall, and is well known for its urban heat island effect (Carpio et al., 2020). The monthly average temperature ranges in Wuhan from 2.7 °C in January to 31.1°C in July. Since the UGS can adjust the surface heat budget, the construction of UGS become a very important task for environmental conservation and management, urban planning, and urban ecological monitoring (Havinga et al., 2020).

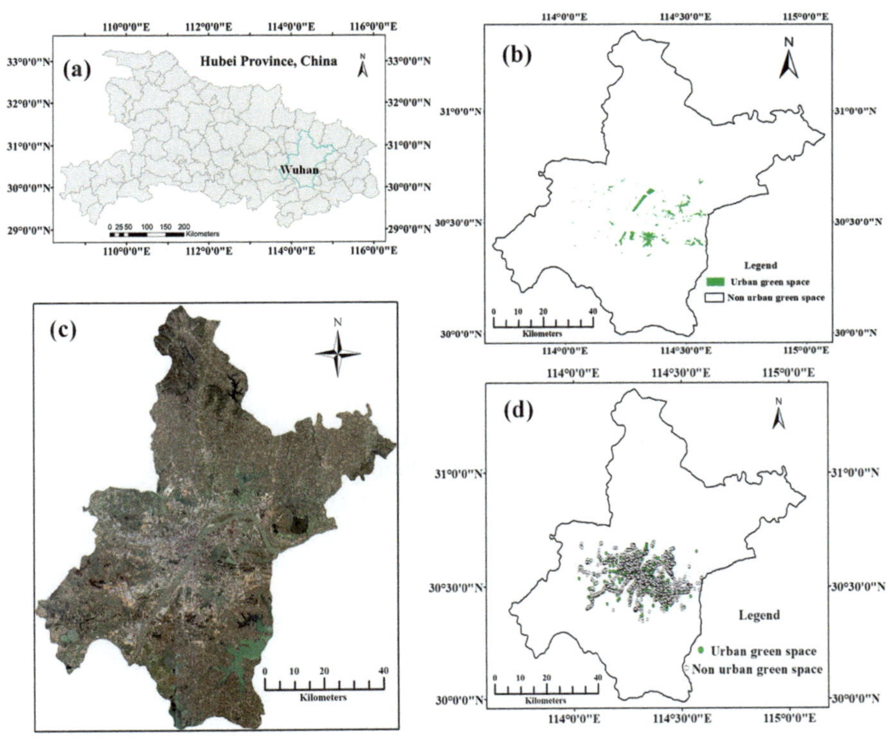

FIGURE 5.4. The study area and datasets: (a) The location of the study area; (b) UGS polygons based on OSM data; (c) The Sentinel-2 imagery of the study area; (d) UGS POIs based on OSM data.

Datasets

In this study, the acquisition date of the OSM data are from 22th August 2020. The dataset can be freely downloaded from https://www.openstreetmap.org.The urban green space polygons were extracted from the OSM data for the study area. The UGS polygons contain of 790 urban green belt, 100 urban wetlands, 10 water, and 150 urban forest polygons. A total of 6500 POI data are obtained in the study area. The 500 UGS POIs were extracted from POI data. The OSM polygon and POI data are show in Figure 5.4(b) and Figure 5.4(d), respectively.

The Sentinel-2 (Sentinel-2A and Sentinel-2B) satellite is dedicated for monitoring agricultures, forests, land-use changes, coastal and inland waters, and urban remote sensing mapping (Wang and Atkinson, 2018). Sentinel-2A was launched on 23th June 2015 and Sentinel-2B on 7th March 2017, both with a revisit period of 5 days. The Sentinel-2 Multispectral Instrument (MSI) comprises two satellites that observe the Earth at 10 m, 20 m, and 60 m spatial resolution (Quintano et al., 2018). Sentinel-2 satellite imagery, which can be freely downloaded (https://scihub.copernicus.eu/), provide abundant spatial-spectral information for urban surface objects (Teodoro and Amaral, 2019).

Sentinel image of the Wuhan region, 22th August 2020, was acquired for testing. Figure 5.4(c) shows the Sentinel-2 image with a spatial resolution of 10 m covering Wuhan. Considering vegetation phenology, a sequence of Sentinel-2 satellite images with a spatial resolution of 10 m was selected from 1st August 2017 to 12th March 2020 for training and was preprocessed by Sentinel Application Platform (SNAP). The proposed neural network was trained and tested with samples from Sentinel-2A images where only the visible and near-infrared bands, three 10 m high-resolution bands (Sentinel-2A band 2, Sentinel-2A band 3, and Sentinel-2A band 8) were used.

MAPPING URBAN GREEN SPACES WITH DIFFERENT METHODS

For UGS mapping, the proposed neural network method was compared with SVM (Kranjčić et al., 2019), DeepLab v3plus (Liu et al., 2019), FCN (Fu et al., 2018). The SVM method is the popular classification algorithm for remote sensing image classification task. The DeepLab v3plus and FCN algorithms are state-of-the-art deep learning for image semantic segmentation task. DeepLab v3plus using an encoder-decoder structure with atrous separable convolution for semantic image segmentation (Liu et al., 2019). The FCN architecture is designed by replacing the fully connected layers by convolutional layers, which enables us to take arbitrary sized images as inputs (Fu et al., 2018). The proposed neural network method was assessed with SVM, DeepLab v3plus, FCN.

The compared methods are trained with the same training dataset. Also, all experimental parameters for deep learning methods are chosen by doing a great deal of experimental and selecting the ones with good performance.

Figure 5.5 shows the results of UGS extraction by the proposed method along with SVM, DeepLab v3plus, and FCN under typical UGS scenes (park green spaces, resident green spaces, and wetland green spaces). From Figure 5.5(c), it is

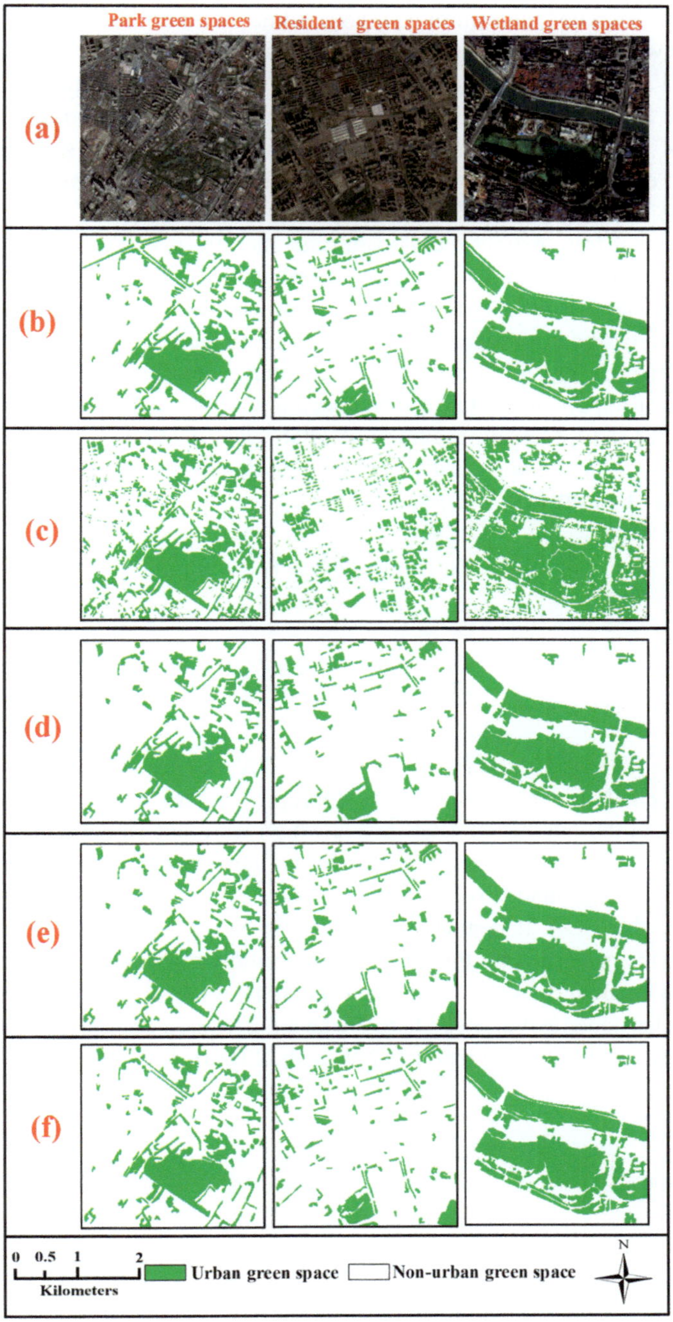

FIGURE 5.5 Extraction results by different methods for the different types of urban green spaces: (a) Original Sentinel-2 images; (b) The ground truth reference; (c) SVM; (d) DeepLab v3plus; (e) FCN; and (f) the proposed method.

found that the SVM algorithm confuses some building shadows and artificial grasslands with UGS. Building shadows are misclassified as UGS in dense building areas by SVM since it cannot extract high-level semantic information. Figure 5.5(d) and Figure 5.5(e) show that the DeepLab v3plus and FCN methods can solve the problem of the confusion between building shadows and urban green spaces, but they fail to extract small urban green spaces, leading to underestimation of the total area of the urban green spaces. This is because the DeepLab v3plus and FCN methods cannot fully extract multi-scale semantic information. From Figure 5.5(f), the proposed method can achieve satisfactorily results and solve the underestimation issue for typical urban green space scenes, which is mainly attributed to multi-scale feature extraction module of the proposed method.

Three metrics are used for the evaluation: overall accuracy (OA), kappa coefficient (Kappa), and edge overall accuracy (EOA). In this paper, OA and Kappa are used to evaluate accuracy of UGS classification. For the urban green space class, OA and Kappa can be defined by Equation (5.8) and Equation (5.9), respectively (Pontius and Millones, 2011).

$$OA = \frac{(TN + TP)}{M} \tag{5.8}$$

$$Kappa = \frac{M \times (TN + TP) - \nabla}{M^2 - \nabla} \tag{5.9}$$

where, $\nabla = (TP + FP) \times (TP + FN) + (FN + TN) \times (FN + TN)$, M is the total number of pixels for the whole image, TP represents the number of true positive, FN represents the number of UGS pixels which were not classified, FP represents the number of false-positive, TN represents the number of background pixels classified as background pixels.

Besides, the EOA (Chen et al., 2018b) is used to evaluate UGS boundary accuracy. This paper defines the buffer area as follows: (1) The boundary of UGS is obtained by manual drawing; (2) The morphological expansion is performed in the UGS boundary obtained in Step (1) to create a buffer area centered on the boundary line and having a radius of 3 pixels (Chen et al., 2018a). The EOA can be defined by Equation (5.10).

$$EOA = \frac{N_B}{N} \tag{5.10}$$

where N represents the total number of pixels in the buffer area with a radius of 3 pixels, N_B represents the number of UGS pixels that were correctly classified in the buffer area within a radius of 3 pixels.

Figure 5.6 shows quantitative evaluation results based on three types of urban green spaces. Overall, higher values of accuracy, Kappa, and edge overall accuracy indicate batter performance of a mapping algorithm. The proposed method produces an average overall accuracy of 0.971, average Kappa of 0.921,

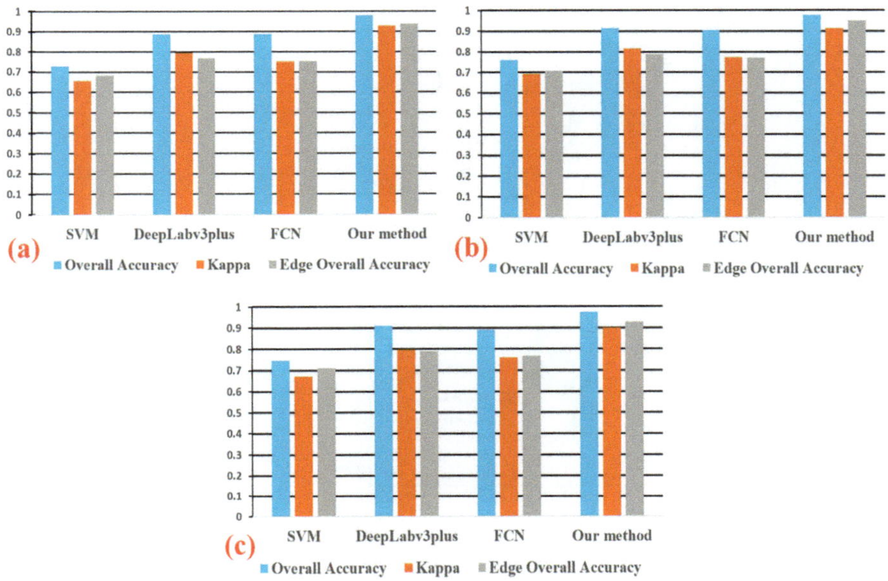

FIGURE 5.6 Quantitative evaluation of the different methods based on three type of urban green spaces: (a) park green spaces; (b) residential green spaces; (c) urban wetlands.

and average edge overall accuracy of 0.931 for the three types of urban green spaces. While for SVM method, the average overall accuracy, average Kappa, and average edge overall accuracy yield 0.718, 0.659, and 0.704, respectively. This poorer performance of SVM is mainly due to its incapable complete separation of building shadows from urban green spaces. From Figure 5.6(b), we can observe that the overall extraction accuracy of SVM method is lowest for residential green spaces. Although the DeepLab v3plus and FCN methods produces best values for overall accuracy and Kappa. However, the edge overall accuracy of DeepLab v3plus and FCN methods are lower for three types of urban green spaces, this is mainly due to the stride convolution and pooling operations can blur UGS boundaries.

Based on visual assessment and quantitative evaluation, the proposed method can reliably extract the different types of UGS under complex background and restrain building shadows and reserve UGS simultaneously.

In addition, we assess the effect of OSM data for UGS mapping of the proposed method with and without OSM data. Figure 5.7 shows the results of mapping using the proposed method with and without OSM data. The proposed method with OSM data can provide more satisfactory result and present a clearer UGS pattern, especially in the residential areas (see Figure 5.7(b)). It benefits from the more prior information from OSM data. However, the proposed method without OSM data fail to detect complete UGS structure. Some building shadows are detected wrongly as UGS (see Figure 5.7(d)). The qualitative results show edge overall accuracy significantly improves by about 0.053 for the entire city with the OSM data. The overall accuracy slightly increases by approximately

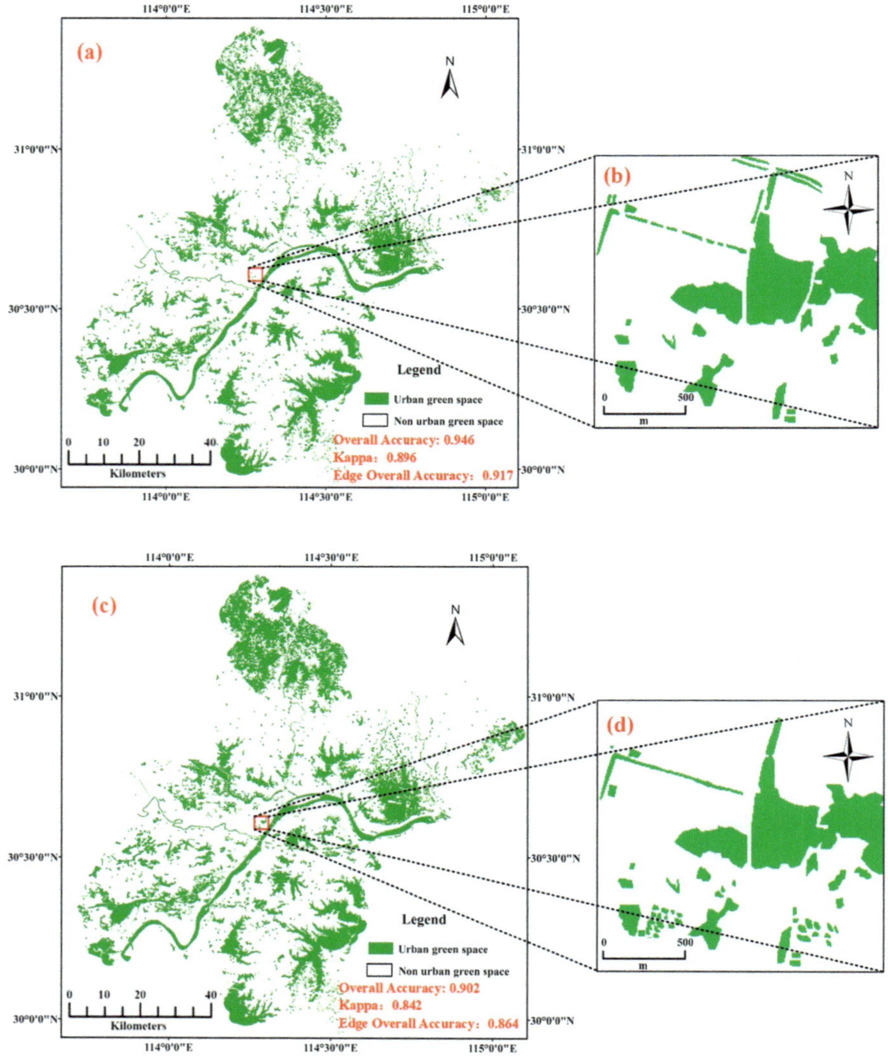

FIGURE 5.7 UGS mapping of the study areas using the proposed method with and without OSM data: (a) The proposed method with OSM data (b) The local zoomed of (a) (c) The proposed method without OSM data (d) The local zoomed of (c).

0.044 when using OSM data. This suggests that OSM data is valuable for large scale mapping of urban green space.

CONCLUSIONS

Accurate mapping of UGS is significant for providing data and information in support of ecological conservation, environmental management, urban planning, sustainable development and urban climate studies. UGS mapping faces an overestimation issue because certain types of objects, such as, building shadow, dark roads, and farmland within a city may possess similar spectral values as urban green spaces. To accurately map the urban green spaces, this study proposes an automatic UGS mapping neural network (UGSM-NN) method that integrates remote sensing imagery with crowdsourced geospatial big data. This proposed method aims to solve the overestimation problem on the UGS mapping when using remote sensing images. The proposed neural network consists of a multi-scale feature extraction module, a multi-modal information fuse module, and a boundary enhancement module. The experimental results show the proposed method can reliably extract different types of UGS including park green spaces, resident green spaces, and wetland green spaces, and present a clear spatial pattern of UGS.

Compared to existing studies, our work is innovative in three aspects. First, it focuses on mapping of UGS at the city scale. Second, it integrates remote sensing images and crowdsourced geospatial big data using a multi-scale feature extraction module and a multi-modal information fuse module. Finally, the boundary enhancement module is used, for the first time, to obtain a clearer UGS map of the large scale. The proposed method reduces commission errors in urban green spaces mapping using remote sensing image, which benefits for the overall urban and UGS mapping, as well as urban ecological monitoring. However, the proposed method has two limitations: 1) The UGSM-NN method was conducted in a single city, and cannot be applied for new regions without validation; and 2) The effect of time gap between POI data and Sentinel-2 images were not considered in this study. In future works, we will explore the performance of the proposed method for mapping at a larger scale while trying to solve these identified weaknesses.

REFERENCES

Abbott, B. W., Bishop, K., Zarnetske, J. P., et al., 2019. Human domination of the global water cycle absent from depictions and perceptions. *Nature Geoscience*, 12(7), 533–540.

Arsanjani, J. J., Helbich, M., Bakillah, M., 2013. Exploiting volunteered geographic information to ease land use mapping of an urban landscape. Proceedings of the International Archives of the Photogrammetry (pp.29–31), Remote Sensing and Spatial Information Sciences.

Astell-Burt, T., Navakatikyan, M. A., Feng, X., 2020. Urban green space, tree canopy and 11-year risk of dementia in a cohort of 109,688 Australians. *Environment International*, 145(12), 106102.

Carpio, M., González, Á., González, M., et al., 2020. Influence of pavements on the urban heat island phenomenon: A scientific evolution analysis. *Energy and Buildings*, 226(7), 110379.

Chen, L. C., Papandreou, G., Kokkinos, I., et al., 2017. Deeplab: Semantic image segmentation with deep convolutional nets, atrous convolution, and fully connected crfs. *IEEE transactions on pattern analysis and machine intelligence*, 40(4), 834–848.

Chen, Y., Fan, R., Bilal, M., et al., 2018a. Multilevel cloud detection for high-resolution remote sensing imagery using multiple convolutional neural networks. *ISPRS International Journal of Geo-Information*, 7(5), 181.

Chen, Y., Fan, R., Yang, X., et al., 2018b. Extraction of urban water bodies from high-resolution remote-sensing imagery using deep learning. *Water*, 10(5), 585.

Chen, Y., Lin, Z., Zhao, X., et al., 2014. Deep learning-based classification of hyperspectral data. *IEEE Journal of Selected topics in applied earth observations and remote sensing*, 7(6), 2094–2107.

Chen, Y., Weng, Q., Tang, L., et al., 2021. Thick Clouds Removing From Multitemporal Landsat Images Using Spatiotemporal Neural Networks. *IEEE Transactions on Geoscience and Remote Sensing*, 59(8) 1–14.

Dai, Z., Guldmann, J. M., Hu, Y., 2018. Spatial regression models of park and land-use impacts on the urban heat island in central Beijing. *Science of the Total Environment*, 626, 1136–1147.

Di, S., Li, Z. L., Tang, R., et al., 2019. Urban green space classification and water consumption analysis with remote-sensing technology: a case study in Beijing, China. *International Journal of Remote Sensing*, 40(6), 1909–1929.

Du, S., Du, S., Liu, B., et al., 2020. Large-scale urban functional zone mapping by integrating remote sensing images and open social data. *GIScience & Remote Sensing*, 57(3), 411–430.

Eigen, D., Fergus, R., 2015. Predicting depth, surface normals and semantic labels with a common multi-scale convolutional architecture. IEEE international conference on computer vision (pp. 2650–2658), IEEE.

Elwood, S., Goodchild, M. F., Sui, D. Z., 2012. Researching volunteered geographic information: Spatial data, geographic research, and new social practice. *Annals of the Association of American geographers*, 102(3), 571–590.

Fonte, C. C., Patriarca, J., Jesus, I., et al., 2020. Automatic Extraction and Filtering of OpenStreetMap Data to Generate Training Datasets for Land Use Land Cover Classification. *Remote Sensing*, 12(20), 3428.

Foody, G. M., 1995. Using prior knowledge in artificial neural network classification with a minimal training set. *Remote Sensing*, 16(2), 301–312.

Fu, X., Qu, H., 2018. Research on semantic segmentation of high-resolution remote sensing image based on full convolutional neural network. 12th International Symposium on Antennas, Propagation and EM Theory (ISAPE) (pp. 1–4). IEEE.

Haklay, M., Weber, P., 2008. Openstreetmap: User-generated street maps. *IEEE Pervasive Computing*, 7(4), 12–18.

Hartling, S., et al., 2019. Urban tree species classification using a WorldView-2/3 and LiDAR data fusion approach and deep learning. *Sensors*, 19(6), 1284.

Havinga, I., Bogaart, P. W., Hein, L., et al., 2020. Defining and spatially modelling cultural ecosystem services using crowdsourced data. *Ecosystem Services*, 43(6), 101091.

He, K., Zhang, X., Ren, S., 2015. Spatial pyramid pooling in deep convolutional networks for visual recognition. *IEEE Transactions on Pattern Analysis and Machine Intelligence*, 37(9), 1904–1916.

Huang, B., Zhao, B., Song, Y., 2018. Urban land-use mapping using a deep convolutional neural network with high spatial resolution multispectral remote sensing imagery. *Remote Sensing of Environment*, 214(9), 73–86.

Islam, M. A., Kalash, M., Rochan, M., et al., 2017. Salient Object Detection using a Context-Aware Refinement Network. British Machine Vision Conference (pp. 1–5), IEEE.

Ji, Y., Zhang, H., Wu, Q. J., et al., 2018. Salient object detection via multi-scale attention CNN. *Neurocomputing*, 322, 130–140.

Johnson, B. A., Iizuka, K., 2016. Integrating OpenStreetMap crowdsourced data and Landsat time-series imagery for rapid land use/land cover (LULC) mapping: Case study of the Laguna de Bay area of the Philippines. *Applied Geography*, 67(2), 140–149.

Kranjčić, N., Medak, D., Župan, R., et al., 2019. Support vector machine accuracy assessment for extracting green urban areas in towns. *Remote Sensing*, 11(6), 655.

Kuang, W., Dou, Y. 2020. Investigating the Patterns and Dynamics of Urban Green Space in China's 70 Major Cities Using Satellite Remote Sensing. *Remote Sensing*, 12(12), 1929.

Li, D., Wu, S., Liang, Z., et al, 2020. The impacts of urbanization and climate change on urban vegetation dynamics in china. *Urban Forestry & Urban Greening*, 54 (10), 126764.

Liang, P., Shi, W., Zhang, X., 2018. Remote sensing image classification based on stacked denoising autoencoder. *Remote Sensing*, 10(1), 16.

Lee, J. S., Grunes, M. R., Ainsworth, T. L., et al., 1999. Unsupervised classification using polarimetric decomposition and the complex Wishart classifier. *IEEE Transactions on Geoscience and Remote Sensing*, 37(5), 2249–2258.

LeCun, Y., Bengio, Y., Hinton, G., et al., 2015. Deep learning. *Nature*, 521(7553), 436–444.

Liu, D., Chen, N., Zhang, X., et al., 2020. Annual large-scale urban land mapping based on landsat time series in google earth engine and openstreetmap data: a case study in the middle yangtze river basin. *ISPRS Journal of Photogrammetry and Remote Sensing*, 159 (1), 337–351.

Liu, W., Yue, A., Shi, W., et al., 2019. An Automatic Extraction Architecture of Urban Green Space Based on DeepLabv3plus Semantic Segmentation Model. 2019 IEEE 4th International Conference on Image, Vision and Computing (ICIVC) (pp. 311–315), IEEE.

Liu, Y., Zhang, Y. M., Zhang, X. Y., et al., 2016. Adaptive spatial pooling for image classification. *Pattern Recognition*, 55(7), 58–67.

Luo, N., Wan, T., Hao, H., et al., 2019. Fusing high-spatial-resolution remotely sensed imagery and openstreetmap data for land cover classification over urban areas. *Remote Sensing*, 11(1), 88.

Ma, L., Cheng, L., Han, W., et al., 2014. Cultivated land information extraction from high-resolution unmanned aerial vehicle imagery data. *Journal of Applied Remote Sensing*, 8(1), 083673.

Peng, C., Li, Y., Jiao, L., et al., 2019. Densely based multi-scale and multi-modal fully convolutional networks for high-resolution remote-sensing image semantic segmentation. *IEEE Journal of Selected Topics in Applied Earth Observations and Remote Sensing*, 12(8), 2612–2626.

Peng, C., Zhang, X., Yu, G., et al., 2017. Large kernel matters--improve semantic segmentation by global convolutional network. In Proceedings of the IEEE conference on computer vision and pattern recognition (pp. 4353–4361), IEEE.

Pontius Jr, R. G., Millones, M., 2011. Death to Kappa: birth of quantity disagreement and allocation disagreement for accuracy assessment. *International Journal of Remote Sensing*, 32(15), 4407–4429.

Qi, C. R., Su, H., Mo, K., et al., 2017. Pointnet: Deep learning on point sets for 3d classification and segmentation. IEEE conference on computer vision and pattern recognition (pp. 652–660), IEEE.

Qin, X., Zhang, Z., Huang, C., et al., 2019. Basnet: Boundary-aware salient object detection. IEEE Conference on Computer Vision and Pattern Recognition (pp. 7479–7489), IEEE.

Quintano, C., Fernández-Manso, A., Fernández-Manso, O., 2018. Combination of Landsat and Sentinel-2 MSI data for initial assessing of burn severity. *International Journal of Applied Earth Observation and Geoinformation*, 64, 221–225.

Reichstein, M., Camps-Valls, G., Stevens, B., et al., 2019. Deep learning and process understanding for data-driven Earth system science. *Nature*, 566(7743), 195–204.

Scherer, D., Müller, A., Behnke, S., 2010. Evaluation of pooling operations in convolutional architectures for object recognition. In International conference on artificial neural networks (pp. 92–101), Springer.

Selway, C. A., Mills, J. G., Weinstein, P., et al., 2020. Transfer of environmental microbes to the skin and respiratory tract of humans after urban green space exposure. *Environment International*, 145(12), 106084.

Studer, L., Wallau, J., Ingold, R., et al., 2020. Effects of Graph Pooling Layers on Classification with Graph Neural Networks. 7th Swiss Conference on Data Science (SDS) (pp. 57–58), IEEE.

Sun, M., Song, Z., Jiang, X., et al., 2017. Learning pooling for convolutional neural network. *Neurocomputing*, 224(2), 96–104.

Tavares, P.A., Beltrão, N., Guimarães, U.S., et al., 2019. Urban Ecosystem Services Quantification through Remote Sensing Approach: A Systematic Review. *Environments* 6(5), 51.

Teodoro, A., Amaral, A., 2019. A statistical and spatial analysis of Portuguese forest fires in summer 2016 considering Landsat 8 and Sentinel 2A data. *Environments*, 6(3), 36.

Teodoro, A. C., Araujo, R., 2016. Comparison of performance of object-based image analysis techniques available in open source software (spring and orfeo toolbox/monteverdi) considering very high spatial resolution data. *Journal of Applied Remote Sensing*, 10(1), 016011.

Tong, Z., Tanaka, G. (2019). Hybrid pooling for enhancement of generalization ability in deep convolutional neural networks. *Neurocomputing*, 333, 76–85.

Van Herzele, A., Wiedemann, T., 2003. A monitoring tool for the provision of accessible and attractive urban green spaces. *Landscape and Urban Planning*, 63(2), 109–126.

Voltersen, M., Berger, C., Hese, S., et al., 2014. Object-based land cover mapping and comprehensive feature calculation for an automated derivation of urban structure types at block level. *Remote Sensing of Environment*, 154(11), 192–201.

Wan, T., Lu, H., Lu, Q., et al., 2017. Classification of high-resolution remote-sensing image using Openstreetmap information. *IEEE Geoscience and Remote Sensing Letters*, 14(12), 2305–2309.

Wang, F., 1990. Fuzzy supervised classification of remote sensing images. *IEEE Transactions on Geoscience and Remote Sensing*, 28(2), 194–201.

Wang, Q., Atkinson, P. M., 2018. Spatio-temporal fusion for daily Sentinel-2 images. *Remote Sensing of Environment*, 204, 31–42.

Wang, Q., Xu, Y., Wang, Y., et al., 2020. Individual and combined impacts of future land-use and climate conditions on extreme hydrological events in a representative basin of the Yangtze River Delta, China. *Atmospheric Research*, 236(5), 104805.

Wang, Z., Ziou, D., Armenakis, C., et al., 2005. A comparative analysis of image fusion methods. *IEEE transactions on Geoscience and Remote Sensing*, 43(6), 1391–1402.

Xu, F., Somers, B., 2021. Unmixing-based sentinel-2 downscaling for urban land cover mapping. *ISPRS Journal of Photogrammetry and Remote Sensing*, 171(1), 133–154.

Xu, Z., Zhou, Y., Wang, S., et al., 2020. A Novel Intelligent Classification Method for Urban Green Space Based on High-Resolution Remote Sensing Images. *Remote Sensing*, 12(22), 3845.

Xiao, Z., Miao, Y., Du, X., et al., 2020. Impacts of regional transport and boundary layer structure on the PM2. 5 pollution in Wuhan, Central China. *Atmospheric Environment*, 230(6), 117508.

Yue, C., Ciais, P., Houghton, R. A., et al., 2020. Contribution of land use to the interannual variability of the land carbon cycle. *Nature Communications*, 11(1), 1–11.

Yao, Y., Li, X., Liu, X., et al., 2017. Sensing spatial distribution of urban land use by integrating points-of-interest and Google Word2Vec model. *International Journal of Geographical Information Science*, 31(4), 825–848.

Zhang, C., Sargent, I., Pan, X., Li, H., et al., 2018. An object-based convolutional neural network (OCNN) for urban land use classification. *Remote Sensing of Environment*, 216(10), 57–70.

Zhang, J., Kerekes, J., 2011. Unsupervised urban land-cover classification using WorldView-2 data and self-organizing maps. IEEE International Geoscience and Remote Sensing Symposium (pp. 150–153). IEEE.

Zhang, T., 2019. Element-wise estimation error of a total variation regularized estimator for change point detection. arXiv preprint arXiv:1901.00914.

Zhao, W., Bo, Y., Chen, J., et al., 2019. Exploring semantic elements for urban scene recognition: Deep integration of high-resolution imagery and OpenStreetMap (OSM). *ISPRS Journal of Photogrammetry and Remote Sensing*, 151(5), 237–250.

Zheng, Y., Iwana, B. K., Uchida, S., 2019. Mining the displacement of max-pooling for text recognition. *Pattern Recognition*, 93 (9), 558–569.

Zou, X., et al., 2017. Tree classification in complex forest point clouds based on deep learning. *IEEE Geoscience and Remote Sensing Letters*, 14(12), 2360–2364.

6 Contrasting Pattern of Urban Expansion and Urban Land Use Intensification of Global Megacities using Nighttime Light Time Series Data

Qiming Zheng and Qihao Weng

INTRODUCTION

The world population, as reported by World Population Prospects 2019, is increasing towards 9.7 billion by 2050 (United Nations, 2019). The ongoing urbanization is projected to account for about 26% (2.5 billion) of the growth, whilst the urban dwellers will increase from 55% to 68% of the overall world's population by 2050 (Seto et al., 2017). This dramatic demographic transition, shifting population from rural to urban, is bringing a substantial, sophisticated and diverse changes to urban land cover and land use. These changes include: 1) urban growth, the conversion in the biophysical property of land cover, where natural landscapes, agricultural or forest lands are taken by built-up areas to fulfill the demand of urbanization (Gong et al., 2012); 2) urban land use intensification, also known as urban renewal, including cleaning out slums inside urban areas, and remodeling and/or reconstructing old infrastructures and buildings (e.g., brown-fields) into new uses (Stokes & Seto, 2019); 3) deurbanization, occurs whereby urban built-up areas are changed back to greens fields for conservation, recreation or other purposes (Satterthwaite et al., 2010); and (iv) urban land use degradation, highly developed urban areas being degraded into less developed areas, or even into vacant lands (Long & Wu, 2016).

The knowledge on the diversity of urban land changes is of high significance for understanding of urban environments and socio-economic changes. It has been well-documented that urban changes triggered a succession of repercussions on

DOI: 10.1201/9781003244561-8

local environment, human health and global climate change (Seto et al., 2012; Song et al.). Monitoring urban land changes in a spatially explicit and temporally intensive manner provides essential information for addressing these environmental issues (Batty, 2008; Weng, 2012). Understanding urban land changes is highly pertinent to the focus of Sustainable Development Goal 11, making cities and human settlement sustainable (Griggs et al., 2013; Guneralp et al., 2017), which requires information on urban land use and land cover, urban form and growth patterns, slums, buildings, transportation and infrastructure, green and public spaces, cultural and natural heritage sites.

The demands for more specific information on urban related uses pose new opportunities and challenges that require developing new geospatial methods to characterize different types of urban land changes. "Daytime" remote sensing (DRS) imagery, like Landsat or MODIS, can only present the biophysical attributes of earth's surface. Previous studies with DRS imagery have focused efforts on tracking a single type of urban land change, i.e., growth of urban extents (Song et al.; Zhu et al., 2019). Such unidimensional measurement largely ignores the other types of urban land changes, such as urban land use intensification and degradation, and deurbanization. Nighttime light (NTL) imagery provides a unique optical remote sensing data and is able to capture artificial light emitted at night with a global coverage (Román et al., 2018; Wang et al., 2021; Zheng et al., 2022).The detected NTL intensity is an integrated indication of both physical and socio-economic attributes of urban areas, so it may provide an alternative approach to disentangle and characterize different types of urban land changes (Stokes & Seto, 2019).

In this study, we disentangle urban land changes in 34 typical global megacities into the following types: urban growth (UG), deurbanization (DU), urban land use intensification (LUI), urban land use degradation (LUD) and no significant change (NC) by using the Visible Infrared Imaging Radiometer Suite (VIIRS) imagery of Suomi-NPP satellite.

STUDY AREA AND DATASETS

We have chosen 34 megacities as our study areas to represent a diverse range of factors, including urbanization status, urban size, climate, and landscape configuration (Figure 6.1). These megacities are typical representatives of current and

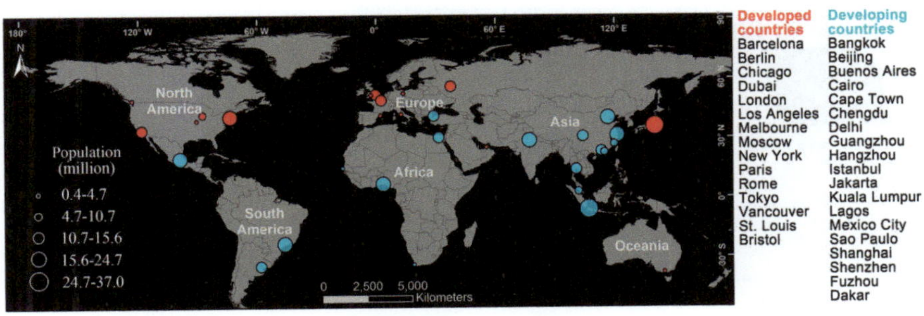

FIGURE 6.1 Study areas: 34 global megacities and the corresponding population.

projected urbanization hotspots, based on population data from 2016 and projections for 2030 (United Nations, 2016). Monitoring urban land changes in these megacities is crucial for understanding global urbanization trends and informing future development planning in less urbanized regions. We intentionally included megacities in tropical and sub-frigid areas, like Jakarta (6°N) and Moscow (55°N), to assess the robustness of our proposed method given different environmental conditions.

We utilized 84 monthly averaged VIIRS time series images, spanning from April 2012 to March 2019, to map urban changes. These images have a 15 arcsec spatial resolution and offer a high dynamic range (14 bit) and sensitive low light detection limit (2×10^{-11} W/cm2/sr) (Elvidge et al., 2017; Zhao et al., 2018). The VIIRS data consists of two datasets: average radiance (avg_rad) and the number of cloud-free observations (cf_cvg) used for calculating the avg_rad. Additionally, we obtained preprocessed data from the Google Earth Engine platform, including MODIS 16-Day NDVI (MOD13A1, 500 m), annual land cover maps (MCD12Q1, 500 m), and the Global Human Settlement Layer (GHSL), a 30-m resolution urban footprint map (Pesaresi et al., 2013).

METHODS

Figure 6.2 depicts the schematic flowchart of how we modeled and classified urban land change of the selected 34 megacities. We first used a Logistic-Harmonic model to fit monthly VIIRS NTL time series and then utilized the temporal features extracted from the fitted time series, together with a Random Forest classifier, to generate monthly urban built-up area maps. At last, we presented a post-classification scheme to categorize urban land change into five types: urban growth, deurbanization, urban land use intensification and degradation, and no change.

TIME SERIES MODELING

We used a non-linear Logistic-Harmonic (LogH) model to fit the monthly VIIRS time series (Eq. 6.1 and Figure 6.3). The model captures the trend of NTL with a logistic term and models seasonality with two harmonic terms (Zheng et al., 2021).

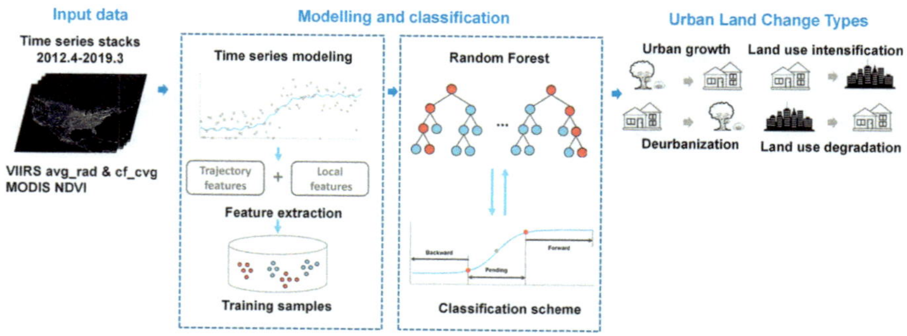

FIGURE 6.2 Methodological framework of this study.

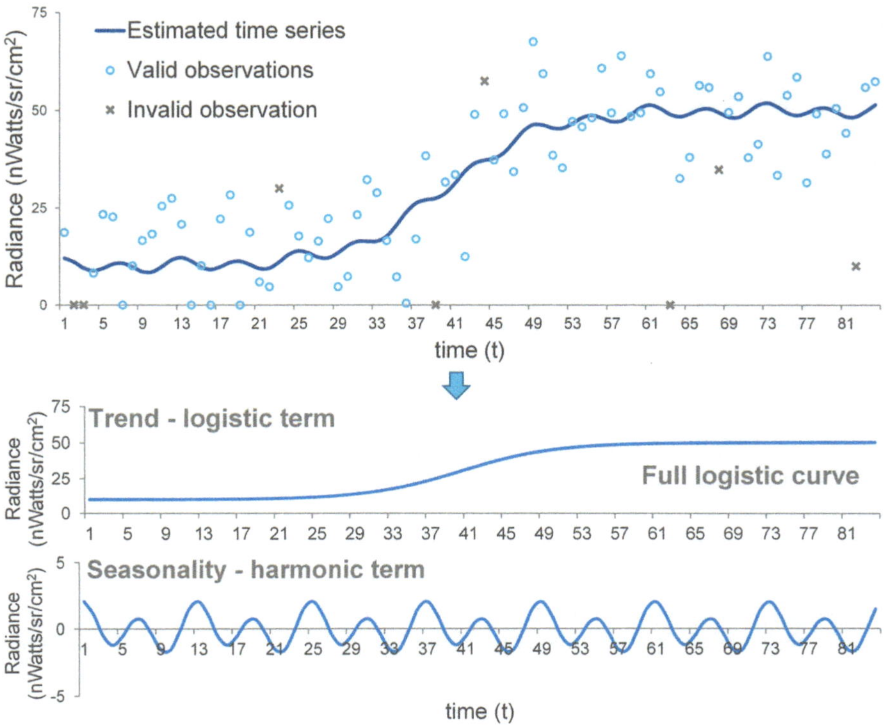

FIGURE 6.3 Illustrative figure of LogH model fitting.

The sigmoid function, known for its superiority in quantifying urban development stages and trajectory types, is suitable for NTL time series analysis and urban land change mapping (Song, 2016). The harmonic terms aim to offset the seasonality noises of NTL time series (Levin, 2017). The non-linear least square model with the Levenberg-Marquardt algorithm was used to estimate model parameters for each VIIRS pixel time series. We followed the approach proposed by Zheng et al. (2021) to screen out the impact from low-quality observations and estimate initial values for model parameters to ensure robust model performance.

$$Y_{log}(t) = \frac{a}{1 + \boldsymbol{e}^{bt+c}} + d + \sum_{k}^{2} \left[f_k \sin\left(\frac{2\pi kt}{12}\right) + g_k \cos\left(\frac{2\pi kt}{12}\right) \right] \tag{6.1}$$

where the estimated Y_{log} is a function of t (the tth month of observation) with the following parameters: a (magnitude factor), b (change rate), c (timing of change), d (either pre-change or post-change radiance value), and f_k and g_k (amplitude factors of seasonality); $\boldsymbol{e}$ is the base of natural logarithm. The number of harmonic term k is set to 2 to account for the annual and semi-annual seasonal variation based on our previous study with the Lomb Scargle Periodogram test (Zheng et al., 2019). If a time series is not changed significantly or has a higher R^2 with a linear trend term, it will be fitted by a Linear-Harmonic model.

Urban Built-Up Area Mapping

We utilized the temporal information from the fitted trend of VIIRS NTL time series as input features, together with a Random Forest classifier, to create monthly urban built-up area maps. The trajectory shape of the trend term can vary significantly throughout the study period (t1 to t84), exhibiting a full logistic curve (Figure 6.4(a)), a partial curve covering a substantial portion (Figure 6.4(d)), or even a small portion (Figure 6.4(b)) of the full curve. However, using the estimated model parameters directly as input features would compromise classification accuracy (Gómez et al., 2016). To ensure robustness and stability, metrics and statistical features derived from time series trajectories were employed (Vogelmann et al., 2016). Our feature space consisted of trajectory features from the estimated trend term and local features from the target dates (Table 6.1). For VIIRS time, series data modeled by the Linear-Harmonic model or when tcp1, tcp2, or tcp3 fell outside the study period (t1 to t84), tcp1, tcp2, and tcp3 were set as t1, t42, and t84, respectively, to guarantee the quality and robustness of the extracted features.

Due to limited data availability, obtaining monthly stable training samples for global megacities with high-resolution data is not feasible. However, the MODIS MCD12Q1 land cover product provides a reliable 500 m built-up area map on a yearly basis since 2000 (Sulla-Menashe et al., 2019). To create stable training samples for each megacity, we randomly selected 500 built-up and 500 non-built-up pixels from the MCD12Q1 dataset. The sample selection followed specific criteria: 1) The selected pixel and its eight neighboring pixels belonged to the same land use/cover type throughout the study period. 2) To prevent spatial autocorrelation, no two samples were adjacent to each other. 3) For each megacity, 100 stable built-up and non-built-up samples were randomly chosen, and these samples were further

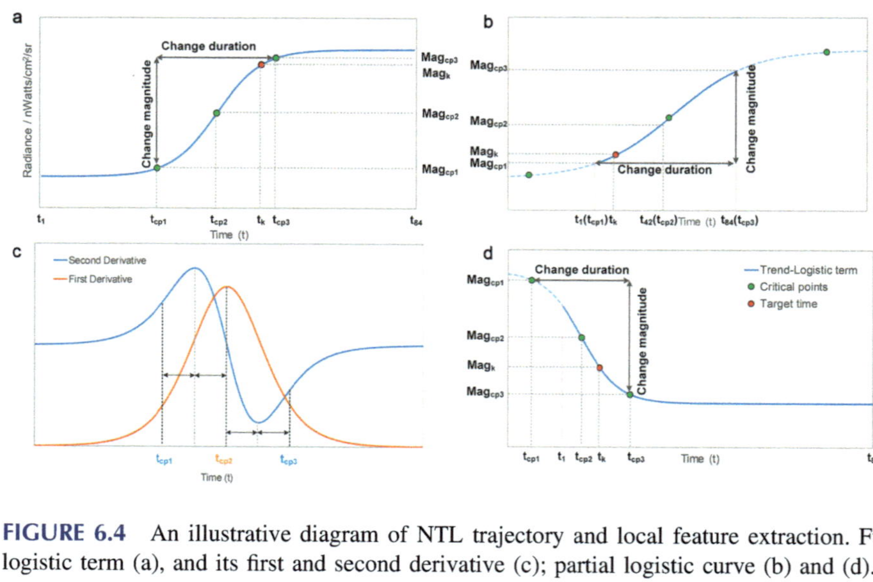

FIGURE 6.4 An illustrative diagram of NTL trajectory and local feature extraction. Full logistic term (a), and its first and second derivative (c); partial logistic curve (b) and (d).

TABLE 6.1
Explanation of Trajectory and Local Features

Trajectory Features	
Critical points 1 (CP1)	The start of change, $t_{CP1} = 2 \times t_{max}\text{-}t_{cp2}$
Critical points 2 (CP2)	The point with largest change rate, t_{cp2}
Critical points 3 (CP3)	The end of change, $t_{CP3} = 2 \times t_{min}\text{-}t_{cp2}$
NTL radiance of CPs	Mag_{cp1}, Mag_{cp2}, Mag_{cp3}
Change magnitude	$Change_{magnitude} = Mag_{cp3} - Mag_{cp1}$
Change duration	$Change_{duration} = \mid t_{cp3} - t_{cp1} \mid$
Change rate	$Change_{rate} = Change_{magnitude}/Change_{duration}$
Seasonality magnitude	$Seasonality_{magnitude} = \Sigma_k^2 \sqrt{f_k^2 + g_k^2}$
Local Features	
Time of target date	t_k
NTL radiance of target month	Mag_{tk}
Maximum NDVI (MOD13A1)	$NDVI_{max}$ is the maximum NDVI value within the 12 closest months of the target month (t_k)

validated using Google Earth images. If more than five inaccurate pixels were identified, all sample pixels were reselected.

At last, with the help of these stable training samples, we trained a Random Forest classifier each month and create monthly urban built-up areas maps. Classification accuracy was evaluated by calculating the out-of-bag accuracy score of the Random Forest (RF) algorithm and the overall accuracy. The reference data used for evaluation was obtained from the corresponding year's MCD12Q1 dataset. Additionally, a reference dataset was generated using GHSL, a medium spatial resolution built-up area map product, for a combined assessment. In this joint assessment, a VIIRS pixel of 500 m resolution was labeled as a built-up area if it contained more than 20% of the built-up area derived from GHSL.

Urban Land Change Classification and Analysis

We proposed a post-classification scheme to further categorize each pixel into five urban land change types: 1) urban land cover changes: urban growth (non-built-up area to built-up area) and deurbanization (built-up area to non-built-up area); 2) urban land use intensity changes: urban land use intensification (significant NTL intensity increase) and urban land use degradation (significant NTL intensity decrease); and 3) no statistically significant change (non-built-up to non-built-up or no significant NTL change) (Table 6.2).

At last, we analyzed the urban land change pattern of 34 selected megacities and compared the pattern across cities from developing countries and developed countries. We calculated the following metrics: 1) percentage of changed pixels

TABLE 6.2
Urban Land Change Types Identification

Pre-Change Stage	Post-Change Stage	Urban Land Change Types
Built-up	Built-up (NTL increasing)	Land use intensification
Non-built-up	Built-up	Urban growth (Land cover change)
Built-up	Non-built-up	Deurbanization (Land cover change)
	Built-up (NTL decreasing)	Land use degradation
Non-built-up	Non-built-up	No change
No significant change in NTL		No change

out of the total built-up area of each city; 2) percentage of each urban land change type of all changed pixels; and 3) average change magnitude of urban land use intensity, measured by changes in NTL radiance.

RESULTS

Time Series Modeling

Figure 6.5 summarizes the fitting performance for all pixels of 34 selected megacities. It presents: 1) the proportion of pixels fitted by the Linear-Harmonic model, the Logistic-Harmonic (LogH) model, and those that do not exhibit significant change (No change); 2) the mean R2 values; and 3) the mean normalized RMSE (nRMSE $\in$ [0, 1]). Of the changed time series, approximately 90% were fitted by the LogH model, which demonstrated superior fitting performance (R2 = 0.51) compared to the Linear-Harmonic model (R2 = 0.30). The mean R2 of the LogH model ranged from 0.34 (Vancouver) to 0.66 (Delhi) across 34 megacities. The mean nRMSE for all cities ranged from 0 to 0.1. The LogH model exhibited a slightly higher mean nRMSE value (0.043) than the Linear-Harmonic model (0.027). This can be attributed to the LogH model capturing higher data variation in the fitted time series compared to the Linear-Harmonic model.

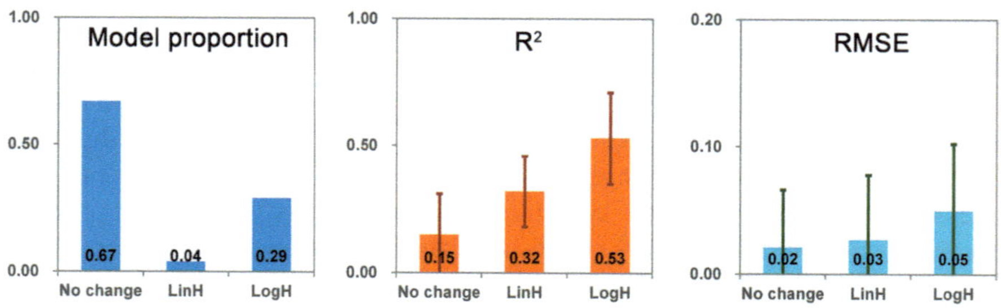

FIGURE 6.5 Results of time series modeling, including the proportion of model types, R2 and RMSE.

Urban Built-Up Area Mapping

Our method exhibited high consistency with MCD12Q1 and slightly lower agreement with GHSL. Discrepancies primarily occurred in scattered and low-density human settlements at urban outskirts and intercity highways. Table 6.3 compares the accuracy of our proposed method (Model 1) with other widely used nighttime light studies (Model 3–5). Temporal feature-based methods (Model 1&2) achieved higher overall accuracy, improving by up to 20%, compared to spectral-based methods (Model 3–5). By incorporating temporal features and NDVI, our proposed method (Model 1) achieved the highest accuracy of 97% (MCD12Q1 data) or 88% (GHSL data). Mapping accuracy varied at the city level, ranging from 93% (MCD12Q1) or 86% (GHSL) in Guangzhou, China, to 98% (MCD12Q1) or 92% (GHSL) in Sao Paolo, Brazil.

Figure 6.6 compares built-up areas identified by our method (Model 1), the NTL thresholding method (Model 4), and the spectral feature-based method (Model 5). Our temporal feature-based approach excels in handling heterogeneous landscapes. In cities dominated by built-up pixels like Shanghai (Figure 6.6(b)) and Melbourne (Figure 6.6(f)), Model 1 outperforms Model 4 by 6–12% (MCD12Q1) or Model 5 by 9–16% (GHSL) in accuracy (Table 6.3). This advantage is even more significant in cities with highly heterogeneous built-up/non-built-up landscapes such as Hangzhou (Figure 6.6(a)) and Bangkok (Figure 6.6(d)), where Model 1 surpasses Model 4 by 10–17% (MCD12Q1) or Model 5 by 11–23% (GHSL) in overall accuracy. This advantage applies to both urban outskirts and inner-city areas. For instance, in the Central Ridge Forest Reserve in downtown Delhi, India (Site 1 in Figure 6.6(b)), surrounded by commercial and public infrastructure, scenic spots, and a busy road, Model 4 and Model 5 fail to distinguish the forest reserve from the surrounding built-up

TABLE 6.3
Accuracy Assessment of Different Models

Methods		Features		References	
Models	**Type**	**Type**	**Feature Sets**	**MCD12Q1**	**GHSL**
1	Classification	Temporal	Temporal features; NDVI	96.9%	88.4%
2	Classification	Temporal	Temporal features	95.0%	87.3%
3	Thresholding	Spectral	VIIRS	72.1%	72.5%
4	Thresholding	Spectral	VANUI	84.6%	77.7%
5	Classification	Spectral	VIIRS; NDVI	85.7%	80.9%

Note: The accuracy yielded with MCD12Q1 indicates the average overall accuracy of all megacities for the 84 months; GHSL evaluates the accuracy of the built-up area map in December 2014 only. **Model 1** is the proposed method. **Model 3–5** are three widely-adopted methods, including thresholding method using raw VIIRS data (Model 3) (Liu et al., 2012) or a NTL index—VANUI* (Model 4) (Zhang et al., 2013), and classification-based method (Model 5) (Dou et al., 2017).

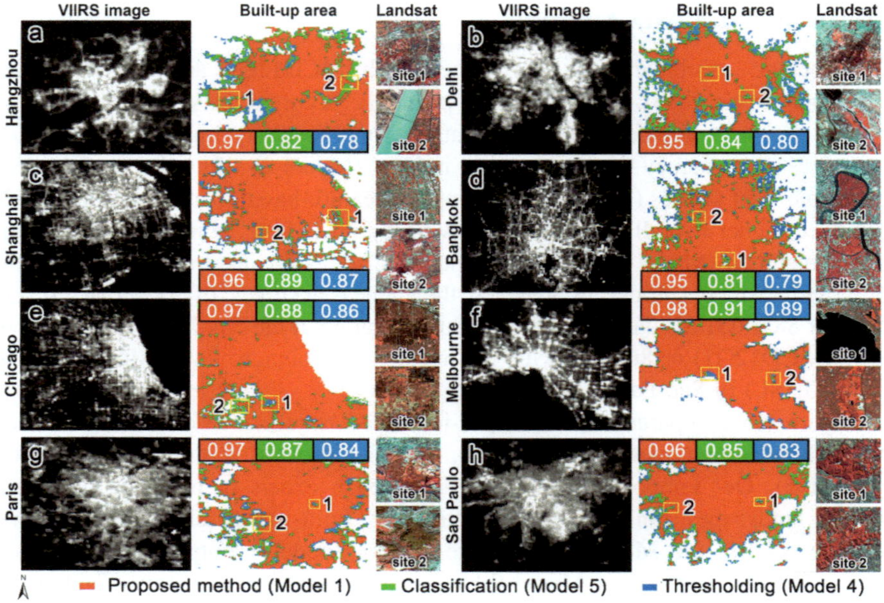

FIGURE 6.6 The left panel is VIIRS image; the middle panel is built-up areas derived by three different types of methods and the corresponding accuracy; the right panel is the zoomed-in Landsat images. (a-h) See text for details.

areas despite a moderate NTL intensity (18 nW/cm2/sr). In contrast, by using temporal features alone (Model 2) or in combination with NDVI (Model 1), a substantial portion of the forest reserve can be successfully identified as non-built-up areas. Another example is the outskirts of Chicago, USA (Site 2 in Figure 6.6(e)), where residential houses coexist with cropland and bare land. Model 1 demonstrates significantly higher accuracy in mapping built-up areas compared to other models.

Urban Land Change Pattern

Using accurate monthly built-up area maps, we categorized urban land change into five types: no change (NC), urban growth (UG), urban land use intensification (LUI), urban land use degradation (LUD), and deurbanization (DU). We analyzed urban land change patterns across 34 megacities (Figure 6.7) and found significant changes in over 53% of urban built-up pixels, varying across cities (11% in Berlin to 59% in Buenos Aires and 89% in Jakarta). Urban land use intensification (84%) and degradation (3%) accounted for the majority of changes, while urban growth represented only 4%. By distinguishing urban core and outskirts, we observed that urban growth, deurbanization, and land use degradation occurred primarily in the outskirts, while land use intensification and no change were evenly distributed.

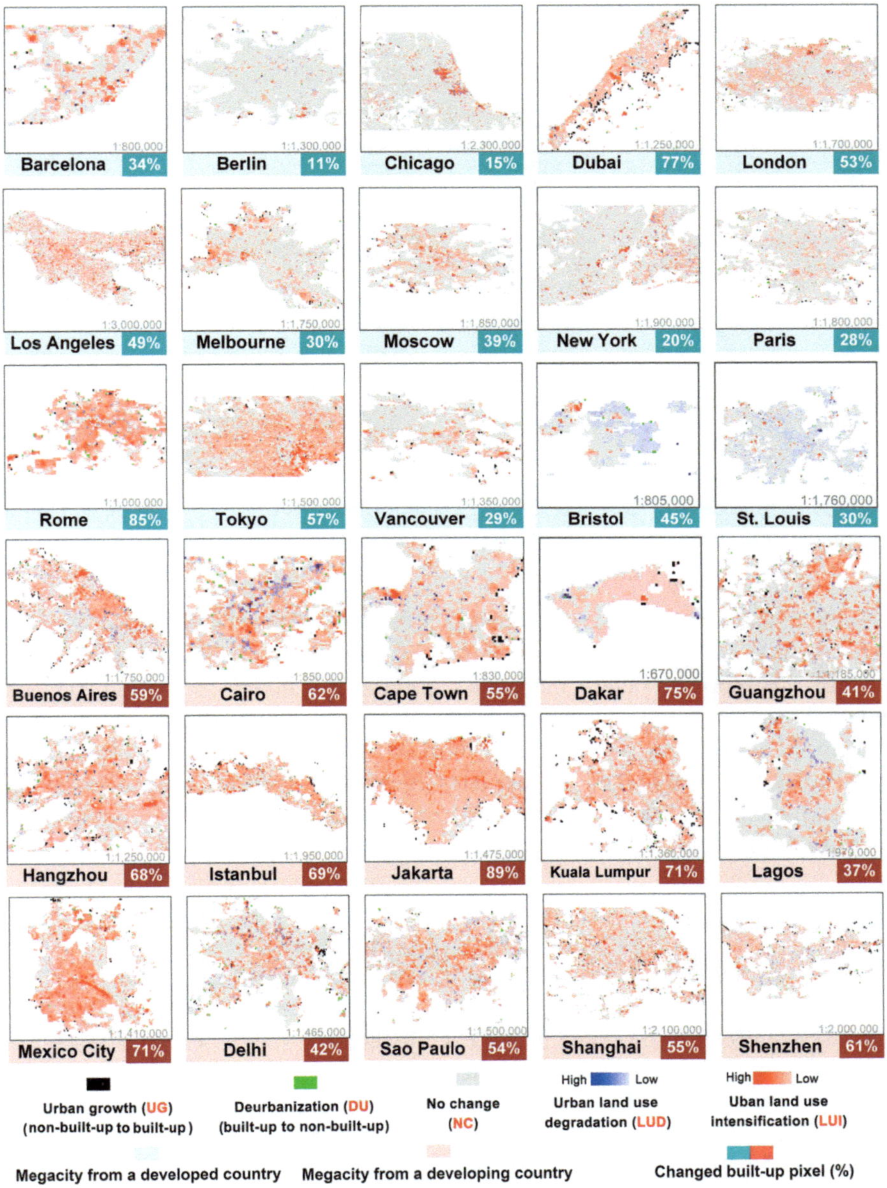

FIGURE 6.7 Urban land change types of 30 megacities. The percentage indicates the proportion of changed pixels in urban built-up areas. For better illustration, we only present the results of 30 out of 34 studied megacities.

Distinct urban change patterns emerged between developed and developing country megacities (Figure 6.8). Developing country megacities were more likely to follow a logistic urban development trajectory (50% compared to 30% in developed countries) and experienced a higher extent of urban land change (60% of pixels

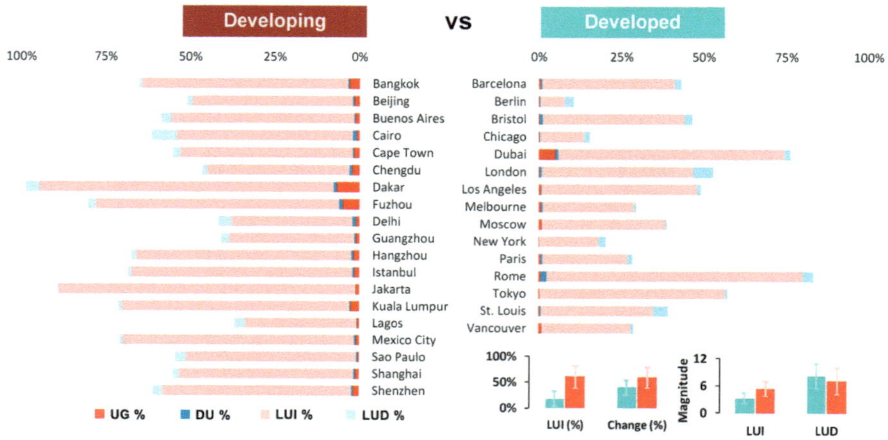

FIGURE 6.8 Urban land change types (%) of 34 megacities and comparison between megacities from developed and developing countries in terms of LUI (%), changed built-up pixels (%) and change magnitude of LUI and LUD.

compared to 39% in developed countries). Coastal megacities in Southeast Asia and Africa, such as Jakarta, Nairobi, and Kuala Lumpur, exhibited notable urban change. Developing country megacities underwent a higher proportion (63%) of urban land use change, including intensification and degradation, compared to developed country megacities (18%). In terms of magnitude, developing country megacities displayed higher land use intensification (5.8 nW/cm^2/sr) and lower land use degradation (−8.2 nW/cm^2/sr) than developed country megacities (3.0 nW/cm^2/sr and −6.5 nW/cm^2/sr, respectively).

DISCUSSIONS

In this study, temporal features derived from monthly VIIRS time series were used for urban built-up area mapping. Compared with traditional spectral-based methods, the advantages of the proposed method lied in the following aspects. First, spectral-based thresholding and classification methods depend only on the spectral value of a pixel, so they cannot effectively distinguish non-built-up areas with moderate or high NTL values from built-up areas, such as Site 1 in Figure 6.6(b). On the contrary, the NTL trajectories of built-up and non-built-up areas could vary significantly and thus are beneficial to differentiate them (Gómez et al., 2016; Zhu, 2017). This explains the high mapping accuracy of our method in non-built-up areas around urban outskirts and inside urban areas, where accommodate substantial moderate and extremely high NTL value pixels (Figure 6.6). Second, fitting time series model allows to isolate "noises" in VIIRS time series, e.g., the seasonal effect and ephemeral signals, so that the temporal pattern can be better modeled. Third, the use of dense monthly VIIRS time series can track gradual changes as well as subtle and abrupt changes. These changes would be buried in annual NTL time series or land cover map products that are released at an annual or 5–10 year interval and with a lag of 1 to 2 years, such as GlobeLand30, GHSL and Global Urban Footprint (Esch et al., 2017; Pesaresi et al., 2013).

This study provides valuable insights into the temporal dynamics of urban land change, including types, spatial extent, intensity, and trend trajectories. These findings are crucial for advancing sustainable urban development goals. Two key implications emerge from our results. Firstly, while previous remote sensing studies have predominantly focused on urban growth, our analysis highlights the significance of land use intensification (84%) in shaping urban landscapes. This suggests a potential shift towards intensified land use in global megacities, necessitating further investigation into its environmental implications, such as impacts on urban ecosystems, local climate, energy consumption, and population distribution. Secondly, a notable distinction in urban land change patterns emerges between megacities in developing and developed countries. This disparity offers valuable insights for urban spatial planning, particularly in developing countries undergoing rapid urbanization, as they can draw lessons from the experiences of developed countries and their urban land trajectories.

Our proposed method has several limitations that should be acknowledged. Firstly, the presence of mixed pixels within the VIIRS imagery poses a challenge in accurately identifying land cover and land use changes within a single pixel. Due to the coarse resolution of VIIRS images, it is difficult to address this issue effectively. Additionally, the relationship between NTL intensity and specific urban land use types varies greatly across different regions, making it challenging to account for this variability. Secondly, our method assumes that there is only one dominant urban change type occurring during the study period within each pixel. However, in rare cases, consecutive urban land changes may occur, such as urban growth followed by urban land use intensification. This violates the premise of our approach and can result in difficulties in time series fitting and subsequent urban change mapping. Nevertheless, such cases were extremely rare in the selected megacities (<0.01%). Thirdly, the spectral range of VIIRS is limited to 500 nm to 900 nm, which means that certain types of light, such as that emitted by LEDs, cannot be detected. The exact contribution and how to compensate for this missing portion of light require further investigation. Lastly, although there are two coarse resolution NTL datasets available, DMSP-OLS and VIIRS, our study exclusively utilized VIIRS time series. This choice is due to the absence of an onboard calibration system in DMSP-OLS and its incompatibility with VIIRS. Despite recent efforts to harmonize DMSP-OLS and VIIRS, consistent time series in radiance values have not yet been achieved. These limitations should be considered in the interpretation and application of our method to ensure the accuracy and reliability of the results. Future research can focus on addressing these challenges to further improve urban land change mapping.

CONCLUSIONS

The study developed an effective method to analyze urban land changes in 30 global megacities. The proposed Logistic-Harmonic model demonstrated robust performance in fitting monthly VIIRS time series, enabling the creation of accurate monthly built-up area maps and classification of urban land change types. The method achieved high overall accuracy when compared to MCD12Q1 (97%) or GHSL (88%), highlighting the superiority of temporal features over spectral features, particularly in heterogeneous regions. Additionally, the method improved temporal consistency and classification efficiency. Urban land changes were observed in 53% of built-up pixels, with urban land use intensification dominating (84%) the total urban changes over other urban land type, such as urban growth (4%). Megacities in developing countries exhibited wider extent and higher magnitude of urban land change, as well as a larger proportion of land use intensification compared to developed countries.

Further investigation is warranted in several areas. Firstly, exploring the potential of extending the Logistic-Harmonic model and temporal feature-based classification method to the upcoming Black Marble Product Suite, a high-quality daily VIIRS dataset with advanced correction processes, could provide valuable insights into urban land changes over time. Secondly, while our method demonstrated robustness against noise and low-quality observations, its applicability to different cities worldwide, and the potential for modeling other types of remote sensing time series data (e.g., reflectance or spectral indices) should be investigated. Additionally, calibrating DMSP-OLS and VIIRS would enable the examination of urban land cover and land use changes over longer time periods, offering further insights into urban dynamics.

REFERENCES

Batty, M. (2008). The size, scale, and shape of cities. *Science*, 319(5864), 769–771. 10.1126/science.1151419

Dou, Y., Liu, Z., He, C., & Yue, H. (2017). Urban Land Extraction Using VIIRS Nighttime Light Data: An Evaluation of Three Popular Methods. *Remote Sensing*, 9(2), 175. https://doi.org/10.3390/rs9020175

Elvidge, C., Baugh, K., Zhizhin, M., Hsu, F. C., & Ghosh, T. (2017). VIIRS night-time lights [1]. *International Journal of Remote Sensing*, 38(21), 5860–5879. 10.1080/01431161.2017.1342050

Esch, T., Heldens, W., Hirner, A., Keil, M., Marconcini, M., Roth, A., … Strano, E. (2017). Breaking new ground in mapping human settlements from space – The Global Urban Footprint. *ISPRS Journal of Photogrammetry and Remote Sensing*, 134, 30–42. 10.1016/j.isprsjprs.2017.10.012

Gómez, C., White, J. C., & Wulder, M. A. (2016). Optical remotely sensed time series data for land cover classification: A review. *ISPRS Journal of Photogrammetry and Remote Sensing*, 116, 55–72. 10.1016/j.isprsjprs.2016.03.008

Gong, P., Liang, S., Carlton, E. J., Jiang, Q., Wu, J., Wang, L., & Remais, J. V. (2012). Urbanisation and health in China. *The Lancet*, 379(9818), 843–852.

Griggs, D., Stafford-Smith, M., Gaffney, O., Rockström, J., Öhman, M. C., Shyamsundar, P., … Noble, I. (2013). Policy: Sustainable development goals for people and planet. *Nature*, 495(7441), 305.

Guneralp, B., Zhou, Y., Urge-Vorsatz, D., Gupta, M., Yu, S., Patel, P. L., … Seto, K. C. (2017). Global scenarios of urban density and its impacts on building energy use through 2050. *Proc Natl Acad Sci U S A*, 114(34), 8945–8950. 10.1073/pnas.1606035114

Levin, N. (2017). The impact of seasonal changes on observed nighttime brightness from 2014 to 2015 monthly VIIRS DNB composites [1]. *Remote Sensing of Environment*, 193, 150–164. 10.1016/j.rse.2017.03.003

Liu, Z., He, C., Zhang, Q., Huang, Q., & Yang, Y. (2012). Extracting the dynamics of urban expansion in China using DMSP-OLS nighttime light data from 1992 to 2008. *Landscape and Urban Planning*, 106(1), 62–72. https://doi.org/10.1016/j.landurbplan.2012.02.013

Long, Y., & Wu, K. (2016). Shrinking cities in a rapidly urbanizing China. *Environment and Planning A*, 48(2), 220–222.

Pesaresi, M., Huadong, G., Blaes, X., Ehrlich, D., Ferri, S., Gueguen, L., … Lu, L. (2013). A global human settlement layer from optical HR/VHR RS data: concept and first results. *IEEE Journal of Selected Topics in Applied Earth Observations and Remote Sensing*, 6(5), 2102–2131.

Román, M. O., Wang, Z., Sun, Q., Kalb, V., Miller, S. D., Molthan, A., … Masuoka, E. J. (2018). NASA's Black Marble nighttime lights product suite. *Remote Sensing Of Environment*, 210, 113–143. 10.1016/j.rse.2018.03.017

Satterthwaite, D., McGranahan, G., & Tacoli, C. (2010). Urbanization and its implications for food and farming. *Philosophical Transactions of the Royal Society B: Biological Sciences*, 365(1554), 2809–2820.

Seto, K. C., Golden, J. S., Alberti, M., & Turner, B. L., 2nd. (2017). Sustainability in an urbanizing planet. *Proceedings of the National Academy of Sciences U S A*, 114(34), 8935–8938. 10.1073/pnas.1606037114

Seto, K. C., Guneralp, B., & Hutyra, L. R. (2012). Global forecasts of urban expansion to 2030 and direct impacts on biodiversity and carbon pools. *Proceedings of the National Academy of Sciences U S A*, 109(40), 16083–16088. 10.1073/pnas.1211658109

Song, X. (2016). Characterizing themagnitude, timing and duration of urban growth from time series of Landsat-based estimates of impervious cover. *Remote Sensing of Environment*, 175, 1–13. 10.1016/j.rse.2015.12.027

Song, X. P., Hansen, M. C., Stehman, S. V., Potapov, P. V., Tyukavina, A., Vermote, E. F., & Townshend, J. R. (2018). Global land change from 1982 to 2016. *Nature*, 560(7720), 639–643.

Stokes, E. C., & Seto, K. C. (2019). Characterizing urban infrastructural transitions for the Sustainable Development Goals using multi-temporal land, population, and nighttime light data. *Remote Sensing Of Environment*, 234. 10.1016/j.rse.2019.111430

Sulla-Menashe, D., Gray, J. M., Abercrombie, S. P., & Friedl, M. A. (2019). Hierarchical mapping of annual global land cover 2001 to present: The MODIS Collection 6 Land Cover product. *Remote Sensing of Environment*, 222, 183–194. 10.1016/j.rse.2018.12.013

United Nations. (2016). The world's cities in 2016, data booklet. In: United Nations.

United Nations. (2019). World Population Prospects 2019, Online Edition. Retrieved from https://population.un.org/wpp/

Vogelmann, J. E., Gallant, A. L., Shi, H., & Zhu, Z. (2016). Perspectives on monitoring gradual change across the continuity of Landsat sensors using time-series data. *Remote Sensing of Environment*, 185, 258–270. 10.1016/j.rse.2016.02.060

Wang, Z., Román, M. O., Kalb, V. L., Miller, S. D., Zhang, J., & Shrestha, R. M. (2021). Quantifying uncertainties in nighttime light retrievals from Suomi-NPP and NOAA-20 VIIRS Day/Night Band data [1]. *Remote Sensing of Environment*, 263, 112557. 10.1016/j.rse.2021.112557

Weng, Q. (2012). Remote sensing of impervious surfaces in the urban areas: Requirements, methods, and trends. *Remote Sensing of Environment*, 117, 34–49. 10.1016/j.rse.2011.02.030

Zhang, Q., Schaaf, C., & Seto, K. C. (2013). The Vegetation Adjusted NTL Urban Index: A new approach to reduce saturation and increase variation in nighttime luminosity. *Remote Sensing of Environment*, 129, 32–41. https://doi.org/10.1016/j.rse.2012.10.022

Zhao, N., Zhang, W., Liu, Y., Samson, E. L., Chen, Y., & Cao, G. (2018). Improving Nighttime Light Imagery With Location-Based Social Media Data [1]. *IEEE Transactions on Geoscience and Remote Sensing*, 1–12. 10.1109/tgrs.2018.2871788

Zheng, Q., Weng, Q., & Wang, K. (2019). Developing a new cross-sensor calibration model for DMSP-OLS and Suomi-NPP VIIRS night-light imageries [1]. *ISPRS Journal of Photogrammetry and Remote Sensing*, 153, 36–47. 10.1016/j.isprsjprs.2019.04.019

Zheng, Q., Weng, Q., & Wang, K. (2021). Characterizing urban land changes of 30 global megacities using nighttime light time series stacks [1]. *ISPRS Journal of Photogrammetry and Remote Sensing*, 173, 10–23. 10.1016/j.isprsjprs.2021.01.002

Zheng, Q., Weng, Q., Zhou, Y., & Dong, B. (2022). Impact of temporal compositing on nighttime light data and its applications [1]. *Remote Sensing of Environment*, 274, 113016. 10.1016/j.rse.2022.113016

Zhu, Z. (2017). Change detection using landsat time series: A review of frequencies, preprocessing, algorithms, and applications. *ISPRS Journal of Photogrammetry and Remote Sensing*, 130, 370–384. 10.1016/j.isprsjprs.2017.06.013

Zhu, Z., Zhou, Y., Seto, K. C., Stokes, E. C., Deng, C., Pickett, S. T. A., & Taubenböck, H. (2019). Understanding an urbanizing planet: Strategic directions for remote sensing. *Remote Sensing of Environment*, 228, 164–182. 10.1016/j.rse.2019.04.020

7 The Potential of Nature-Based Solutions in Urban Heat Mitigation and Building Energy Savings

Siqi Jia, Qihao Weng, and Yuhong Wang

INTRODUCTION

Given the frequency and severity of heatwaves and Urban heat islands (UHI), there has been a growing awareness of the UHI and urban thermal environment (Ketterer & Matzarakis, 2015; Kim & Brown, 2021; Su et al., 2022). The increasing urban heat not only traps atmospheric pollutants, deteriorates the conditions of living environments, and increases energy consumption (Herath et al., 2018), but it also affects health, leisure activities, and well-being of urban dwellers (Doick et al., 2014; Feyisa et al., 2014; Halder et al., 2021). With more than half the world's population currently residing in cities and this number expected to rise to around 70% by 2050 (Profiroiu et al., 2020; Ritchie & Roser, 2018), combatting the adverse effects of urban heat has become a pressing global issue for both the environment and human beings. Consequently, it is crucial to develop effective heat mitigation solutions to alleviate these challenges (Aram et al., 2019; Wong et al., 2021).

Nature-based solutions (NBS) are increasingly recognized as effective measures to address the challenges of UHI and heatwaves by providing cooling services (Lafortezza et al., 2018; Su et al., 2022), and therefore help to increase energy efficiency. It can foster and simplify implementation actions in urban landscapes by taking into account the service provided by nature (Kabisch et al., 2017). Moreover, it is highlighted that NBS can be cost-effective and that benefits range from environmental protection to creating jobs that stimulate innovation for a green economy. Examples of NBS include provision of urban greenspace such as parks, street trees, and green infrastructure (e.g., green roof, green façade) that may ameliorate high temperature in cities or regulate air and water flows (Gill et al., 2007; Yamamoto, 2006). In this research, the performance of three commonly used NBS in urban heat mitigation and building

DOI: 10.1201/9781003244561-9

energy savings are introduced, including green roof, evaporative pavement, and street trees. Numerous studies have shown that NBS are effective in heat mitigation and energy use reduction (A. Niachou, 2001; Castleton et al., 2010; Romeo & Zinzi, 2013; Zheng & Weng, 2020), while their efficiency is significantly influenced by local climate conditions and building characteristics. However, most of those studies have only considered a few selected buildings in case studies or have focused primarily on the summer season (Costanzo et al., 2016; He et al., 2020), leading to a lack of research on how variations in ambient climate and built environment influence the energy-saving potential of these technologies.

Building energy consumption is greatly susceptible to meteorological feedback, which refers to the dynamic interactions between the building and the surrounding environment, particularly local weather conditions (Li et al., 2019). The heating, cooling, and ventilation requirements of a building are influenced by meteorological feedback, which can have a significant impact on energy consumption (Wong et al., 2011; Yu & Hien, 2006). Accurate prediction and modeling of meteorological feedback is crucial for optimizing building energy performance and reducing energy consumption. However, most existing building energy simulation models used Typical Meteorological Year (TMY) weather files to represent historical and current weather conditions (Fumo, 2014; Hosseini et al., 2020), while failed to incorporate ambient micro-scale climatic conditions and ensure accurate results in energy simulations.

Overall, a comprehensive study that compares these typical NBS in different climate settings is called for. To improve the model accuracy, this study proposes a new integrated approach to study the potential of building energy reductions by different NBS. This approach combines climate change modeling and building energy simulation to quantify the effect of climate change on building energy demand. The research objectives include: 1) to evaluate the effectiveness of NBS in urban heat mitigation; 2) to evaluate the effectiveness of NBS in building energy savings; 3) to identify the influencing factors affecting the performance of typical NBS in different urban built environments and climates. The results of this study can help policymakers and urban planners make informed decisions about the implementation of NBS in their respective cities, leading to greater sustainability and resilience of urban areas in the face of climate change. Furthermore, the study contributes to advancing the scientific understanding of the role of NBS in urban planning and highlights the importance of developing solutions that are tailored to specific local conditions.

NATURE-BASED SOLUTIONS FOR URBAN HEAT MITIGATION AND BUILDING ENERGY SAVINGS

In this research, the performance of three commonly used NBS in urban environment was evaluated, including green roof, evaporative pavement, and street trees. This section covers the mechanism, effectiveness, advantages, and limitations of each NBS.

Green Roofs

A green roof is a layered system that includes a waterproofing membrane, growing medium, and vegetation layer (Castleton et al., 2010). It has become an increasingly popular design strategy for improving urban environments due to its numerous environmental benefits, including reducing stormwater flow, improving air quality, and mitigating the Urban Heat Island (UHI) effect (Barrio, 1998; Berardi, 2016). Green roofs are widely recognized to mitigate increased urban temperatures, particularly in hot climates, where they can reduce temperatures to more comfortable levels (Razzaghmanesh et al., 2016). During winter conditions, they offer effective thermal protection by providing high insulation levels that strangle heat flux (He et al., 2017; Theodosiou, 2003). Green roofs can also lower cooling energy costs by 32% to 100% (Susca et al., 2011), while reducing heating energy demands by 23% (Zhao et al., 2015).

Green roofs can be classified into two types based on their purpose and characteristics: extensive green roofs (EGR) and intensive green roofs (IGR) (Townshend, 2007). EGR is covered with grass only and is designed to be self-sustaining. In contrast, IGR is usually associated with rooftop gardens and requires skilled labor, irrigation, and constant maintenance due to the deeper soil depth needed for growing trees and other vegetation (Bianchini & Hewage, 2012). IGR typically has thicker growing media and higher canopy density, resulting in higher levels of evapotranspiration and shading than EGR. This results in lower outdoor surface temperatures and better cooling effects (Castleton et al., 2010; Silva et al., 2016). However, IGR requires higher capital costs and technical expertise for installation and maintenance compared to EGR, which offers more economic benefits (Townshend, 2007). Figure 7.1 shows the appearance of both types of green roofs.

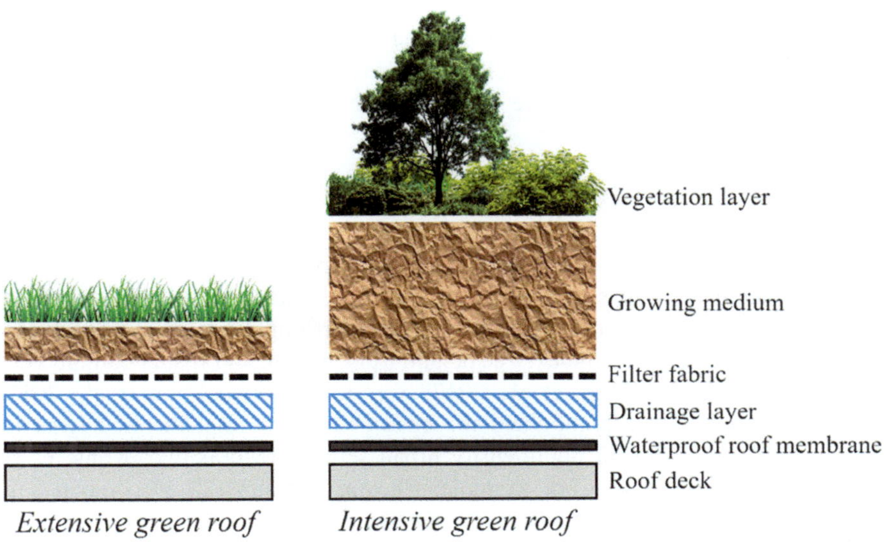

FIGURE 7.1 Extensive green roof and intensive green roof.

Nevertheless, although urban heat mitigation through roof-level strategies has often been investigated, the effectiveness of heat mitigation strategies in improving the pedestrian-level thermal environment has been questioned. For instance, because of the height of roofs, some researchers believed that the pedestrian-level cooling effects of green roofs are negligible (Chen et al., 2009; Zölch et al., 2016). In contrast, the street-level heat mitigation strategies are believed to be more effective in improving pedestrian's thermal comfort through shading, evaporation, and evapotranspiration (Morakinyo, Dahanayake, et al., 2017; Taleghani, 2018).

Evaporative and Water-Retentive Pavements

Evaporative and water-retentive pavements allow water to pass through their material voids, thereby facilitating its penetration into the soil. These pavements rely on evaporation of the water for the maintenance of a desirable thermal environment. Higher moisture contents and increased watering can help keep the pavement surface cooler. To achieve their water-retentive and permeable properties, these pavements eliminate finer particles from their structure or replace them with more porous materials such as pozzolans, slag, fly ash, or silica fumes (Mohajerani et al., 2017). The porosity of these materials imparts a lower albedo compared to impermeable counterparts, and enhances convective fluxes to the atmosphere by increasing the effective surface area via higher void content (Haselbach, 2009; Santamouris, 2013). According to ITO (ITO, 2006), water-retentive pavements can reduce surface temperatures by up to 16°C and air temperatures by approximately 1°C, contributing to the UHI mitigation. Another study on surface heat budgets demonstrated that adding water-retaining materials to asphalt and cement concrete surfaces significantly reduced sensible heat flux at rates of 150 W/m^2 and 100 W/m^2, respectively (Takebayashi & Moriyama, 2012). The effectiveness of evaporative and water-retentive pavements may be limited by climatic conditions (Santamouris, 2013). They are more suited to warm and humid climates where rainwater is a significant factor in cooling pavement surfaces. Figure 7.2 shows the progress of how water-retentive pavements cool the environment.

Street Trees

Urban areas can benefit from large tree or shrub-type greenspaces as they provide notable cooling through shade, evapotranspiration, and increased albedo (Wong et al., 2021). The cooling effect and thermal comfort improvement of trees depend on factors such as leaf area index, tree height, trunk height, and spatial configuration of trees (Morakinyo, Kong, et al., 2017; Tan et al., 2016; Zhou et al., 2017). Numerous studies have highlighted the effectiveness of implementing trees for outdoor heat mitigation in high-density urban settings (Morakinyo, Kong, et al., 2017; Thom et al., 2016). For example, Cheung and Jim (2018) compared the cooling effects of a tree and a concrete shelter, identifying that the tree provides significantly better cooling effects than the shelter with the mean daytime air temperature reduction at 0.6 °C and the maximum daytime air temperature

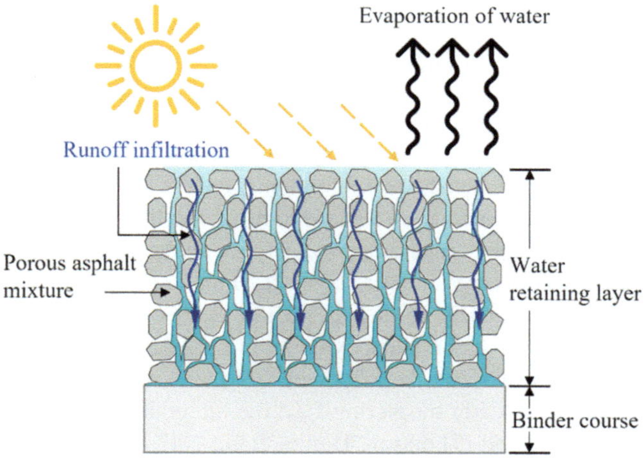

FIGURE 7.2 Water-retentive pavement.

reduction at 2.1 °C (Cheung & Jim, 2018a). Another study used high-resolution satellite data of 293 European cities to investigate the effectiveness of urban trees in mitigating urban heat. Results showed that the surface temperature observed for urban trees is on average 0–4 K lower in Southern European regions and 8–12 K lower in Central Europe (Schwaab et al., 2021). Increasing tree canopy in open and exposed streets should be given priority to achieve the greatest improvement in outdoor thermal comfort (Thom et al., 2016). However, the cooling effect of small greenspace with few trees or shrubs and grass-type greenspace has been questioned. Some studies have suggested that these types of greenspaces are more susceptible to urban and anthropogenic influences, which may increase sensible heat gain instead of providing cooling (Wong et al., 2021). Figure 7.3 shows an example of the typical design of street trees.

FIGURE 7.3 Street trees. Source: Google Map, captured at Nathan Road, Kowloon, Hong Kong.

METHODOLOGY

Assessment of Thermal Comfort

Mean Radiant Temperature

To objectively assess the thermal comfort level of a space, four crucial meteorological variables that govern the human energy balance and outdoor thermal comfort are typically utilized. These include air temperature (T_a), relative humidity (rH), wind speed (V_a) and mean radiant temperature (T_{mrt}). Particularly, T_{mrt} is a critical metric for human thermal comfort assessment (Mayer & Höppe, 1987; Thorsson, Lindberg, et al., 2007) for it is a quantity linking the radiant energy transfer between a human body and surroundings to which the body exposed (ASHRAE, 2013). While all other meteorological parameters can be obtained from on-site sensors, there are two common methods for directly measuring T_{mrt}: the integral radiation method and the globe thermometer method (Thorsson, Honjo, et al., 2007).

The integral radiation method, which was proposed by Höppe in 1992 (Höppe, 1992), has been widely recognized as the most accurate T_{mrt} measurement method (Thorsson, Lindberg, et al., 2007; VDI, 1998; Walikewitz et al., 2015). It is an experimental method that involves the measurement of short-wave and long-wave radiant flux densities emitted from the entire three-dimensional environment (Höppe, 1992; Manavvi & Rajasekar, 2020). However, it is difficult to undertake multi-point observations by this method in outdoor experiments with respect to the high cost and complexity of measurement setup. Therefore, the observations using the integral radiation method are often used as a reference to compare or modify the results of other methods (Chen et al., 2014; Crank et al., 2020; de Lieto Vollaro et al., 2013). Another commonly used method to obtain an estimate of T_{mrt} directly is the globe thermometer method. The black-globe thermometer method has been in use since it was first developed in the 1930s and then quickly gained attention due to its simplicity (Bedford & Warner, 1934). Referring to Kuehn et al. (Kuehn et al., 1970), assuming the globe thermometer is in equilibrium, the globe temperature, which is expressed as the weighted average of radiant and ambient temperatures, reflects the convective and radiative heat exchange around the black-globe thermometer. The globe thermometer method was initially developed to measure indoor radiant temperature. For outdoor use, an empirical equation is required to predict T_{mrt} based on additional parameters including the wind speed and air temperature (Eq. 7.1):

$$T_{mrt} = \left[(T_g + 273.15)^4 + \frac{1.1 \times 10^8 V_a^{0.6}}{\varepsilon D^{0.4}} \times (T_g - T_a) \right]^{1/4} - 273.15 \qquad (7.1)$$

where T_g is the globe temperature (°C), V_a is the air velocity (m/s), T_a is the air temperature (°C), D is the globe diameter (mm), and ε is the globe emissivity. However, the globe thermometer method has some shortcomings in outdoor applications. For example, previous studies revealed the unsuitability of the globe thermometer for T_{mrt} measurement in an outdoor situation, given the constantly changing radiative fluxes and air velocity may prolong the time to reach equilibrium (Spagnolo & de Dear, 2003).

Outdoor Thermal Comfort Indices

Researchers have dedicated substantial effort towards developing metrics for quantifying human thermal comfort (Herrmann & Matzarakis, 2010; Höppe, 1999; Makaremi et al., 2012). Human thermal comfort is described as a state of mind that reflects satisfaction or dissatisfaction with the surrounding environment (Abdel-Ghany et al., 2013). An individual's thermal state is affected by micro-climatic conditions such as wind speed, temperature, humidity, and the positions of sun and shade (Cheng et al., 2011; Comfort, 2003; Lai et al., 2014). To assess outdoor thermal comfort, several indices integrating environmental factors and the energy balance of the human body have been developed (Xi et al., 2012). Some examples of these indices are the Predicted Mean Vote (PMV), Physiological Equivalent Temperature (PET), and Universal Thermal Climate Index (UTCI) (Fang et al., 2019; Lai et al., 2014; Provencal et al., 2016).

Although all of these indices have been applied to assess human responses to the environment, the use of steady-state indices (e.g., PMV) may lead to estimation errors for pedestrians in an outdoor environment where dynamic thermal adaptation and transient process are involved. For instance, studies by Nikolopoulou et al. (Nikolopoulou et al., 2001) and Lai et al. (Lai et al., 2014) found significant discrepancies between a person's subjective thermal satisfaction and objective thermal states described by PMV. Hence, recent studies commonly use PET and UTCI as outdoor thermal indices (Cheung & Jim, 2018b). Both indices simulate dynamic heat transport within the body and the heat exchange between the body and its environment (Fiala et al., 1999, 2001). Compared to the two-node models used for PET, multi-node models used for UTCI provide more detailed simulations of the human body, predicting both overall (whole-body) and local (multi-node) physiological responses (Fang et al., 2019; Jia & Wang, 2021; Rakha, 2015). Therefore, this study employed UTCI to represent outdoor thermal comfort.

Microclimatic Modeling

ENVI-met

For simulating the meteorological parameters and universal thermal indices, ENVI-met has been applied to account for the thermal exchange of every location in the environment with fine resolutions. ENVI-met is a Computational Fluid Dynamics (CFD) based microclimate and local air quality model. ENVI-met is capable of calculating both surface temperature and air temperature of different heights within the interval times for 24 to 48 hours (Hien et al., 2012). Key input parameters include weather conditions, initial soil wetness and temperature profiles, structures, the physical properties of urban surfaces, and plants (Berardi, 2016). ENVI-met can simulate the long-wave and short-wave radiative exchange within the plants, the plant canopy effects on convective heat transfer, evapotranspiration from soil and plants, heat conduction in the soil layer, and moisture-dependent thermal properties (Sailor, 2008; Yang et al., 2019; Zölch et al., 2016). ENVI-met consists of a one-dimensional (1D) boundary sub-model which is used to calculate the inflow profile and top boundary, a three-dimensional (3D) atmospheric sub-model which is the main model, and a 3D/1D soil sub-model (Huttner, 2012). Flow simulation in the atmospheric model is based on the Reynold-averaged Navier-Stokes (RANS) equations (Acero & Arrizabalaga, 2016), and Yamada and Mellor E-ε (Yamada & Mellor, 1975) is chosen as the turbulence model. Further details of ENVI-met can be found in the literature (Ali-Toudert & Mayer, 2006; Bruse & Fleer, 1998).

Scenario Development

The computational domain in this study is set as $x \times y \times z$ = 250 m × 250 m × 200 m. Both the buildings and streets in the study area are carefully modeled. Key input parameters include weather conditions, initial soil wetness and temperature profiles, structures, the physical properties of urban surfaces, and plants (Berardi, 2016). Three types of land cover are considered for the walkways: soil, permeable surface using grasscrete paver, and asphalt pavement. Default albedo values are used for each type of land cover. The buildings (height: 30 m) and trees (height: 5 m, canopy width: 5 m) are distributed on these land cover types. Additional simulation details are shown in Table 7.1.

The base scenario represents the current outdoor microclimate, urban layout, and surface conditions of the study area. Additional simulation scenarios are developed for evaluating the influences of different types of heat mitigation strategies on outdoor microclimate and thermal comfort. The details of each simulation scenario are shown in Table 7.2, and the layouts of scenarios are displayed in Figure 7.4. The influences of four types of heat mitigation strategies (EGR, IGR, ST, and PP) and the reference (REF) on outdoor microclimate and thermal comfort are evaluated based on the outputs of ENVI-met, including surface air temperature (T_a), mean radiant temperature (T_{mrt}), relative humidity (rH), and wind speed (V_a).

TABLE 7.1
ENVI-met Simulation Details

Start and Duration of the Simulation	
Start date	1st August 2020
Start time	0:00 am
Total simulation time period (h)	24
Domain size and grid resolution	
Size of computational domain (*dx*, *dy*, *dz*)	250 m × 250 m × 200 m
Grid resolution (*dx*, *dy*, *dz*)	2 m × 2 m × 2 m
Initial meteorological conditions	
Air temperature range (°*C*)	22–34
Wind speed at the height of 10 m (m/s)	2.5
Wind direction (deg)	0°
Roughness length at the measurement site (m)	0.1
Relative humidity (%)	60 –70

TABLE 7.2
Description of Simulation Scenarios

Scenarios	Details
Extensive green roof (EGR)	Buildings are installed with extensive green roofs (covered by grass)
Intensive green roof (IGR)	Buildings are installed with intensive green roofs (covered by shrubs and small trees)
Permeable pavement (PP)	The pavement is covered by the permeable surface using grasscrete paver
Street tree (ST)	More street trees are added on vacant spaces along the sidewalks

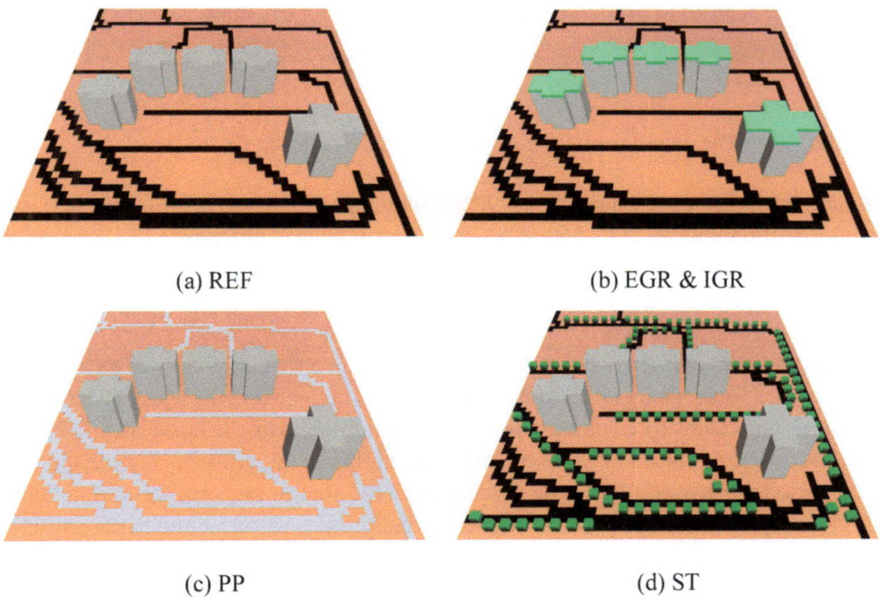

FIGURE 7.4 The layouts of simulation scenarios: (a) reference (REF); (b) extensive green roof (EGR) & intensive green roof (IGR); (c) permeable pavement (PP); and (d) street trees (ST).

Validation of the Meteorological Outputs

The general validity of the ENVI-met model was evaluated from an experimental site in Hong Kong (Figure 7.5(a)). The experimental area is 100 m × 100 m with a grid resolution of 0.5 m. Two sets of experiments were conducted under sunny conditions on different dates and seasons (October 29–31, 2020; April 3–5, 2021) at the same site. The diurnal profiles obtained from the three-day observations were used to validate the model.

To measure and record the background weather parameters, a Kestrel 5400 weather station (Nielsen-Kellerman Co. USA) was set up at a height of 1.5 m above the ground level (Figure 7.5(b)). The following parameters were measured: air temperature (T_a, range: −29 to 70 °C), globe temperature (T_g, range: −29 to 70 °C),

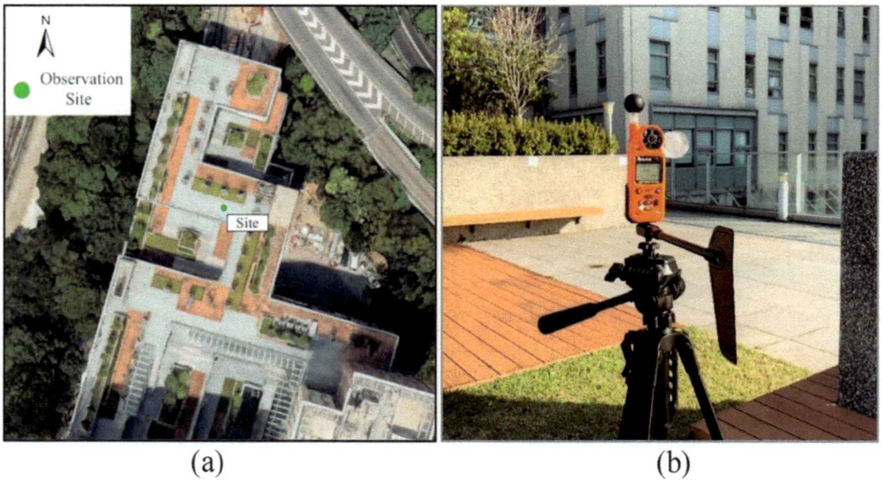

FIGURE 7.5 The experimental site (a) and experimental setup (b).

relative humidity (rH, range: 10 to 90%), wind speed (V_a, range: 0 to 40 m/s), and wind direction (range: 0 to 360 °C). The mean radiant temperature (T_{mrt}) based on the globe thermometer method (Chen et al., 2014; Mehrotra et al., 2019; Standard & ISO, 1998; Tan et al., 2013) was calculated using the globe temperature, air temperature, and wind speed data. In the second experiment, another Kestrel weather station was installed at the inlet boundary of the modeled area to measure the inlet wind speed. The global radiation data was collected with a CMP 3 Kipp & Zonen pyranometer to calculate the solar radiation adjustment factor in ENVI-met. The hourly mean values of air temperature and relative humidity during the modeled time period were used for simple forcing in ENVI-met to perform the simulations.

The outputs were compared with the diurnal profiles of measured meteorological variables. Table 7.3 provides two error terms obtained from ENVI-met and field measurements: the mean absolute error (MAE), and the error range. The relatively low MAE values in both experiments indicate that the prediction results for the four meteorological parameters are partially accurate. ENVI-met was then applied to the scenario analysis in this study.

TABLE 7.3
Estimation Errors of Meteorological Parameters in Two Experiments

	Autumn (October 2020)		Spring (April 2021)	
	MAE	Range	MAE	Range
T_a (°C)	0.82	−1.70 ~ 1.12	1.15	−3.79 ~ 0.76
T_{mrt} (°C)	1.01	−4.28 ~ 2.61	1.96	−5.79 ~ 3.87
rH (%)	3.40	−3.91 ~ 10.19	2.84	−2.51 ~ 12.22
V_a (m/s)	0.11	−0.19 ~ 0.21	0.17	−0.41 ~ 0.17

BUILDING ENERGY SIMULATION

EnergyPlus

This research proposes an integrated modeling approach that combines ENVI-met and a building energy simulation program (i.e., EnergyPlus). Figure 7.6 illustrates the development of the integrated modeling approach. The model first simulates microclimatic conditions in various urban built environments under different climates using ENVI-met. For the model inputs, the climate conditions incorporate the climates of six evaluated cities, while the built environments incorporate the typical urban building types based on the local climate zone (LCZ) system. Case studies were conducted in six selected cities in different climatic zones around the world, including Cairo, Hong Kong, Seoul, London, Los Angeles, and Sao Paulo. The forcing weather data required for the simulations was obtained from

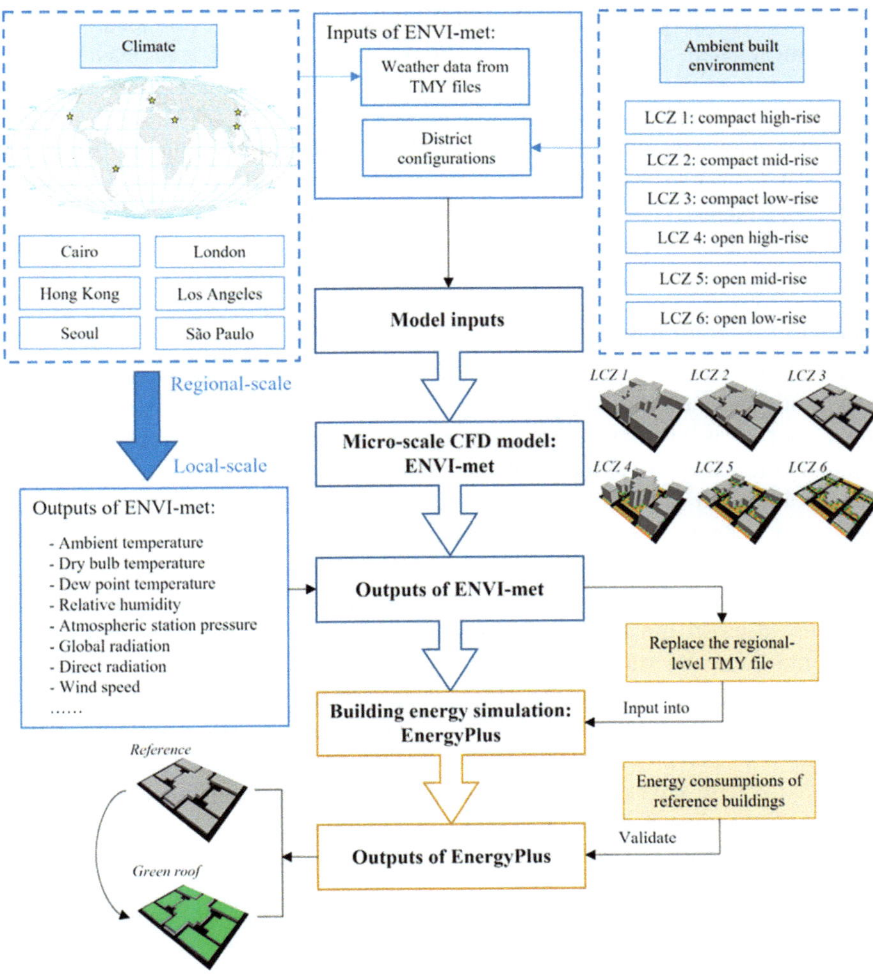

FIGURE 7.6 Development of the integrated modeling approach.

EnergyPlus Weather (EPW) files for each city. ENVI-met can generate much more detailed meteorological outputs (e.g., ambient temperature, relative humidity, wind speed, and reflectance) compared to the region-level EPW files. The outputs of ENVI-met are then used to configure EnergyPlus by modifying the TMY weather inputs. Based on the micro-scale environmental parameters, the detailed building energy use is simulated for different modeled environments. EnergyPlus is an open-source simulation program used for estimating building energy use, including heating, cooling, ventilation, and lighting (Crawley et al., 2001). It enables users to model buildings with mechanical and electrical systems that have integrated thermal controls and perform simulations based on real building descriptions (Morakinyo, Dahanayake, et al., 2017). EnergyPlus has been widely used to help optimize building energy performance and improve indoor thermal comfort while reducing energy consumption (Fumo et al., 2010).

The integrated modeling framework described in this research offers a powerful tool for conducting detailed observations and simulations of local physical environments and meteorological conditions. This approach can lead to more accurate estimations of building energy use compared to traditional modeling methods. In this study, the framework is then applied to a set of scenarios after validation. The scenario development involved modeling the building energy use with or without energy-saving solutions in different ambient climates and urban built environments. By comparing the results of these simulations with those of the baseline scenarios, we are able to determine the effectiveness of cool and green roofs for reducing building energy use in different environments.

Scenario Development

mong the three evaluated NBS (i.e., green roof, permeable pavement, and street trees), only the building energy-saving performance of roof-level strategies (i.e., green roof) was evaluated. For scenarios incorporating the effect of ambient climate, the distinct climate of six typical cities around the world was incorporated into scenarios evaluating the energy-saving performance of green roofs. For scenarios incorporating the effect of urban built environment, six scenarios incorporating different urban built environments were constructed according to the LCZ system (Stewart & Oke, 2012; Zheng et al., 2018). LCZs are being used in the numerous studies focusing on temperature, rainfall, and other environmental variables (Das & Das, 2020; Patel et al., 2023; Yang et al., 2021; Yang et al., 2022). In this study, each scenario corresponds to a different LCZ, including: 1) Scenario a (LCZ 1, compact high-rise environment); 2) Scenario b (LCZ 2, compact mid-rise environment); 3) Scenario c (LCZ 3, compact low-rise environment); 4) Scenario d (LCZ 4, open high-rise environment); 5) Scenario e (LCZ 5, open mid-rise environment); and 6) Scenario f (LCZ 6, open low-rise environment). Table 7.4 presents geometric and surface cover properties for each LCZ.

The layout corresponding to each LCZ was modeled in ENVI-met. The computational domain in this study was set as X × Y = 200 m × 225 m, with grid cells of 5 m. Two types of land cover – natural surfaces (covered by soil) and concrete pavement – were considered to characterize the pervious and

TABLE 7.4
Geometric and Surface Cover Properties for Each LCZ

LCZ	Description	Sky View Factor	Aspect Ratio	Building Surface Fraction (%)	Impervious Surface Fraction (%)	Pervious Surface Fraction (%)	Height (m) of Roughness Elements
LCZ 1	Compact high-rise	0.2–0.4	>2	40–60	40–60	<10	>35
LCZ 2	Compact mid-rise	0.3–0.6	0.75–2	40–70	30–50	<20	10–25
LCZ 3	Compact low-rise	0.2–0.6	0.75–1.5	40–70	20–50	<30	3–10
LCZ 4	Open high-rise	0.5–0.7	0.75–1.-25	20–40	30–40	30–40	>25
LCZ 5	Open mid-rise	0.5–0.8	0.3–0.75	20–40	30–50	20–40	10–25
LCZ 6	Open low-rise	0.6–0.9	0.3–0.75	20–40	20–50	30–60	3–10

Source: Stewart & Oke (Stewart & Oke, 2012).

impervious surfaces within the modeled space, respectively. Default albedo values were used for the different types of land cover. For modeled spaces of compact urban densities corresponding to LCZs 1 to 3, the building surface fraction was 60%, while the impervious surface fraction was 40%. A total of seven buildings were uniformly distributed on the concrete surfaces, and there was no previous surface among these three modeled spaces. For LCZs 4 to 6, six buildings were included in the modeled environment, with a total building surface fraction of 30%. The impervious and pervious surface fractions were 40% and 30%, respectively. Trees (height: 10 m) and shrubs (height: 0.5 m) were uniformly distributed on the natural surfaces with a canopy width of 10 m. Assuming an average floor height of 3 m, the building heights for all buildings within the modeled space ranged from 30 to 60 m (10–20 story height) for LCZs 1 and 4, 15 – 24 m (5–8 story height) for LCZs 2 and 5, and 3 – 9 m (1–3 story height) for LCZs 3 and 6.

Validation of the Building Energy Use

The proposed methodology was validated using a building energy dataset obtained from the U.S. Department of Energy's Building Technologies Office (Luo et al., 2022). The target building is a medium-sized office building (i.e., Building 59), located in Berkeley, California. The building has four floors (only the top two floors are the office floors) and a conditioned space of 10,400 m^2. The building's roof is composed of a white single-ply PVC roofing membrane installed on a concrete roof deck. Figure 7.7 depicts the appearance and layout of the model building.

The dataset comprises hourly data of whole-building and end-use energy consumption, HVAC system operating conditions, outdoor and indoor

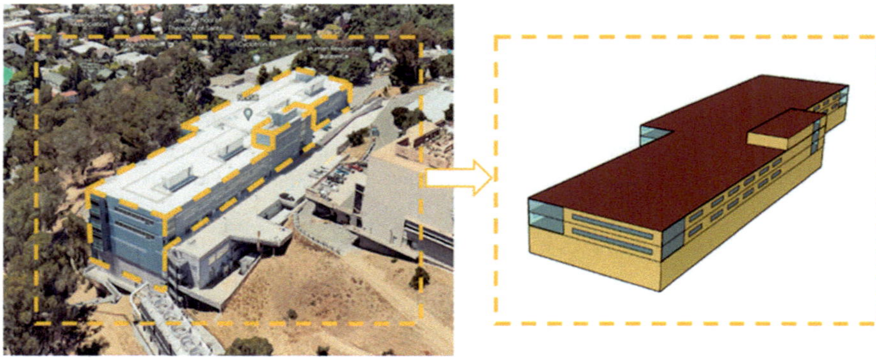

FIGURE 7.7 Appearance and layout of the model building.

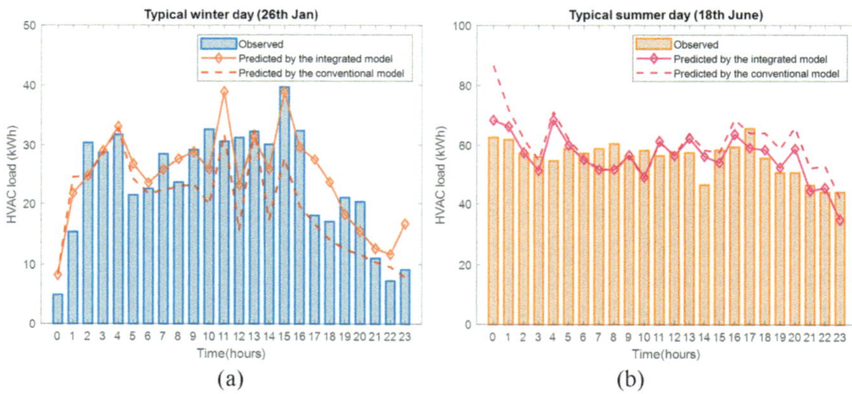

FIGURE 7.8 Accuracy of the hourly energy use simulation: (a) on a typical winter day; (b) on a typical summer day.

environmental parameters, and occupant data. The dataset covers a period of three years from 2018 to 2020. To validate the accuracy of the proposed integrated method, we first simulated the diurnal profiles of energy consumption in the reference office building for a typical winter day and summer day. We then compared the simulated hourly profiles with the observed hourly building energy use on those two days. After validation, the proposed method was used to simulate building consumption in various scenarios. The diurnal profiles of energy consumption in the reference office building were simulated on two typical days: the day with the minimum daily ambient temperature (January 26, and the day with the maximum temperature (June 18). Figure 7.8 shows the accuracy of the simulated hourly building use by the integrated modeling approach on both days, along with the building energy use predicted by the conventional method with inputting the regional weather data as a comparison.

The proposed method using micro-level weather data is capable of accurately simulating building energy use at a high temporal resolution. On a typical winter day (Figure 7.8(a)), the proposed method reduces the root-mean-square error

(RMSE) from 7.25 kWh (1.27%) with the conventional method to 4.96 kWh (0.87%). On a typical summer day (Figure 7.8(b)), the proposed method also outperforms the conventional method with an RMSE of 6.08 kWh (0.46%), which is 3.09 kWh lower than that of the conventional method (0.69%). This method has further potential applications for building energy simulations in various climates and built environments after further validation.

RESULTS

Performance of Nature-Based Solutions in Urban Heat Mitigation

The heat mitigation potentials of the different types of heat mitigation strategies are analyzed. Figure 7.9 presents the diurnal air temperature, mean radiant temperature, and UTCI at the pedestrian-level (at a height of 1.5 m) of the walkways for the five simulation scenarios (one reference and four NBS). Note that the temperature shown in the figure is the average temperature of all the points on the walkways for each simulation scenario.

The figure reveals the effectiveness of different types of heat mitigation strategies at improving the thermal environment, especially between 13:00 and 17:00 when it is the hottest. All evaluated NBS can cause some reductions in air temperature. Compared with the reference, street trees cause a maximum reduction in air temperature and mean radiant temperature by 0.27 °C and 4.23 °C, respectively. According to Figure 7.9(c), street trees also result in the lowest values of UTCI during the daytime, with a maximum UTCI reduction of 0.8817 °C at 11:00. The UTCI values of all other heat mitigation strategies are only slightly lower than those of the reference. At night, there is almost no difference in UTCI values among all of the scenarios, except for the slightly increased values under the ST scenario. Notably, the thermal environment may affect air conditioning use in the neighborhood, as lower environmental temperatures may be associated with less heat emission from air conditioning. This factor, however, is not incorporated into estimates of UTCI values. Next, the distributions of UTCI reductions compared to the reference are shown in Figure 7.10. UTCI reduction value is calculated by subtracting the UTCI value of the simulation scenario from the reference value at the same position. All of the values in the figure are obtained from the simulation results at 16:00 when the reference's UTCI reaches a maximum.

Figure 7.10 reveals that there is an interactive effect between the location and type of heat mitigation strategy. For the roof-level NBS (EGR and IGR, see Figure 7.10(a) and (b)), the effects of green roofs on improving pedestrian-level thermal comfort are more noticeable along the E–W directions of buildings, and the thermal comfort improvement is more obvious in the S-W corner of the study area. In the PP scenario (Figure 7.10(d)), the biggest UTCI reductions occur along E-W oriented walkways in the southern part of the study area. The figure also suggests that adding street trees causes the most significant reductions in UTCI over the entire neighborhood. The improvements to thermal comfort are remarkable in some street tree locations, corresponding to the dark blue pixels shown in Figure 7.10(c).

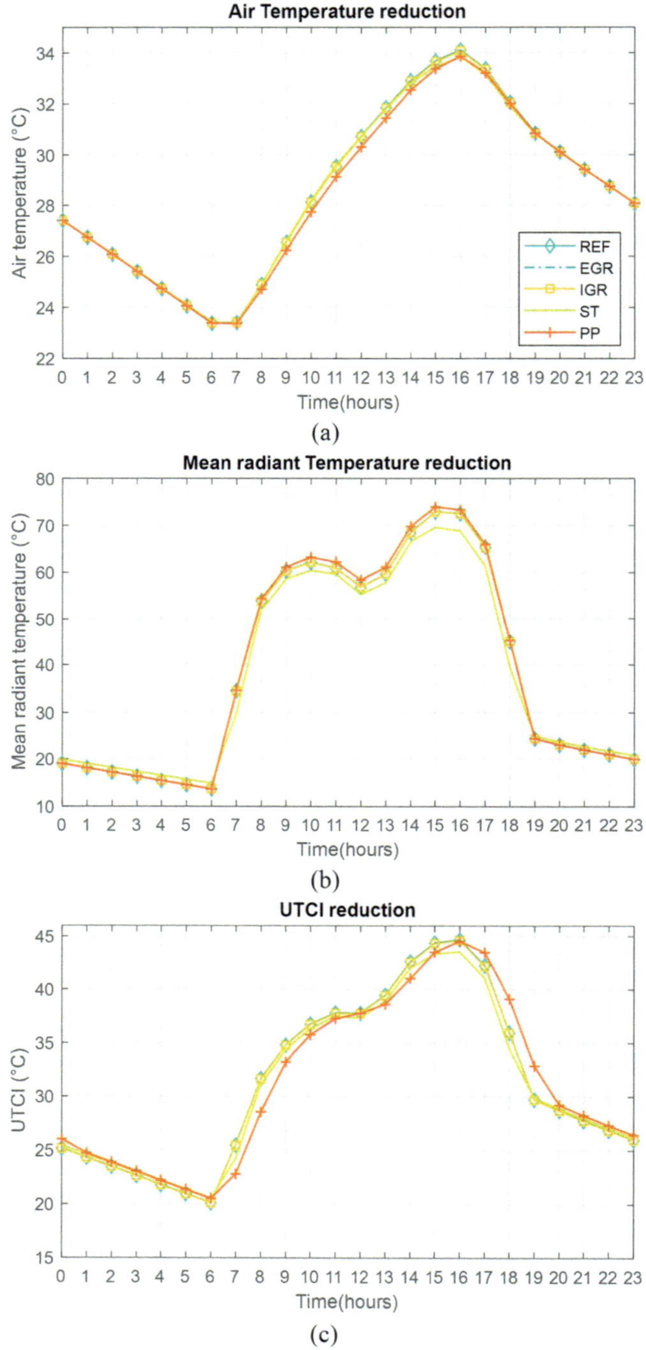

FIGURE 7.9 Diurnal profiles of the performance of NBS in heat mitigation: (a) air temperature reduction; (b) mean radiant temperature reduction; (c) UTCI reduction.

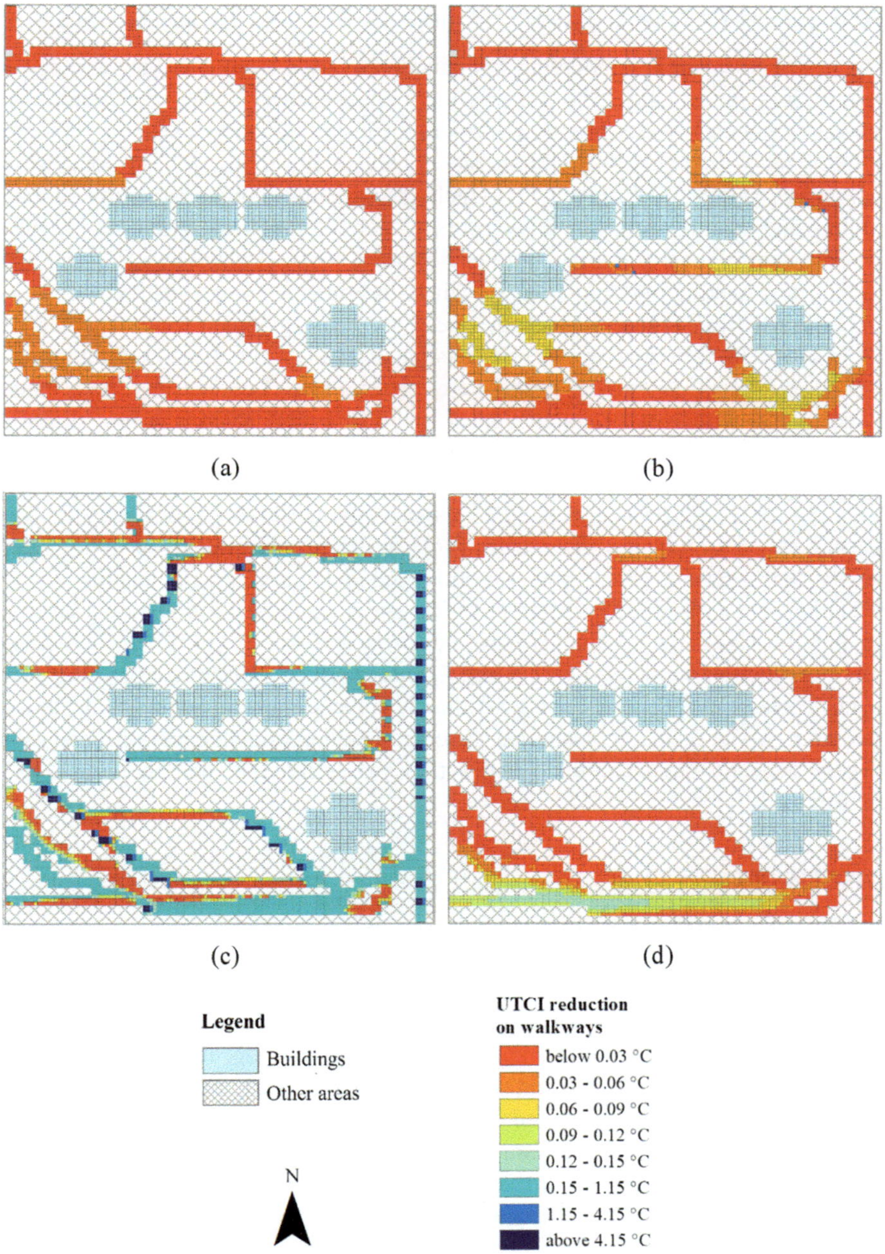

FIGURE 7.10 Distributions of UTCI reductions for all NBS: (a) EGR; (b) IGR; (c) ST; (d) PP.

Performance of Nature-Based Solutions in Building Energy Savings

The proposed integrated modeling framework was used to simulate the energy use of buildings in six cities with varying local climates over a one-year period. We compared the energy consumption of buildings installing green roofs to a reference building using concrete roofs without any energy-saving strategies. The comparison results are shown in Figure 7.11, which displays the annual reduction of both cooling and heating energy by green roofs in each city. Figure 7.11(a) and (b) show the effects of climate and urban built environment (characterized by LCZs) on both cooling and heating building energy savings by green roofs, respectively.

Observed from Figure 7.11(a), the mean reduction of green roofs ranged from 63.38% to 83.21% among the six evaluated cities. The best cooling energy reduction effects for green roofs were observed in London and Sao Paulo, respectively, with mean reduction rates of 78.63% and 83.21% across different built environments. The heating energy reduction effect of green roofs was relatively significant in Sao Paulo, Cairo, and Los Angeles with mean reduction rates of 87.92%, 85.65%, and 80.37%, respectively. Observed from Figure 7.11(b), urban built environment plays a crucial role in determining the green roof energy-saving performance. Significant variations in both cooling and heating energy reductions were observed in different LCZs, with mean reduction rate ranging from 38.38% to 100% for cooling energy reduction and 47.4% to 100% for heating energy reduction. Notably, LCZs 3 and 6, characterized by low-rise buildings, can achieve nearly zero annual energy demands for cooling or heating by installing green roofs. The results indicate that the energy-saving effectiveness of green roofs largely depends on the climate zones they are applied in and the urban built environment.

DISCUSSION

The Effectiveness of NBS in Cooling the Environment and Improving Thermal Comfort

The simulation results show the effectiveness of different NBS towards cooling the environment and improving thermal comfort. Trees and vegetation seem to be the key strategy to reduce temperature and improve thermal comfort in hot summer conditions through their combined effects of shading and evapotranspiration (Ali-Toudert & Mayer, 2007; Makaremi et al., 2012; Wang & Akbari, 2016). At night, the mean radiant temperature of the street tree (ST) scenario is higher than that of the reference. This may be explained by the subdued evapotranspiration of vegetation at nighttime (Peng et al., 2020). Although many existing studies conclude that green roofs have no cooling effect on pedestrian-level microclimate (Chen et al., 2009; Zölch et al., 2016), this study finds that there is some cooling brought about by roof-level heat mitigation strategies, especially by the intensive green roofs (IGR). The more voluminous biomass and extended canopy areas of IGR may provide some passive cooling at the pedestrian-level through evapotranspiration (Peng & Jim, 2013).

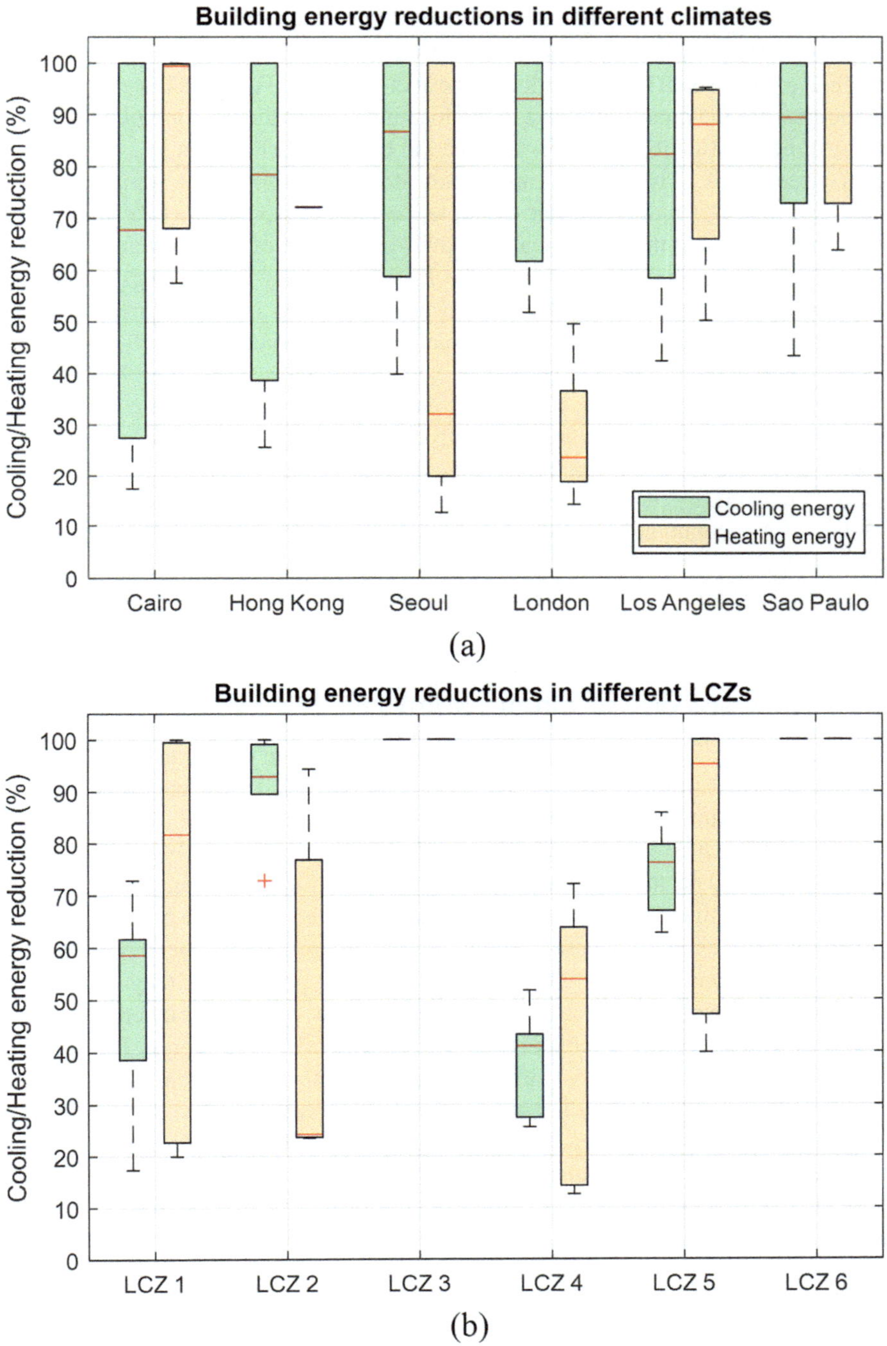

FIGURE 7.11 Annual cooling/heating energy reductions by green roofs: (a) in different climates: (b) in different LCZs.

Among all evaluated types of NBS, street trees cause the biggest reduction in average UTCI, with a maximum reduction of 0.8817 °C at 11:00, while the extensive green roof (EGR) has the least effect on UTCI. For the ST scenario, UTCI reductions are the most significant at the locations where the trees are planted. Other types of heat mitigation strategies also lead to UTCI reductions but to a lesser extent than street trees. These other mitigation strategies also show some locational variation in UTCI reduction. For example, permeable pavement (PP) shows more noticeable effects at the paths located in the south region of the study area, and IGR causes more obvious changes along the E–W directions of buildings. Therefore, location effects may be considered when planning green infrastructure.

Although adding street trees is beneficial, its implementation in well-established, densely populated urban areas may face some challenges. First, there is often less land space than is required to plant a tree and have it develop a mature root system. Second, as a city with a large number of tropical cyclones, falling trees often disrupt traffic and pose a public hazard in Hong Kong. These challenges necessitate the proper choosing of trees and planting strategies. In addition to street trees, some studies also recommend the use of protrusions from façades or buildings located next to the streets to reduce thermal stress exposure and improve walkability (Rodríguez-Algeciras et al., 2017). Such features are not evaluated in this study due to the difficulty of retrofitting existing buildings.

The Effectiveness of Green Roofs in Reducing Building Energy Use

In hot and sunny regions and relatively cold regions, green roofs can reliably reduce energy demand throughout the year. This finding is consistent with existing literature (Costanzo et al., 2016; Silva et al., 2016; Yang et al., 2018; Zinzi & Agnoli, 2012). Due to the extra insulation soil layer of green roofs, which tends to keep heat stored in the buildings, green roofs can provide heating energy-saving benefits as well in cities with colder climates. However, in some warm regions with cloudy or rainy climates, where greenery has less cooling effect due to reduced rates of evaporation and transpiration of plants, the use of green roofs may be carefully considered. Under this case, cool/reflective pavements with higher albedo (e.g., cool roofs) maybe adopted for reducing cooling energy use of buildings. The kinds of strategies reduce surface and air temperature in the surrounding environment mainly by increasing the percentage of reflected solar radiation due to an increase of reflectivity (Mohajerani et al., 2017; Wang et al., 2021). In addition, since urban built environment can significantly affect the performance of green roofs, the possible overshadowing effects of adjacent buildings should be carefully considered and modeled. Some studies also recommend conducting solar simulation before implementing these mitigation strategies (Wong et al., 2021).

When deciding on the implementation of green roofs in the long-term development of buildings, it is also important to consider not only their effectiveness in reducing energy consumption but also their economic and environmental benefits. For example, compared to cool roofs, green roofs provide additional benefits such as reducing urban heat island effects, improving air quality, increasing water-permeable surface, enhancing stormwater management, and improving the durability of roof materials (Hamel et al., 2021; Meng et al., 2022; Wong et al., 2021). Especially in high-urbanized regions, the application of green roofs can bring more ecosystem service benefits to both urban environments and humans.

CONCLUSION

Given the growing issue of urban heat islands (UHI) and increasing global building energy demand, this study aimed to assess the effectiveness of various nature-based solutions (NBS), including green roofs, permeable pavements, and street trees, in mitigating UHI, improving thermal comfort, and reducing building energy consumption. To improve modeling accuracy, this study developed an integrated modeling approach that combines climate change modeling and building energy simulation to quantify the impact of microclimate change on building energy demand. Through scenario analysis, the impacts of each NBS scenario on air temperature (T_a) reduction, mean radiant temperature (T_{mrt}) reduction, and Universal Thermal Climate Index (UTCI) reduction were analyzed. Our results show that ST scenario was the most effective in reducing T_a, T_{mrt}, and UTCI. However, the effectiveness of each type of NBS varied depending on the location within the study area, highlighting the importance of considering urban forms in infrastructure planning. To further investigate the potential for NBS to reduce building energy consumption, we evaluated the energy-saving performance of green roofs across different ambient climates and urban environments in six global cities: Cairo, Hong Kong, Seoul, London, Los Angeles, and Sao Paulo.

In summary, this study provides important insights into the effects of local ambient climate and urban built environment on the performance of typical NBS in UHI mitigation and building energy savings, which can inform the selection and design of appropriate NBS for different urban contexts to cope with climate change. Limitations of the study however include the fact that this paper largely depends on computer simulations, while the large-scale and long-term experimental data for energy use in buildings are yet to be carried out globally for different NBS. Future works should incorporate this experimental data to gain a better understanding of the holistic potential of NBS under different ambient environments. By incorporating wider study areas, more specific advice can be provided for cooling the urban environment, reducing building energy use and greenhouse gas emissions globally.

REFERENCES

Abdel-Ghany, A. M., Al-Helal, I. M., & Shady, M. R. (2013). Human Thermal Comfort and Heat Stress in an Outdoor Urban Arid Environment: A Case Study. *Advances in Meteorology*, 2013, 1–7. 10.1155/2013/693541

Acero, J. A., & Arrizabalaga, J. (2016). Evaluating the performance of ENVI-met model in diurnal cycles for different meteorological conditions. *Theoretical and Applied Climatology*, 131(1–2), 455–469. 10.1007/s00704-016-1971-y

Ali-Toudert, F., & Mayer, H. (2006). Numerical study on the effects of aspect ratio and orientation of an urban street canyon on outdoor thermal comfort in hot and dry climate. *Building and Environment*, 41(2), 94–108. 10.1016/j.buildenv.2005.01.013

Ali-Toudert, F., & Mayer, H. (2007). Effects of asymmetry, galleries, overhanging façades and vegetation on thermal comfort in urban street canyons. *Solar Energy*, 81(6), 742–754. 10.1016/j.solener.2006.10.007

Aram, F., Higueras Garcia, E., Solgi, E., & Mansournia, S. (2019). Urban green space cooling effect in cities. *Heliyon*, 5(4), e01339. 10.1016/j.heliyon.2019.e01339

ASHRAE, F. (2013). *Fundamentals handbook*. IP Edition, 21.

Barrio, D. (1998). Analysis of the green roofs cooling potential in buildings.

Bedford, T., & Warner, C. (1934). The globe thermometer in studies of heating and ventilation. *Epidemiology & Infection*, 34(4), 458–473.

Berardi, U. (2016). The outdoor microclimate benefits and energy saving resulting from green roofs retrofits. *Energy and Buildings*, 121, 217–229. 10.1016/j.enbuild.2016.03.021

Bianchini, F., & Hewage, K. (2012). How "green" are the green roofs? Lifecycle analysis of green roof materials. *Building and Environment*, 48, 57–65. 10.1016/j.buildenv.2011.08.019

Bruse, M., & Fleer, H. (1998). Simulating surface–plant–air interactions inside urban environments with a three dimensional numerical model. *Environmental Modelling & Software*, 13(3-4), 373–384.

Castleton, H. F., Stovin, V., Beck, S. B. M., & Davison, J. B. (2010). Green roofs; building energy savings and the potential for retrofit. *Energy and Buildings*, 42(10), 1582–1591. 10.1016/j.enbuild.2010.05.004

Chen, H., Ooka, R., Huang, H., & Tsuchiya, T. (2009). Study on mitigation measures for outdoor thermal environment on present urban blocks in Tokyo using coupled simulation. *Building and Environment*, 44(11), 2290–2299. 10.1016/j.buildenv.2009.03.012

Chen, Y.-C., Lin, T.-P., & Matzarakis, A. (2014). Comparison of mean radiant temperature from field experiment and modelling: a case study in Freiburg, Germany. *Theoretical and Applied Climatology*, 118(3), 535–551. 10.1007/s00704-013-1081-z

Cheng, V., Ng, E., Chan, C., & Givoni, B. (2011). Outdoor thermal comfort study in a sub-tropical climate: a longitudinal study based in Hong Kong. *International Journal of Biometeorology*, 56(1), 43–56. 10.1007/s00484-010-0396-z

Cheung, P. K., & Jim, C. Y. (2018a). Comparing the cooling effects of a tree and a concrete shelter using PET and UTCI. *Building and Environment*, 130, 49–61. 10.1016/j.buildenv.2017.12.013

Cheung, P. K., & Jim, C. Y. (2018b). Subjective outdoor thermal comfort and urban green space usage in humid-subtropical Hong Kong. *Energy and Buildings*, 173, 150–162. 10.1016/j.enbuild.2018.05.029

Comfort, A. S. o. C. E. T. C. o. O. H. (2003). Outdoor Human Comfort and Its Assessment: State of the Art.

Costanzo, V., Evola, G., & Marletta, L. (2016). Energy savings in buildings or UHI mitigation? Comparison between green roofs and cool roofs. *Energy and Buildings*, 114, 247–255. 10.1016/j.enbuild.2015.04.053

Crank, P. J., Middel, A., Wagner, M., Hoots, D., Smith, M., & Brazel, A. (2020). Validation of seasonal mean radiant temperature simulations in hot arid urban climates. *Science of The Total Environment*, 749. 10.1016/j.scitotenv.2020.141392

Crawley, D. B., Lawrie, L. K., Winkelmann, F. C., Buhl, W. F., Huang, Y. J., Pedersen, C. O., Strand, R. K., Liesen, R. J., Fisher, D. E., & Witte, M. J. (2001). EnergyPlus: creating a new-generation building energy simulation program. *Energy and Buildings*, 33(4), 319–331.

Das, M., & Das, A. (2020). Assessing the relationship between local climatic zones (LCZs) and land surface temperature (LST)–A case study of Sriniketan-Santiniketan Planning Area (SSPA), West Bengal, India. *Urban Climate*, 32, 100591.

de Lieto Vollaro, R., Vallati, A., & Bottillo, S. (2013). Differents Methods to Estimate the Mean Radiant Temperature in an Urban Canyon. *Advanced Materials Research*, 650, 647–651. 10.4028/www.scientific.net/AMR.650.647

Doick, K. J., Peace, A., & Hutchings, T. R. (2014). The role of one large greenspace in mitigating London's nocturnal urban heat island. *Science of The Total Environment*, 493, 662–671. 10.1016/j.scitotenv.2014.06.048

Fang, Z., Feng, X., Liu, J., Lin, Z., Mak, C. M., Niu, J., Tse, K.-T., & Xu, X. (2019). Investigation into the differences among several outdoor thermal comfort indices against field survey in subtropics. *Sustainable Cities and Society*, 44, 676–690. 10.1016/j.scs.2018.10.022

Feyisa, G. L., Dons, K., & Meilby, H. (2014). Efficiency of parks in mitigating urban heat island effect: An example from Addis Ababa. *Landscape and Urban Planning*, 123, 87–95. 10.1016/j.landurbplan.2013.12.008

Fiala, D., Lomas, K. J., & Stohrer, M. (1999). A computer model of human thermoregulation for a wide range of environmental conditions: the passive system. *Journal of Applied Physiology*, 87(5), 1957–1972.

Fiala, D., Lomas, K. J., & Stohrer, M. (2001). Computer prediction of human thermoregulatory and temperature responses to a wide range of environmental conditions. *International Journal of Biometeorology*, 45(3), 143–159.

Fumo, N. (2014). A review on the basics of building energy estimation. *Renewable and Sustainable Energy Reviews*, 31, 53–60. 10.1016/j.rser.2013.11.040

Fumo, N., Mago, P., & Luck, R. (2010). Methodology to estimate building energy consumption using EnergyPlus Benchmark Models. *Energy and Buildings*, 42(12), 2331–2337. 10.1016/j.enbuild.2010.07.027

Gill, S. E., Handley, J. F., Ennos, A. R., & Pauleit, S. (2007). Adapting cities for climate change: the role of the green infrastructure. *Built Environment*, 33(1), 115–133.

Halder, B., Bandyopadhyay, J., & Banik, P. (2021). Monitoring the effect of urban development on urban heat island based on remote sensing and geo-spatial approach in Kolkata and adjacent areas, India. *Sustainable Cities and Society*, 74. 10.1016/j.scs.2021.103186

Hamel, P., Guerry, A. D., Polasky, S., Han, B., Douglass, J. A., Hamann, M., Janke, B., Kuiper, J. J., Levrel, H., Liu, H., Lonsdorf, E., McDonald, R. I., Nootenboom, C., Ouyang, Z., Remme, R. P., Sharp, R. P., Tardieu, L., Viguié, V., Xu, D., Zheng, H., & Daily, G. C. (2021). Mapping the benefits of nature in cities with the InVEST software. *npj Urban Sustainability*, 1(1). 10.1038/s42949-021-00027-9

Haselbach, L. (2009). Pervious concrete and mitigation of the urban heat island effect.

He, Y., Yu, H., Ozaki, A., & Dong, N. (2020). Thermal and energy performance of green roof and cool roof: A comparison study in Shanghai area. *Journal of Cleaner Production*, 267. 10.1016/j.jclepro.2020.122205

He, Y., Yu, H., Ozaki, A., Dong, N., & Zheng, S. (2017). Influence of plant and soil layer on energy balance and thermal performance of green roof system. *Energy*, 141, 1285–1299. 10.1016/j.energy.2017.08.064

Herath, H. M. P. I. K., Halwatura, R. U., & Jayasinghe, G. Y. (2018). Evaluation of green infrastructure effects on tropical Sri Lankan urban context as an urban heat island adaptation strategy. *Urban Forestry & Urban Greening*, 29, 212–222. 10.1016/j.ufug.2017.11.013

Herrmann, J., & Matzarakis, A. (2010). Influence of mean radiant temperature on thermal comfort of humans in idealized urban environments. *Berichte des Meteorologischen Instituts der Albert-Ludwigs-Universität Freiburg*, 522.

Hien, W. N., Ignatius, M., Eliza, A., Jusuf, S. K., & Samsudin, R. (2012). Comparison of STEVE and ENVI-met as temperature prediction models for Singapore context.

International Journal of Sustainable Building Technology and Urban Development, 3(3), 197–209. 10.1080/2093761x.2012.720224

Höppe, P. (1992). A new procedure to determine the mean radiant temperature outdoors. *Wetter und Leben*, 44, 147–151.

Höppe, P. (1999). The physiological equivalent temperature–a universal index for the biometeorological assessment of the thermal environment. *International Journal of Biometeorology*, 43(2), 71–75.

Hosseini, M., Bigtashi, A., & Lee, B. (2020). A systematic approach in constructing typical meteorological year weather files using machine learning. *Energy and Buildings*, 226. 10.1016/j.enbuild.2020.110375

Huttner, S. (2012). *Further development and application of the 3D microclimate simulation ENVI-met*. Mainz: Johannes Gutenberg-Universitat in Mainz, 147.

ITO, M. (2006). Study on pavement technologies to mitigate the heat island effect and their effectiveness.

Jia, S., & Wang, Y. (2021). Effect of heat mitigation strategies on thermal environment, thermal comfort, and walkability: A case study in Hong Kong. *Building and Environment*, 201. 10.1016/j.buildenv.2021.107988

Kabisch, N., Korn, H., Stadler, J., & Bonn, A. (2017). *Nature-based solutions to climate change adaptation in urban areas: Linkages between science, policy and practice*. Springer Nature.

Ketterer, C., & Matzarakis, A. (2015). Comparison of different methods for the assessment of the urban heat island in Stuttgart, Germany. *International Journal of Biometeorology*, 59(9), 1299–1309. 10.1007/s00484-014-0940-3

Kim, S. W., & Brown, R. D. (2021). Urban heat island (UHI) variations within a city boundary: A systematic literature review. *Renewable and Sustainable Energy Reviews*, 148. 10.1016/j.rser.2021.111256

Kuehn, L., Stubbs, R., & Weaver, R. (1970). Theory of the globe thermometer. *Journal of Applied Physiology*, 29(5), 750–757.

Lafortezza, R., Chen, J., van den Bosch, C. K., & Randrup, T. B. (2018). Nature-based solutions for resilient landscapes and cities. *Environmental Research*, 165, 431–441. 10.1016/j.envres.2017.11.038

Lai, D., Guo, D., Hou, Y., Lin, C., & Chen, Q. (2014). Studies of outdoor thermal comfort in northern China. *Building and Environment*, 77, 110–118. 10.1016/j.buildenv.2014.03.026

Li, X., Zhou, Y., Yu, S., Jia, G., Li, H., & Li, W. (2019). Urban heat island impacts on building energy consumption: A review of approaches and findings. *Energy*, 174, 407–419. 10.1016/j.energy.2019.02.183

Luo, N., Wang, Z., Blum, D., Weyandt, C., Bourassa, N., Piette, M. A., & Hong, T. (2022). A three-year dataset supporting research on building energy management and occupancy analytics. *Scientific Data*, 9(1), 156. 10.1038/s41597-022-01257-x

Makaremi, N., Salleh, E., Jaafar, M. Z., & GhaffarianHoseini, A. (2012). Thermal comfort conditions of shaded outdoor spaces in hot and humid climate of Malaysia. *Building and Environment*, 48, 7–14. 10.1016/j.buildenv.2011.07.024

Manavvi, S., & Rajasekar, E. (2020). Estimating outdoor mean radiant temperature in a humid subtropical climate. *Building and Environment*, 171. 10.1016/j.buildenv.2020.106658

Mayer, H., & Höppe, P. (1987). Thermal comfort of man in different urban environments. *Theoretical and Applied Climatology*, 38(1), 43–49.

Mehrotra, S., Bardhan, R., & Ramamritham, K. (2019). Outdoor thermal performance of heterogeneous urban environment: An indicator-based approach for climate-sensitive planning. *Science of The Total Environment*, 669, 872–886. 10.1016/j.scitotenv.2019.03.152

Meng, L., Zhou, Y., Roman, M. O., Stokes, E. C., Wang, Z., Asrar, G. R., Mao, J., Richardson, A. D., Gu, L., & Wang, Y. (2022). Artificial light at night: an underappreciated effect on phenology of deciduous woody plants. *PNAS Nexus*, 1(2), pgac046. 10.1093/pnasnexus/pgac046

Mohajerani, A., Bakaric, J., & Jeffrey-Bailey, T. (2017). The urban heat island effect, its causes, and mitigation, with reference to the thermal properties of asphalt concrete. *Journal of Environmental Management*, 197, 522–538. 10.1016/j.jenvman.2017.03.095

Morakinyo, T. E., Dahanayake, K. W. D. K. C., Ng, E., & Chow, C. L. (2017). Temperature and cooling demand reduction by green-roof types in different climates and urban densities: A co-simulation parametric study. *Energy and Buildings*, 145, 226–237. 10.1016/j.enbuild.2017.03.066

Morakinyo, T. E., Kong, L., Lau, K. K.-L., Yuan, C., & Ng, E. (2017). A study on the impact of shadow-cast and tree species on in-canyon and neighborhood's thermal comfort. *Building and Environment*, 115, 1–17. 10.1016/j.buildenv.2017.01.005

Nikolopoulou, M., Baker, N., & Steemers, K. (2001). Thermal comfort in outdoor urban spaces: understanding the human parameter. *Solar Energy*, 70(3), 227–235.

Niachou, A., Papakonstantinou, K., Santamouris, M., Tsangrassoulis, A., Mihalakakou, G. (2001). Analysis of the green roof thermal properties and investigation of its energy performance. *Energy and Buildings*, 33, 719–729.

Patel, P., Kalyanam, R., He, L., Aliaga, D., & Niyogi, D. (2023). Deep learning-based urban morphology for city-scale environmental modeling. *PNAS Nexus*, 2(3), pgad027. 10.1093/pnasnexus/pgad027

Peng, L., & Jim, C. (2013). Green-Roof Effects on Neighborhood Microclimate and Human Thermal Sensation. *Energies*, 6(2), 598–618. 10.3390/en6020598

Peng, L. L. H., Jiang, Z., Yang, X., He, Y., Xu, T., & Chen, S. S. (2020). Cooling effects of block-scale facade greening and their relationship with urban form. *Building and Environment*, 169. 10.1016/j.buildenv.2019.106552

Profiroiu, C. M., Bodislav, D. A., Burlacu, S., & Rădulescu, C. V. (2020). Challenges of sustainable urban development in the context of population Growth. *European Journal of Sustainable Development*, 9(3), 51- 51.

Provencal, S., Bergeron, O., Leduc, R., & Barrette, N. (2016). Thermal comfort in Quebec City, Canada: sensitivity analysis of the UTCI and other popular thermal comfort indices in a mid-latitude continental city. *International Journal of Biometeorology*, 60(4), 591–603. 10.1007/s00484-015-1054-2

Rakha, T. (2015). Towards comfortable and walkable cities: spatially resolved outdoor thermal comfort analysis linked to travel survey-based human activity schedules. Massachusetts Institute of Technology.

Razzaghmanesh, M., Beecham, S., & Salemi, T. (2016). The role of green roofs in mitigating Urban Heat Island effects in the metropolitan area of Adelaide, South Australia. *Urban Forestry & Urban Greening*, 15, 89–102. 10.1016/j.ufug.2015.11.013

Ritchie, H., & Roser, M. (2018). Urbanization. Our world in data.

Rodríguez-Algeciras, J., Tablada, A., & Matzarakis, A. (2017). Effect of asymmetrical street canyons on pedestrian thermal comfort in warm-humid climate of Cuba. *Theoretical and Applied Climatology*, 133(3-4), 663–679. 10.1007/s00704-017-2204-8

Romeo, C., & Zinzi, M. (2013). Impact of a cool roof application on the energy and comfort performance in an existing non-residential building. A Sicilian case study. *Energy and Buildings*, 67, 647–657. 10.1016/j.enbuild.2011.07.023

Sailor, D. J. (2008). A green roof model for building energy simulation programs. *Energy and Buildings*, 40(8), 1466–1478. 10.1016/j.enbuild.2008.02.001

Santamouris, M. (2013). Using cool pavements as a mitigation strategy to fight urban heat island—A review of the actual developments. *Renewable and Sustainable Energy Reviews*, 26, 224–240. 10.1016/j.rser.2013.05.047

Schwaab, J., Meier, R., Mussetti, G., Seneviratne, S., Burgi, C., & Davin, E. L. (2021). The role of urban trees in reducing land surface temperatures in European cities. *Nature Communication*, 12(1), 6763. 10.1038/s41467-021-26768-w

Silva, C. M., Gomes, M. G., & Silva, M. (2016). Green roofs energy performance in Mediterranean climate. *Energy and Buildings*, 116, 318–325. 10.1016/j.enbuild.2016.01.012

Spagnolo, J., & de Dear, R. (2003). A field study of thermal comfort in outdoor and semi-outdoor environments in subtropical Sydney Australia. *Building and Environment*, 38(5), 721–738. 10.1016/s0360-1323(02)00209-3

Standard, I., & ISO, B. (1998). Ergonomics of the thermal environment—instruments for measuring physical quantities.

Stewart, I. D., & Oke, T. R. (2012). Local climate zones for urban temperature studies. *Bulletin of the American Meteorological Society*, 93(12), 1879–1900.

Su, W., Zhang, L., & Chang, Q. (2022). Nature-based solutions for urban heat mitigation in historical and cultural block: The case of Beijing Old City. *Building and Environment*, 225. 10.1016/j.buildenv.2022.109600

Susca, T., Gaffin, S. R., & Dell'osso, G. R. (2011). Positive effects of vegetation: urban heat island and green roofs. *Environmental Pollution*, 159(8-9), 2119–2126. 10.1016/j.envpol.2011.03.007

Takebayashi, H., & Moriyama, M. (2012). Study on Surface Heat Budget of Various Pavements for Urban Heat Island Mitigation. *Advances in Materials Science and Engineering*, 2012, 1–11. 10.1155/2012/523051

Taleghani, M. (2018). Outdoor thermal comfort by different heat mitigation strategies- A review. *Renewable and Sustainable Energy Reviews*, 81, 2011–2018. 10.1016/j.rser.2017.06.010

Tan, C. L., Wong, N. H., & Jusuf, S. K. (2013). Outdoor mean radiant temperature estimation in the tropical urban environment. *Building and Environment*, 64, 118–129. 10.1016/j.buildenv.2013.03.012

Tan, Z., Lau, K. K.-L., & Ng, E. (2016). Urban tree design approaches for mitigating daytime urban heat island effects in a high-density urban environment. *Energy and Buildings*, 114, 265–274. 10.1016/j.enbuild.2015.06.031

Theodosiou, T. G. (2003). Summer period analysis of the performance of a planted roof as a passive cooling technique. *Energy and Buildings*, 35(9), 909–917. 10.1016/s0378-7788(03)00023-9

Thom, J. K., Coutts, A. M., Broadbent, A. M., & Tapper, N. J. (2016). The influence of increasing tree cover on mean radiant temperature across a mixed development suburb in Adelaide, Australia. *Urban Forestry & Urban Greening*, 20, 233–242. 10.1016/j.ufug.2016.08.016

Thorsson, S., Honjo, T., Lindberg, F., Eliasson, I., & Lim, E.-M. (2007). Thermal comfort and outdoor activity in Japanese urban public places. *Environment and Behavior*, 39(5), 660–684.

Thorsson, S., Lindberg, F., Eliasson, I., & Holmer, B. (2007). Different methods for estimating the mean radiant temperature in an outdoor urban setting. *International Journal of Climatology*, 27(14), 1983–1993. 10.1002/joc.1537

Townshend, D. (2007). Study on green roof application in Hong Kong. *Architectural services department*.

VDI. (1998). *Methods for the human-biometerological assessment of climate and air Hygiene for Urban and Regional Planning*. In: Beuth Berlin.

Walikewitz, N., Jänicke, B., Langner, M., Meier, F., & Endlicher, W. (2015). The difference between the mean radiant temperature and the air temperature within indoor environments: A case study during summer conditions. *Building and Environment*, 84, 151–161. 10.1016/j.buildenv.2014.11.004

Wang, C., Wang, Z.-H., Kaloush, K. E., & Shacat, J. (2021). Cool pavements for urban heat island mitigation: A synthetic review. *Renewable and Sustainable Energy Reviews*, 146. 10.1016/j.rser.2021.111171

Wang, Y., & Akbari, H. (2016). Analysis of urban heat island phenomenon and mitigation solutions evaluation for Montreal. *Sustainable Cities and Society*, 26, 438–446. 10.1016/j.scs.2016.04.015

Wong, N. H., Jusuf, S. K., Syafii, N. I., Chen, Y., Hajadi, N., Sathyanarayanan, H., & Manickavasagam, Y. V. (2011). Evaluation of the impact of the surrounding urban morphology on building energy consumption. *Solar Energy*, 85(1), 57–71. 10.1016/j.solener.2010.11.002

Wong, N. H., Tan, C. L., Kolokotsa, D. D., & Takebayashi, H. (2021). Greenery as a mitigation and adaptation strategy to urban heat. *Nature Reviews Earth & Environment*, 2(3), 166–181. 10.1038/s43017-020-00129-5

Xi, T., Li, Q., Mochida, A., & Meng, Q. (2012). Study on the outdoor thermal environment and thermal comfort around campus clusters in subtropical urban areas. *Building and Environment*, 52, 162–170.

Yamada, T., & Mellor, G. (1975). A simulation of the Wangara atmospheric boundary layer data. *Journal of Atmospheric Sciences*, 32(12), 2309–2329.

Yamamoto, Y. (2006). Measures to mitigate urban heat islands (1349-3663).

Yang, J., Mohan Kumar, D. l., Pyrgou, A., Chong, A., Santamouris, M., Kolokotsa, D., & Lee, S. E. (2018). Green and cool roofs' urban heat island mitigation potential in tropical climate. *Solar Energy*, 173, 597–609. 10.1016/j.solener.2018.08.006

Yang, J., Ren, J., Sun, D., Xiao, X., Xia, J. C., Jin, C., & Li, X. (2021). Understanding land surface temperature impact factors based on local climate zones. *Sustainable Cities and Society*, 69, 102818.

Yang, J., Xin, J., Zhang, Y., Xiao, X., & Xia, J. C. (2022). Contributions of sea–land breeze and local climate zones to daytime and nighttime heat island intensity. *NPJ Urban Sustainability*, 2(1), 12.

Yang, Y., Gatto, E., Gao, Z., Buccolieri, R., Morakinyo, T. E., & Lan, H. (2019). The "plant evaluation model" for the assessment of the impact of vegetation on outdoor microclimate in the urban environment. *Building and Environment*, 159. 10.1016/j.buildenv.2019.05.029

Yu, C., & Hien, W. N. (2006). Thermal benefits of city parks. *Energy and Buildings*, 38(2), 105–120. 10.1016/j.enbuild.2005.04.003

Zhao, M., Srebric, J., Berghage, R. D., & Dressler, K. A. (2015). Accumulated snow layer influence on the heat transfer process through green roof assemblies. *Building and Environment*, 87, 82–91. 10.1016/j.buildenv.2014.12.018

Zheng, Y., Ren, C., Xu, Y., Wang, R., Ho, J., Lau, K., & Ng, E. (2018). GIS-based mapping of Local Climate Zone in the high-density city of Hong Kong. *Urban Climate*, 24, 419–448. 10.1016/j.uclim.2017.05.008

Zheng, Y., & Weng, Q. (2020). Modeling the Effect of Green Roof Systems and Photovoltaic Panels for Building Energy Savings to Mitigate Climate Change. *Remote Sensing*, 12(15). 10.3390/rs12152402

Zhou, W., Wang, J., & Cadenasso, M. L. (2017). Effects of the spatial configuration of trees on urban heat mitigation: A comparative study. *Remote Sensing of Environment*, 195, 1–12. 10.1016/j.rse.2017.03.043

Zinzi, M., & Agnoli, S. (2012). Cool and green roofs. An energy and comfort comparison between passive cooling and mitigation urban heat island techniques for residential buildings in the Mediterranean region. *Energy and Buildings*, 55, 66–76. 10.1016/j.enbuild.2011.09.024

Zölch, T., Maderspacher, J., Wamsler, C., & Pauleit, S. (2016). Using green infrastructure for urban climate-proofing: An evaluation of heat mitigation measures at the micro-scale. *Urban Forestry & Urban Greening*, 20, 305–316. 10.1016/j.ufug.2016.09.011

Part III

Geospatial Techniques for Resilient Cities

8 GeoAI for High-Resolution Urban Air Temperature Estimation and Urban Heat Island Monitoring

Cheolhee Yoo and Qihao Weng

INTRODUCTION

Although urban areas constitute merely 3% of the Earth's total surface, they accommodate more than half of the global population today (Seto et al., 2012). This urban population is projected to increase to 2.5 billion by 2050 (Angel et al., 2011). Urban areas are expanding rapidly, replacing existing natural land cover and vegetation with impervious surfaces (Foley et al., 2005). The significant solar absorption and high thermal capacity of these impervious surfaces warm the core areas of cities more than their peripheries, a phenomenon known as the Urban Heat Island (UHI) effect. The UHI effect is well-known to exacerbate environmental issues such as air pollution and drought (Stone, 2005; Im & Eltahir, 2018). More importantly, its negative impact on human health, including issues like heatstroke and mental health, has been extensively revealed, making the issue of UHIs in densely populated cities a growing concern (Basu & Samet, 2002; Tzoulas et al., 2007).

In order to measure the intensity of UHIs, it is first necessary to delineate the urban area (Voogt & Oke, 2003). A common method used in previous studies involves clustering urban areas using impervious surface data and defining the extent of neighboring areas through a buffering approach (Yang et al., 2003; Lu et al., 2011). However, this method struggles to reflect the heterogeneous surface characteristics created by diverse elements within the city, such as buildings and trees, as it represents the urban area as a single cluster. To address this issue, a classification system known as the Local Climate Zone (LCZ) has been proposed (Stewart & Oke, 2012). This standard system divides urban areas into 17 classes based on building characteristics (height, density) and vegetation types. Typically, studies define the UHI intensity as the difference between the average temperature of each LCZ class within the city and the average temperature of a base class, a Low

DOI: 10.1201/9781003244561-11

Plant (LCZD) (Alexander & Mills, 2014). This method allows for the differentiation of thermal environmental characteristics by different urban forms, thus enabling a more nuanced analysis of UHI intensity. Consequently, LCZ is widely utilized in current research studies on urban thermal environmental analysis (Quan & Bansal, 2021; Yang et al., 2019; Eldesoky et al., 2021).

Alongside the delineation of urban areas, another consideration is how urban temperatures are monitored. A common approach is to use satellite-based land surface temperature (LST) data (Chang et al., 2021; Yoo et al., 2022). Many Satellite-based LSTs are public and offer broad coverage to capture comprehensive thermal environmental phenomena within and around cities (Zhou et al., 2018). However, these LST data present several challenges in urban studies. Firstly, multispectral sensor-based satellite sensors cannot obtain values beneath clouds, posing significant limitations, particularly in cities located in humid climate regions that are heavily affected by cloud cover (Zeng et al., 2018; Yoo et al., 2020a).

Another challenge is the resolution trade-off (Xia et al., 2019). Typically, satellite data with higher spatial resolution have lower temporal resolution. For example, LST data based on the Moderate Resolution Imaging Spectroradiometer (MODIS) sensor, which is capable of monitoring urban areas and is mounted on the Aqua and Terra satellites, can obtain LST data four times daily, but its spatial resolution of 1 km limits detailed monitoring of urban temperature distributions. In contrast, Landsat-based thermal data, known to be effective for detailed spatial analysis of urban thermal environments due to its relatively high-spatial resolution (~100 m), has a significant limitation: its temporal resolution is approximately 16 days, precluding frequent temperature monitoring at multiple time points.

Furthermore, satellite-based LST data, which measures surface temperature, typically differs from the air temperature that humans can perceive, rendering it somewhat insufficient for direct use in thermal environment studies related to human exposure to heat risk (Yoo et al., 2018). Specifically, in cities, the spatial distribution of LST and air temperature can significantly vary and their patterns are also different and dynamic over time (Cao et al., 2021), complicating the use of satellite-based surface temperature data in urban thermal environment studies.

Alternatively, another typical method of monitoring urban temperatures is to acquire temperature data based on observation stations, which measure the temperature at a height of approximately 1.5 m above the ground in city-installed weather stations (Meier et al., 2017). These well-maintained stations provide high temporal resolution temperature data, making them effective for analyzing diurnal temperature changes (Waqas & Athar, 2018). However, a major issue with using observation station-based air temperature to monitor the urban thermal environment is the number of stations (Shi et al., 2015). While these stations require consistent and sophisticated management to provide high-quality temperature data, most reliable stations are managed by national institutions, meaning there are typically insufficient numbers of these stations to monitor the detailed temperature distribution within a city accurately.

To seamlessly monitor the temperature distribution in cities using observation station-based air temperature data, various preliminary studies have been conducted, initially involving spatial modeling based on interpolation (Hart et al., 2009; Bárdossy et al., 2021). The air temperature data from observation stations are used to produce a temperature map of the city, primarily through interpolation techniques such as Kriging or IDW. However, these simple interpolation techniques are known to be more suitable for natural-type land surface regions with relatively homogeneous spatial distributions of temperature (Ashiq et al., 2010). Given the complex surface characteristics of cities, their temperature distribution is markedly heterogeneous, which presents limitations for the application of interpolation techniques. To overcome this, more complex statistical models have been used (Ho et al., 2014; Yoo et al., 2018). These models use input predictors reflecting thermal characteristics, and modeling is conducted with the measured air temperature from the observation stations as the target. In particular, various machine learning techniques are being used today for urban atmospheric temperature mapping (Wang et al., 2023). For example, Ho et al. (2014) used the Random Forest machine learning technique to produce a daily maximum air temperature for Vancouver, Canada, with a resolution of 100 m and 16 days, using Landsat-based LST as input data.

Furthermore, dos Santos (2020) used the Gradient Boosting machine learning technique to map the daily air temperature of London with a 1 km resolution. Venter et al. (2020) used the Random Forest machine learning technique to produce daily maximum and minimum temperatures for Oslo, Norway, with a 30 m resolution. Although these techniques, based on machine learning, mapped air temperatures with relatively high accuracy, most have a daily temporal resolution, which still presents limitations for monitoring the urban thermal environment at high spatiotemporal resolution.

Hrisko et al. (2020) produced hourly air temperature based on a Neural Network machine learning technique using GOES-16 LST from a geostationary satellite as input data for major cities in the United States. However, the spatial resolution was as low as 2 km. To date, there have been few studies that have been able to analyze the detailed temperature distribution of cities with a high hourly temporal resolution and discuss the spatiotemporal patterns of UHI phenomena in-depth.

In this chapter, we have devised a research method to map air temperature with a high spatiotemporal resolution of hourly 250 m for Seoul, South Korea. In particular, we propose an effective method using multi-task machine learning techniques to map the temperature of various time frames with high accuracy simultaneously. We will compare and analyze the spatial patterns of air temperature by LCZ class using the created hourly 250 m air temperature and simulate the diurnal cycle by calculating the intensity of the UHI effect.

STUDY AREA AND DATA

STUDY AREA

The city chosen for this study is Seoul, the capital city of South Korea, located in the northwestern part of the country. Geographically divided by the Han River into northern and southern regions, Seoul is encircled by four prominent mountains.

The city center is approximately 60 km from South Korea's western coastline and covers an area of around 605 square kilometers. Seoul is characterized by its high population density, with a populace of approximately 9.98 million as of 2022.

According to the Köppen–Geiger climate classification system, Seoul is characterized by a humid subtropical climate. Its summers are marked by elevated temperatures, high humidity, and substantial rainfall, primarily due to the influence of the East Asian Monsoons. Seoul is widely acknowledged as a city susceptible to heatwaves and the UHI phenomenon. Consequently, monitoring urban temperatures, particularly during the summer season, is essential. However, the effectiveness of this activity is significantly constrained by the prevalent cloud cover, which restricts the applicability of satellite-based LST. Given these conditions, Seoul emerges as an appropriate region for the high-resolution temperature mapping that constitutes the goal of the current study.

The research period for this study was from 2018 to 2020. For analytical purposes, the study duration was segregated into two distinct seasons: the cold season (November–April) and the warm season (May–October).

Satellite and In-Situ Meteorological Data

This study involved the use of data obtained from MODIS Terra and Aqua satellites. The wide application of MODIS data in observing atmospheric, oceanic, and terrestrial phenomena is well-recognized in the scientific community. For this study, specific MODIS products, LST and Reflectance—corresponding to the h28v05 region for Seoul—were procured for the period of 2018 to 2020 from the Reverb ECHO portal (http://reverb.echo.nasa.gov). We utilized daily LST data products MYD11A1 (from Aqua) and MOD11A1 (from Terra), both of which have a spatial resolution of 1 km. These LST data were generated using a generalized split-window algorithm, as outlined in the studies by Wan and Dozier (1996). The Terra MODIS satellite provides LST data collected at 10:30 am solar local time (daytime) and 10:30 pm (nighttime). In contrast, Aqua MODIS offers LST data collected at local solar times of 1:30 pm (daytime) and 1:30 am (nighttime). For this study, both daytime and nighttime LST data from Terra, along with the nighttime LST data from Aqua, were collected.

MODIS Reflectance data in the Red and Near-Infrared (NIR) bands were obtained from a MOD09GQ product featuring a spatial resolution of 250 meters. Digital surface model (DSM) data, at a spatial resolution of 30 meters, was sourced from the Advanced Land Observing Satellite (ALOS) Global Digital Surface Model. The study also utilized the Global Man-made Impervious Surface (GMIS) dataset, acquired at an equivalent 30-meter resolution. To procure high-resolution LST data for the input parameter for air temperature modeling, a total of 32 clear-sky daytime scenes from Landsat 8 LST and three nighttime scenes from ASTER LST were collected. The full list of satellite-based data used is described in Table 8.1.

Hourly air temperature readings from 2018 to 2020 were collated from a variety of weather stations situated across Seoul. These encompassed 43 Automatic Weather Stations (AWS) and Automated Surface Observing Systems (ASOS) managed by the Korea Meteorological Administration, 37 stations operated by

TABLE 8.1
Description of Satellite-Based Data Products with Their Source and Resolution Used in This Study

Products	Source	Resolution
Surface Reflectance (Red/NIR)	Clear-sky MODIS MOD09GQ	250 m daily
Terra Daytime LST (TD)	Clear-sky MODIS MOD11A1 (from Terra)	1 km daily, 10:30 AM
Aqua Nighttime LST (AN)	Clear-sky MODIS MYD11A1 (from Aqua)	1 km daily, 01:30 AM
Terra Nighttime LST (TN)	Clear-sky MODIS MOD11A1 (from Terra)	1 km daily, 10:30 PM
Digital Surface Model (DSM)	ALOS Global Digital Surface Model	30 m
Impervious surface (IMP)	Impervious percentage from Global Man-made Impervious Surface	30 m
Landsat daytime HLST	Landsat8 Level2 LST (32 clear-sky scenes)	100 m 16-day
ASTER nighttime HLST	ASTER Level2 LST (3 clear-sky scenes)	90 m 16-day

a Korean public institution, and three stations maintained by the Forest Meteorological Observatory. The cumulative data from these sources served as the reference dataset for this study. The spatial distribution of these weather stations is illustrated in Figure 8.1.

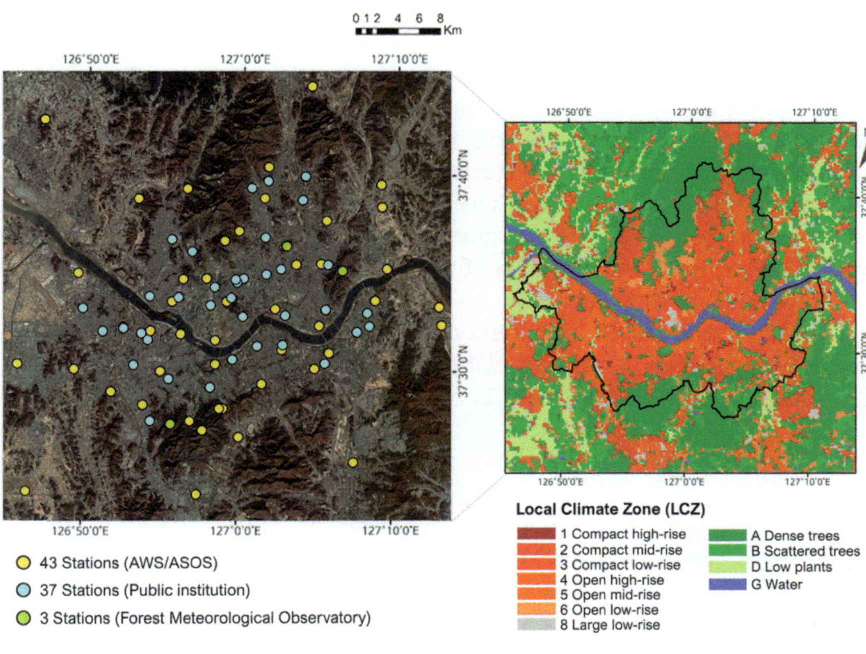

FIGURE 8.1 Study areas with the location of weather stations (left) and Local Climate Zones (LCZ) in Seoul (right).

Local Climate Zone Map

This study adopted the LCZ framework to examine the UHI phenomenon in Seoul at varying temporal intervals, such as hourly increments. A comprehensive LCZ map of Seoul, at a resolution of 50 meters, was developed by integrating several data sources, including Sentinel-2A reflectance, Landsat 8 LST, and building datasets. This LCZ map was generated based on the Convolutional Neural Network (CNN) deep learning method, as outlined by Yoo et al. (2020b). Upon map production, the LCZs were resampled to enhance their resolution to 250 meters, using a majority voting approach. The LCZ framework comprises seventeen distinct categories: ten urban (LCZ1 through LCZ10) and seven natural (LCZA through LCZG). The comprehensive list of LCZ classes is depicted in the legend of Figure 8.1.

METHODS

Variable Processing

The 250 m MODIS Red and Near-Infrared (NIR) reflectances were used as input variables. In addition, the Normalized Difference Vegetation Index (NDVI) was calculated by measuring the difference between the visible red and near-infrared bands. The 30 m Digital Surface Model (DSM) was bilinearly resampled to 250 m, and the slope was computed from the DSM. The 30 m impervious surface percentage (Imp) data, high-resolution Landsat 8, and ASTER clear-sky LSTs (HLST) were also resampled to 250 m.

The Euclidean distance from the centroid of the study area point (DCP), a representation of the urban thermal heat phenomena concentrated in the city center, was computed at a resolution of 250 m. Daily solar radiation was estimated using the ArcGIS Solar Analyst tool (Esri). The incoming solar radiation for each pixel was computed by inputting the elevation and the day of the year (DOY) into ArcGIS. For this study, DOY values were transformed to a range from –1 to 1 over a one-year period, using a sine function to account for seasonality (i.e., setting the peak of summer as 1 and the depth of winter as –1), and 250 m pixel-level DOY were used as input variables.

The study used MODIS 1 km Terra daytime (TD) and Aqua daytime (AD) LST, as well as Aqua night (AN) LST. All the input variables and their corresponding acronyms are presented in Table 8.2.

Air Temperature Estimation

The method for estimating 250 m air temperature consisted of two steps. The first step involved downscaling the MODIS 1 km LST to 250 m LST using a Random Forest (RF) model with various input parameters (also called kernels in downscaling). The second step entailed estimating air temperature at an hourly resolution of 250 m, based on a multi-task machine learning model. The flow chart, encapsulating step 1 and step 2, is illustrated in Figure 8.2.

TABLE 8.2
Variable Used for High Spatiotemporal Air Temperature Mapping in This Study

Acronym	Variable	Resolution
Red	Band1 from MODIS Reflectance	250 m daily
NIR	Band2 from MODIS Reflectance	250 m daily
NDVI	Normalized Difference Vegetation Index from MODIS Reflectance	250 m daily
DSM	ALOS Global Digital Surface Model	250 m (Resampled)
Slope	Slope by DSM	250 m
Aspect	Aspect by DSM	250 m
IMP	The impervious surface percentage from GMIS	250 m (Resampled)
HLST	Clear-sky daytime (Landsat8) / nighttime (ASTER) high-resolution LST	250 m
DCP	Euclidean Distance from the Centroid of the study area Point	250 m
DOY	Sine function transformed Day of Year	250 m
SOLAR	Area Solar radiation within a day	250 m
TD, AD LST	MODIS 1 km Terra Daytime and Aqua Daytime LST	1 km
AN LST	MODIS 1 km Aqua Nighttime LST	1 km

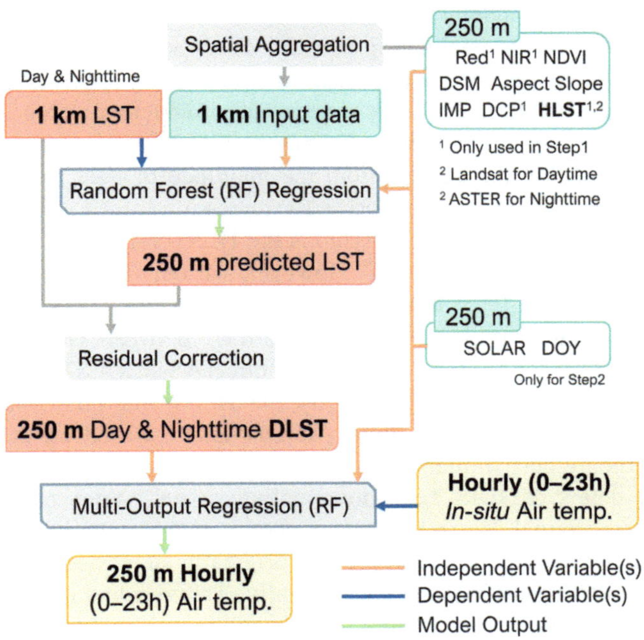

FIGURE 8.2 Flowchart of the study, divided into two main parts: Land Surface Temperature (LST) downscaling (Step 1) and air temperature estimation (Step 2).

Step 1: LST Downscaling

The first step involves downscaling 1 km MODIS LST to a resolution of 250 m. Initially, high-resolution input data, encompassing Red, NIR, NDVI, DSM, Slope, IMP, DSM, and HLST, were resampled to 1 km, matching the resolution and grids of MODIS LST. Subsequently, a downscaling model was established separately for each TD, AD, and AN MODIS LST using a Random Forest classifier. Random Forest is a machine learning algorithm rooted in the principles of Classification and Regression Trees (CART), as proposed by Breiman (2001). This algorithm utilizes a recursive binary split approach to navigate the nodes within a tree structure until terminal nodes are reached. To accomplish this, RF generates a large array of CARTs (usually between 500–1000 trees). These trees are constructed using subsets randomly selected through a bootstrapping method from training samples for a tree and predictor variables at each tree node. This bootstrapping-based randomization strategy effectively addresses the issues intrinsic to the CART method, including its sensitivity to training samples and a propensity for overfitting. RF concludes its decision-making by assimilating the outputs from the multitude of trees through an averaging process for regression.

We constructed a model for each date of each LST image using the Random Forest (RF) machine learning algorithm. This algorithm, based on Classification and Regression Trees (CART) as delineated by Breiman (2001), employs a recursive binary split methodology to navigate through the nodes within a tree structure until it reaches terminal nodes. To accomplish this, RF generates a large array of CARTs, typically between 500–1000 trees. These trees are constructed using subsets that are randomly selected through a bootstrapping method from both training samples for a tree and predictor variables at each tree node. This bootstrapping-based randomization approach effectively addresses the issues inherent to the CART method, such as its sensitivity to training samples and tendency towards overfitting (Belgiu & Drăguţ, 2016). RF concludes its decision-making by integrating the outputs from the multitude of trees through either an averaging or weighted averaging process for regression.

A critical input parameter in this step is the HLST variable, which is directly linked with the target MODIS LST. However, we only had 32 daytime scenes for Landsat LSTs and three nighttime scenes for ASTER LSTs. We calculated the Structural Similarity Index Measure (SSIM) between each MODIS LST and all 32 Landsat LSTs (for daytime) and three ASTER LSTs (for nighttime) after normalization. This was to identify the HLST that could best represent the similar LST pattern of the target MODIS LSTs. Based on the SSIM, the HLST with the highest similarity was selected and used as an input variable for each target date's MODIS LST downscaling model. After constructing the RF model, the original 250 m input data were inputted to produce the predicted 250 m LST. A residual correction process originally suggested by Kustas et al., 2003 was applied to produce the DLST finally. This residual correction process generally uses the following steps: 1) aggregating the simulated high-resolution LST (LSThigh) to the original MODIS LST scale, 2) subtracting the aggregated LSThigh (LSTag) from original MODIS LST (called residuals; Δ), 3) resampling the residuals to LSThigh scale and adding these residuals to the LSThigh for generating MODIS DLST.

Step 2: Air Temperature Estimation

Upon generating the 250 m daytime and nighttime DLST in step 1, step 2 entails the estimation of hourly high-spatial resolution air temperatures through a multi-output regression based on RF. To mitigate the issues of cloud contamination in LST, three combinations of different input variables were constructed (Figure 8.3). Combination1 incorporates AN and TD as input variables alongside auxiliary variables (i.e., NDVI, DSM, Slope, Imp, SOLAR, and DOY). Combination 2 employs TD and AN along with auxiliary variables. Finally, combination 3 simultaneously uses AN, TD, and TN, along with auxiliary variables. The final air temperature map is then produced by merging the three combinations. When TD, AN, and TN are all under clear-sky conditions, combination 3 is used. However, if only TD and AN are under clear-sky conditions, combination 1 is applied. Alternatively, if only TD and TN are under clear-sky conditions, combination 2 is used. If all three LSTs are under cloudy conditions, that date is deemed unsuitable for generating an air temperature map. In step 2, the Leave-One-Station Out Cross-Validation (LOOCV) is repeated for weather stations. The models' performance is evaluated using the coefficient of determination (R^2) and the Root Mean Square Error (RMSE).

Urban Heat Islands Monitoring

Upon the completion of the 250 m hourly air temperature mapping, the UHI phenomena in Seoul were evaluated using an LCZ classification map. The UHI intensity was determined by calculating the difference in air temperatures between regions classified as LCZ2 (compact mid-rise) and LCZD (low plants). The diurnal cycle of the UHI was subsequently examined separately for both the warm and cold

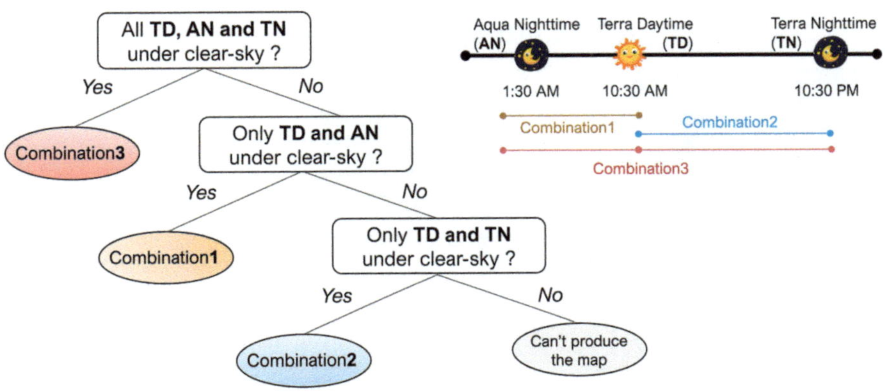

FIGURE 8.3 Combination rule for merging three combinations to create the final air temperature map.

seasons. For a more detailed identification of the UHI, UHI intensity was also calculated based on the difference between LCZ6 (open low-rise) and LCZD.

RESULTS AND DISCUSSION

LST Downscaling Results

Figure 8.4 represents the distribution of the average normalized LST over three time slots (1:30 AM, 10:30 AM, 10:30 PM) during the study period, obtained from 1 km MODIS LST and downscaled to 250 m LST (DLST), along with the average variable importance and error bar from Random Forest modeling. The 250 m

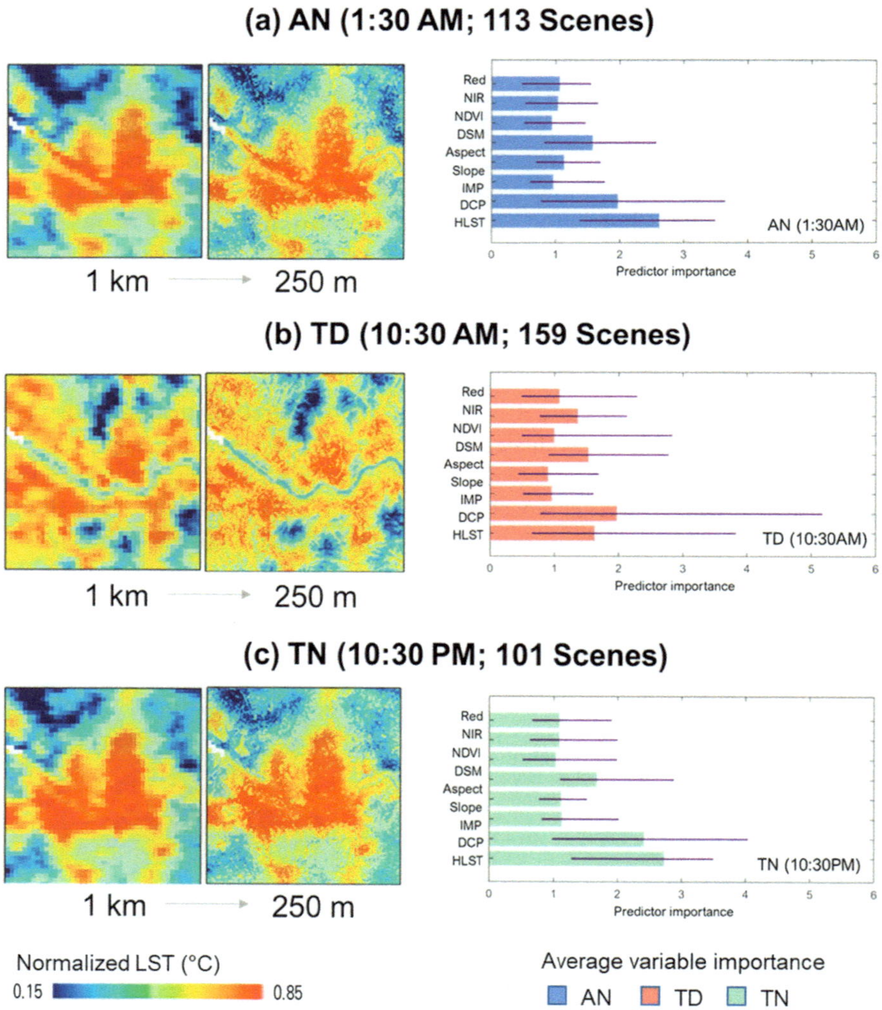

FIGURE 8.4 Averages of normalized 1 km LST and downscaled LST (DLST) with Random Forest (RF) variable importance for AN (a), TD (b), and TN (c) separately.

DLST, downscaled from 1 km, exhibits a similar spatial pattern of overall temperature, but the detailed urban heat environment pattern appears more pronounced. The downscaled technique based on the Random Forest used in this study has been sufficiently validated for its accuracy in previous studies (Hutengs & Vohland, 2016; Ebrahimy et al., 2021).

Upon examining the variable importance, the most important factor in the downscaling model of 1:30 AM AN turned out to be HLST. This suggests that high-resolution LST from other dates similar to the target date can contribute to greater accuracy in downscaling nighttime surface temperatures due to their similar spatial distributions. The second most important variable was DCP, indicating that the distance from the urban center emerged as a significant variable. This implies that urban heat island phenomena exhibit a dominant pattern within the city. For instance, as seen in Figure 8.1, the center of Seoul distinctly shows higher temperatures compared to the surrounding areas. This phenomenon has led to the DCP variable becoming a crucial factor in the detailed representation of temperature patterns. The DSM appeared as the third most important factor in AN downscaling. This interpretation suggests that, given Seoul's geographical features surrounded by mountains, elevation data has been significantly used to simulate the distribution of lower temperatures in the peripheral regions.

At 10:30 AM (TD), the most crucial variable importance was revealed to be DCP rather than HLST. It is apparent that the rank of HLST's importance is lower during the daytime (TD) than at night (AN), which can be attributed to the larger daily variation in the spatial distribution of temperature due to the influence of solar incoming (Luintel et al., 2019). As a result, the number of HLST images with similar thermal environment distributions to the target date LST and the similarity of their distributions appear relatively less significant during daytime. In contrast, DCP and DSM still emerge as the variables with the highest importance. During the day, high temperatures can be observed in urban centers, particularly in the northern region of the river. On the other hand, the surrounding mountainous areas exhibit lower temperatures due to the influence of high altitudes, even during the day.

Lastly, at 10:30 PM (TN), a similar pattern of variable importance to the nighttime (AN) is observed, with HLST having the highest importance, followed by DCP and DSM. The similarity in importance between AN and TN can be attributed to the reduced influence of solar incoming at night, leading to comparable spatial temperature patterns. In summary, HLST, DCP, and DSM demonstrate high importance across all three time periods; however, the importance of HLST is observed to be higher at night compared to during the day.

Hourly Air Temperature Estimation Accuracy Assessment

The Leave-One-Out Cross-Validation (LOOCV) results for each hour of the three combinations are compared in Figure 8.5 All three combinations exhibit high temporal performance (R^2) for all hours and an acceptable error of approximately 1.0 °C RMSE. For combination 1, the early-night (0–3 h) temporal performances are superior to those of late-night (20–23 h). This could be because the input nighttime DLST is captured at early-night (AN; 1:30 AM). Conversely,

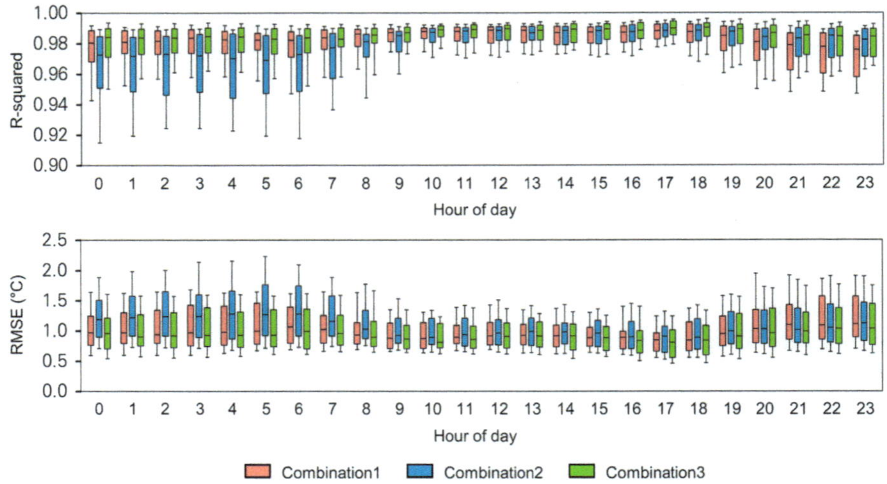

FIGURE 8.5 Leave-One-Out Cross-Validation (LOOCV) results for all hours for the three combinations with a unified sample size according to Combination 3.

combination 2 displays better temporal performance during late-night than early-night, but the difference is significantly larger than that of combination 1. This might be due to the incoming solar heat flows moving from surface to air, making it more challenging to estimate early-night air temperature from late-night LST (Chen et al., 2019).

Combination 3 demonstrates the best performance in terms of R^2 and RMSE among the three combinations, suggesting that using multiple LSTs could produce better air temperature estimation results. These findings are consistent with previous studies, which indicate that incorporating LSTs from multiple time periods improves the accuracy of air temperature estimation (Yoo et al., 2018). Additionally, the R^2 values for daytime are relatively higher than those for nighttime in Figure 8.5. This can be attributed to the larger variation range of temperature during daytime due to the influence of solar incoming, leading to a higher presence of extreme temperature values. Consequently, a high correlation is observed when these values are accurately simulated.

Figure 8.6 presents the all-hour average RMSE distribution based on LOOCV for each station by merging the three combinations. Most stations exhibit a low RMSE below 1.2 °C, particularly in urban core areas. The stations with the highest error are located in mountainous regions, where it is likely challenging to estimate the lower temperature values due to high altitudes. As more stations are established in mountainous areas in the future, the accuracy in these regions is expected to improve.

Nonetheless, when compared to the accuracy of previous studies, the RMSE demonstrates high performance. This is because the range of air temperature estimation error in benchmark studies lies between 1–2 K RMSE (Venter et al., 2020; Hrisko et al., 2020; Ho et al., 2014), indicating that the accuracy of the present study is quite encouraging. In summary, the LOOCV validation results imply that the hourly air temperatures are well estimated within urban areas, and the

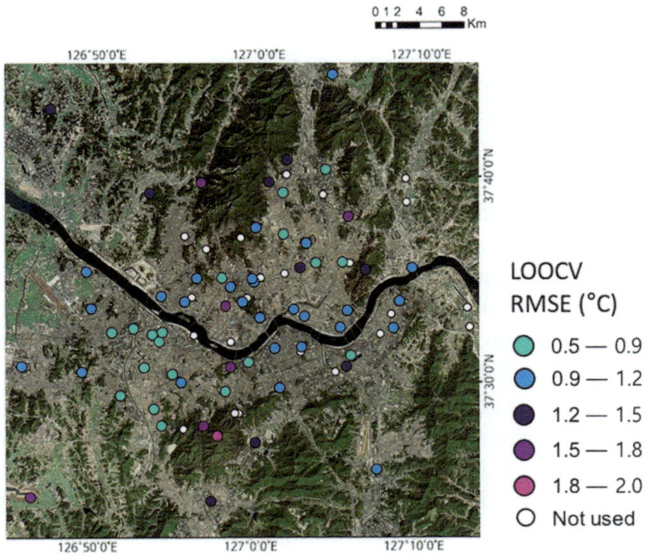

FIGURE 8.6 Average hourly Root Mean Square Error (RMSE) for each station from LOOCV validation results for the three combinations.

entire distribution of air temperatures in the study regions can be represented by the constructed model.

Hourly Air Temperature Mapping Analysis

Figure 8.7 illustrates the distribution of average normalized air temperature during four representative time periods (5 AM and 10 AM, 3 PM, and 8 PM) for both

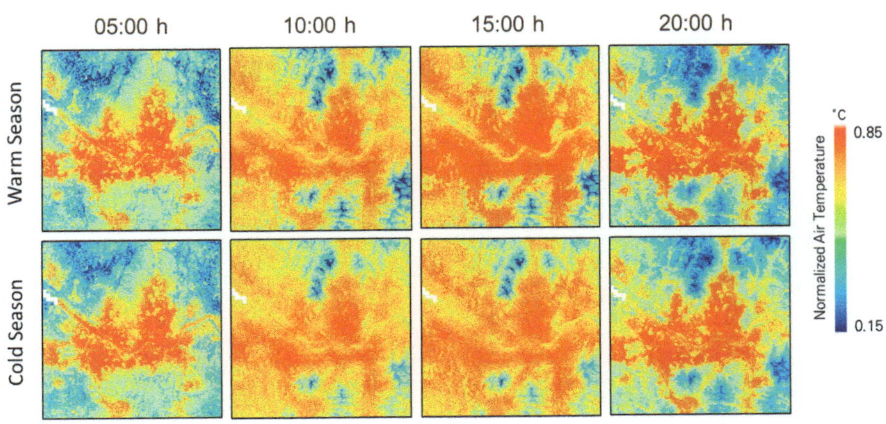

FIGURE 8.7 Final hourly 250 m air temperatures mapped and averaged for warm and cold seasons after normalization, with representative times at 5-hour intervals. Only 95% clear-sky days used: 53 days for Warm Season (May-Oct) and 44 days for Cold Season (Nov-Apr).

warm and cold seasons. For the warm season, a phenomenon of higher temperatures in the city center can be observed at 5 AM, early in the morning. A distinct temperature gradient is evident, with temperatures decreasing as one moves away from the city center. This is a result of the UHI effect, which is more pronounced during early morning hours when the influence of the sun is reduced. Vegetated areas experience a rapid drop in temperature, while impervious surfaces and human anthropogenic heat sources in urban areas retain heat (Sturiale & Scuderi, 2019).

At 10 AM, after the sun has risen, a different temperature distribution can be observed at 5 AM. Low plants and scattered tree areas around the city are heated up due to the influence of solar incoming radiation. At 3 PM, which is typically the time of day when the maximum air temperature is observed, the entire study area, excluding high-altitude mountainous regions, exhibits high temperatures with relatively low variation compared to nighttime. This temperature pattern changes at 8 PM, when the temperature in peripheral areas rapidly decreases, accentuating the higher temperatures in the urban center. A comparison of the temperature patterns at 5 AM and 8 PM reveals that the temperatures in the city center are generally higher than those in the surrounding areas at early morning and late-night. The study confirms that the UHI phenomenon is more pronounced at night than during the day in Seoul, which is consistent with previous studies on cities with a prominent nocturnal UHI effect (Hu & Xue., 2016; Yoo et al., 2022). In the cold season, similar temporal changes in temperature patterns are observed, although the intensity of warming in the city center is less pronounced than in the warm season. The temperature difference between the city center and the surrounding areas is smaller during the daytime (e.g., 10 AM and 3 PM) and also in the cold season.

By plotting the distribution of temperatures for each LCZ during the four time periods (Figure 8.8), distinct UHI patterns can be observed for both warm and cold seasons at 5 AM. Compact built-up types (LCZ1-3) exhibit the highest

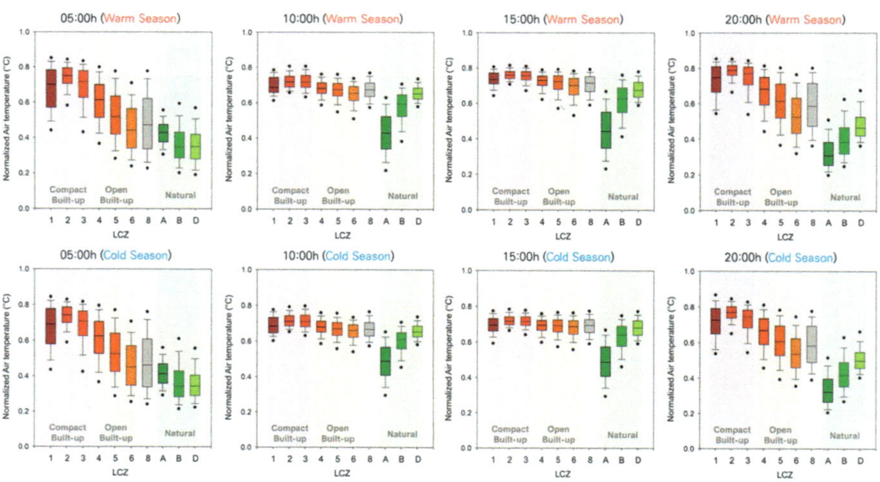

FIGURE 8.8 Boxplots of temperature distribution for hourly 250 m air temperatures using LCZ for each time, shown separately for warm (upper) and cold (lower) seasons.

temperatures, followed by open built-up types (LCZ4-6), and natural types (LCZA, B, and D). Among LCZ1-3, LCZ2 is slightly warmer than LCZ1, likely reflecting the shading effect of tall buildings (Zhou & Chen, 2018). The temperature difference due to height is more pronounced between open built-up types than compact built-up types at night, with lower heights (from LCZ4 to LCZ6) corresponding to lower average temperatures.

In contrast, during the daytime (10 AM and 3 PM), the temperature distribution among most classes does not exhibit significant variation, excluding the mountainous area of LCZA. The temperature difference between compact and open built-up types is not substantial, and natural types exhibit only slightly lower temperatures than urban types. Notably, during the cold season, the low plant natural type (LCZD) exhibits higher temperatures than the open built-up classes at 10 AM.

As night falls, substantial temperature variations are observed between urban and natural types, with a clear distribution pattern within open built-up types: higher building height correspond to higher average temperatures. The study results demonstrate that using LCZs enables a more detailed investigation of the distribution of thermal environments according to different building and vegetation types within a city, and the results suggest distinct differences in these patterns over time.

Diurnal Cycle of UHI Intensity

Figure 8.9 displays the diurnal cycle of UHI intensity. For the difference between LCZ2 and LCZD, distinct curved cycles appear, with high UHI intensity in early nighttime that decreases during daytime and increases again in late-nighttime. The high UHI intensity at nighttime occurs because energy absorbed by man-made materials during the day is released slowly at night, resulting in the heating of air (Arnfield, 2003). Comparing the warm and cold seasons in Figure 8.9(a), higher UHI

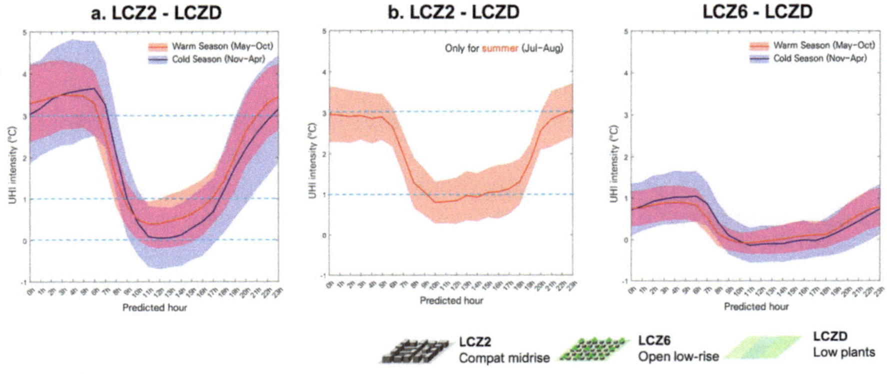

FIGURE 8.9 Diurnal cycle of Urban Heat Island (UHI) intensity between LCZ2 and LCZD for warm and cold seasons (a), summer season only (b), and the difference between LCZ6 and LCZD (c). Only 95% clear-sky days used: 53 days for Warm Season (May-Oct) and 44 days for Cold Season (Nov-Apr).

intensity is observed in early nighttime during the cold season, while daytime UHI is higher in the warm season than in the cold season. Notably, negative UHI intensity appears in the daytime during the cold season (i.e., UCI).

Focusing on the warm season (Figure 8.9(b)), daytime UHI intensity is higher than in other seasons. This result supports previous studies that revealed an association between heatwaves and UHI, where hot summer heatwaves could intensify UHI phenomena (Founda & Santamouris, 2017). In the case of calculating UHI intensity using LCZ6 and LCZD, a distinct UHI cycle between daytime and nighttime is not observed. This suggests that there are different patterns of UHI for each LCZ class, so using detailed air temperatures and LCZs can aid in in-depth UHI assessments.

Novelty and Limitation

The novelty of this study lies in the construction of a two-step air temperature mapping framework, which enables the development of high spatiotemporal urban air temperature data. By performing satellite-based LST downscaling in step 1, we overcame the issue of low spatial resolution in satellite-based LST – mostly about 1 km – by downscaling the MODIS daily from 1 km LST to 250 m. We also addressed the main limitation of low temporal resolution in studies estimating air temperature from satellite-based LST by employing multi-task learning techniques to simultaneously estimate air temperatures at multiple times, resulting in hourly air temperature data. Notably, the validation results showed stable estimation performance (RMSE about 1.2 K) across all time frames, and the mapping analysis confirmed the ability to simulate time-specific urban temperature patterns accurately. Multi-task learning is particularly effective for hourly air temperature mapping, as it allows for simultaneous estimation of similar outputs, enabling efficient and rapid output production. The ability to quickly and effectively estimate the air temperature for all 24 hours is a significant novelty. This method can be particularly useful for real-time operating systems that require rapid production of air temperature data, enabling effective air temperature mapping.

In this study, we used the produced hourly urban air temperature map and LCZs to conduct an in-depth analysis of thermal environment patterns (e.g., UHI) based on urban building and vegetation types. The results from the study area of Seoul revealed that the difference in temperature distribution between LCZ classes is greater at night than during the day, with a clear difference in temperature according to height in open-type LCZs at night. Moreover, the diurnal cycle patterns of UHI intensity were identified using LCZs, with distinct curved patterns showing high intensity in the early morning, lower intensity during the day, and decreasing intensity at night. Particularly, we found that slight cooling heat island phenomena appeared during the cold season. This study highlights the benefits of using LCZs in contrast to previous research, which mostly analyzed specific time frames in line with satellite image capture times by directly mapping hourly air temperature and examining time patterns and diurnal cycles for each LCZ class.

A major limitation of this study is its reliance on satellite LST data. Although we proposed a method to analyze the thermal environment of the target date using past HLSTs to overcome issues such as cloud coverage in conventional satellite images, the downscaling accuracy may decrease if there are no similar HLSTs for the target date. Additionally, despite proposing a combination method to address cloud-related issues, there are still dates when mapping could not be performed due to the influence of clouds. We expect that utilizing LST data with gap-filled cloud coverage could further improve this issue (Shiff et al., 2021). As this study is a case study targeting Seoul, the thermal environment patterns revealed, such as high UHI intensity at nighttime compared to daytime, may not be generalized to all cities. Different patterns may emerge depending on the city's climate condition and geographical characteristics. As more cities acquire sufficient and reliable observation station data, we anticipate that more interesting analyzes of city-specific thermal environment patterns will become possible.

CONCLUSION

Urban areas have been experiencing an increase in temperature due to rapid urbanization and climate change, resulting in the phenomenon known as the UHI effect. Notably, the thermal environment within cities exhibits diverse patterns depending on the types of buildings and vegetation, and the distribution and intensity of UHI vary over time. To analyze these patterns, high spatiotemporal resolution temperature data is required, which can monitor extensive urban areas. However, satellite imagery providing LST faces limitations due to the trade-off between spatial and temporal resolutions and cloud contamination issues. Additionally, 1.5 m ground-above air temperature data from observation stations is insufficient for monitoring detailed urban thermal environment distributions due to the limited number of stations within cities.

In this study, we propose a novel two-step algorithm for mapping hourly urban air temperature at a 250 m spatial resolution in Seoul. In step 1, we downscaled the 1 km MODIS LST data to 250 m using a machine learning Random Forest model, employing high-resolution satellite LST data from sources such as ASTER and Landsat as primary input data for downscaling. In step 2, we estimated hourly air temperature from the downscaled 250 m LST data for multiple time periods. To address the cloud contamination issue in satellite-based LST, we proposed a method for combining LST data from various time periods. We also suggested an efficient method for simultaneously calculating 24-hour hourly air temperature using a Random Forest-based multi-task learning estimation model.

The results of the proposed air temperature estimation technique showed high estimation accuracy, with an average RMSE of approximately 1.0 °C across all 24 hours. The generated air temperature maps effectively simulated detailed urban temperature patterns and their variations across different times. By analyzing the thermal environment distribution of urban types using LCZs, we found that in Seoul, higher building density and taller buildings within open built-up type LCZ classes were associated with higher temperatures and distinct UHI patterns during the night. In contrast, during the day, most regions within the city exhibited relatively lower UHI effects due to sufficient incoming solar heat-up. The diurnal cycle of UHI intensity displayed a curved pattern, with higher intensity at night, lower intensity during the day, and increasing intensity again at night. However, the magnitude of this cycle varied depending on which LCZ class was used as a basis for calculating intensity. The hourly air temperature maps produced in this study can be considered valuable resources for enhancing the understanding of urban thermal environments in both spatial and temporal dimensions. In particular, the multi-task learning technique employed for estimating multiple time periods simultaneously is expected to be effectively utilized in practical applications.

REFERENCES

Alexander, P. J., & Mills, G. (2014). Local climate classification and Dublin's urban heat island. *Atmosphere*, 5(4), 755–774.

Angel, S., Parent, J., Civco, D. L., Blei, A., & Potere, D. (2011). The dimensions of global urban expansion: Estimates and projections for all countries, 2000–2050. *Progress in Planning*, 75(2), 53–107.

Arnfield, A. J. (2003). Two decades of urban climate research: A review of turbulence, exchanges of energy and water, and the urban heat island. *International Journal of Climatology: a Journal of the Royal Meteorological Society*, 23(1), 1–26.

Ashiq, M. W., Zhao, C., Ni, J., & Akhtar, M. (2010). GIS-based high-resolution spatial interpolation of precipitation in mountain–plain areas of Upper Pakistan for regional climate change impact studies. *Theoretical and Applied Climatology*, 99, 239–253.

Bárdossy, A., Seidel, J., & El Hachem, A. (2021). The use of personal weather station observations to improve precipitation estimation and interpolation. *Hydrology and Earth System Sciences*, 25(2), 583–601.

Basu, R., & Samet, J. M. (2002). Relation between elevated ambient temperature and mortality: a review of the epidemiologic evidence. *Epidemiologic Reviews*, 24(2), 190–202.

Belgiu, M., & Drăguţ, L. (2016). Random forest in remote sensing: A review of applications and future directions. *ISPRS Journal of Photogrammetry and Remote Sensing*, 114, 24–31.

Breiman, L. (2001). Random forests. *Machine Learning*, 45, 5–32.

Cao, J., Zhou, W., Zheng, Z., Ren, T., & Wang, W. (2021). Within-city spatial and temporal heterogeneity of air temperature and its relationship with land surface temperature. *Landscape and Urban Planning*, 206, 103979.

Chang, Y., Xiao, J., Li, X., Frolking, S., Zhou, D., Schneider, A., … & Wu, Y. (2021). Exploring diurnal cycles of surface urban heat island intensity in Boston with land surface temperature data derived from GOES-R geostationary satellites. *Science of the Total Environment*, 763, 144224.

Chen, J., Chu, R., Wang, H., Zhang, L., Chen, X., & Du, Y. (2019). Alleviating urban heat island effect using high-conductivity permeable concrete pavement. *Journal of Cleaner Production*, 237, 117722.

dos Santos, R. S. (2020). Estimating spatio-temporal air temperature in London (UK) using machine learning and earth observation satellite data. *International Journal of Applied Earth Observation and Geoinformation*, 88, 102066.

Ebrahimy, H., Aghighi, H., Azadbakht, M., Amani, M., Mahdavi, S., & Matkan, A. A. (2021). Downscaling MODIS land surface temperature product using an adaptive random forest regression method and Google Earth Engine for a 19-years spatio-temporal trend analysis over Iran. *IEEE Journal of Selected Topics in Applied Earth Observations and Remote Sensing*, 14, 2103–2112.

Eldesoky, A. H., Gil, J., & Pont, M. B. (2021). The suitability of the urban local climate zone classification scheme for surface temperature studies in distinct macroclimate regions. *Urban Climate*, 37, 100823.

Foley, J. A., DeFries, R., Asner, G. P., Barford, C., Bonan, G., Carpenter, S. R., … & Snyder, P. K. (2005). Global consequences of land use. *Science*, 309(5734), 570–574.

Founda, D., & Santamouris, M. (2017). Synergies between Urban Heat Island and Heat Waves in Athens (Greece), during an extremely hot summer (2012). *Scientific Reports*, 7(1), 10973.

Hart, Q. J., Brugnach, M., Temesgen, B., Rueda, C., Ustin, S. L., & Frame, K. (2009). Daily reference evapotranspiration for California using satellite imagery and weather station measurement interpolation. *Civil Engineering and Environmental Systems*, 26(1), 19–33.

Ho, H. C., Knudby, A., Sirovyak, P., Xu, Y., Hodul, M., & Henderson, S. B. (2014). Mapping maximum urban air temperature on hot summer days. *Remote Sensing of Environment*, 154, 38–45.

Hrisko, J., Ramamurthy, P., Yu, Y., Yu, P., & Melecio-Vázquez, D. (2020). Urban air temperature model using GOES-16 LST and a diurnal regressive neural network algorithm. *Remote Sensing of Environment*, 237, 111495.

Hu, X. M., & Xue, M. (2016). Influence of synoptic sea-breeze fronts on the urban heat island intensity in Dallas–Fort Worth, Texas. *Monthly Weather Review*, 144(4), 1487–1507.

Hutengs, C., & Vohland, M. (2016). Downscaling land surface temperatures at regional scales with random forest regression. *Remote Sensing of Environment*, 178, 127–141.

Im, E. S., & Eltahir, E. A. (2018). Urbanization-induced drought: Anthropogenic land-use change impacts on the regional climate. *Journal of Climate*, 31(24), 10077–10094. doi: 10.1175/JCLI-D-18-0090.1

Kustas, W. P., Norman, J. M., Anderson, M. C., & French, A. N. (2003). Estimating subpixel surface temperatures and energy fluxes from the vegetation index–radiometric temperature relationship. *Remote Sensing of Environment*, 85(4), 429–440.

Lu, D., Moran, E., & Hetrick, S. (2011). Detection of impervious surface change with multitemporal Landsat images in an urban–rural frontier. *ISPRS Journal of Photogrammetry and Remote Sensing*, 66(3), 298–306.

Luintel, N., Ma, W., Ma, Y., Wang, B., & Subba, S. (2019). Spatial and temporal variation of daytime and nighttime MODIS land surface temperature across Nepal. *Atmospheric and Oceanic Science Letters*, 12(5), 305–312.

Meier, F., Fenner, D., Grassmann, T., Otto, M., & Scherer, D. (2017). Crowdsourcing air temperature from citizen weather stations for urban climate research. *Urban Climate*, 19, 170–191.

Quan, S. J., & Bansal, P. (2021). A systematic review of GIS-based local climate zone mapping studies. *Building and Environment*, 196, 107791.

Seto, K. C., Güneralp, B., & Hutyra, L. R. (2012). Global forecasts of urban expansion to 2030 and direct impacts on biodiversity and carbon pools. *Proceedings of the National Academy of Sciences*, 109(40), 16083–16088.

Shi, T., Huang, Y., Wang, H., Shi, C. E., & Yang, Y. J. (2015). Influence of urbanization on the thermal environment of meteorological station: Satellite-observed evidence. *Advances in Climate Change Research*, 6(1), 7–15.

Shiff, S., Helman, D., & Lensky, I. M. (2021). Worldwide continuous gap-filled MODIS land surface temperature dataset. *Scientific Data*, 8(1), 74.

Stewart, I. D., & Oke, T. R. (2012). Local Climate Zones for urban temperature studies. *Bulletin of the American Meteorological Society*, 93(12), 1879–1900. doi: 10.1175/BAMS-D-11-00019.1

Stone Jr, B. (2005). Urban heat and air pollution: An emerging role for planners in the climate change debate. *Journal of the American Planning Association*, 71(1), 13–25.

Sturiale, L., & Scuderi, A. (2019). The role of green infrastructures in urban planning for climate change adaptation. *Climate*, 7(10), 119.

Tzoulas, K., Korpela, K., Venn, S., Yli-Pelkonen, V., Kaźmierczak, A., Niemela, J., & James, P. (2007). Promoting ecosystem and human health in urban areas using Green Infrastructure: A literature review. *Landscape and Urban Planning*, 81(3), 167–178.

Venter, Z. S., Brousse, O., Esau, I., & Meier, F. (2020). Hyperlocal mapping of urban air temperature using remote sensing and crowdsourced weather data. *Remote Sensing of Environment*, 242, 111791.

Voogt, J. A., & Oke, T. R. (2003). Thermal remote sensing of urban climates. *Remote Sensing of Environment*, 86(3), 370–384.

Wan, Z., & Dozier, J. (1996). A generalized split-window algorithm for retrieving land-surface temperature from space. *IEEE Transactions on Geoscience and Remote Sensing*, 34(4), 892–905.

Wang, H., Yang, J., Chen, G., Ren, C., & Zhang, J. (2023). Machine learning applications on air temperature prediction in the urban canopy layer: A critical review of 2011–2022. *Urban Climate*, 49, 101499.

Waqas, A., & Athar, H. (2018). Observed diurnal temperature range variations and its association with observed cloud cover in northern Pakistan. *International Journal of Climatology*, 38(8), 3323–3336.

Xia, H., Chen, Y., Li, Y., & Quan, J. (2019). Combining kernel-driven and fusion-based methods to generate daily high-spatial-resolution land surface temperatures. *Remote Sensing of Environment*, 224, 259–274.

Yang, J., Jin, S., Xiao, X., Jin, C., Xia, J. C., Li, X., & Wang, S. (2019). Local climate zone ventilation and urban land surface temperatures: Towards a performance-based and wind-sensitive planning proposal in megacities. *Sustainable Cities and Society*, 47, 101487.

Yang, L., Xian, G., Klaver, J. M., & Deal, B. (2003). Urban land-cover change detection through sub-pixel imperviousness mapping using remotely sensed data. *Photogrammetric Engineering & Remote Sensing*, 69(9), 1003–1010.

Yoo, C., Im, J., Cho, D., Lee, Y., Bae, D., & Sismanidis, P. (2022). Downscaling MODIS nighttime land surface temperatures in urban areas using ASTER thermal data through local linear forest. *International Journal of Applied Earth Observation and Geoinformation*, 110, 102827.

Yoo, C., Im, J., Cho, D., Yokoya, N., Xia, J., & Bechtel, B. (2020a). Estimation of all-weather 1 km MODIS land surface temperature for humid summer days. *Remote Sensing*, 12(9), 1398.

Yoo, C., Im, J., Park, S., & Quackenbush, L. J. (2018). Estimation of daily maximum and minimum air temperatures in urban landscapes using MODIS time series satellite data. *ISPRS Journal of Photogrammetry and Remote Sensing*, 137, 149–162.

Yoo, C., Lee, Y., Cho, D., Im, J., & Han, D. (2020b). Improving local climate zone classification using incomplete building data and Sentinel 2 images based on convolutional neural networks. *Remote Sensing*, 12(21), 3552.

Zeng, C., Long, D., Shen, H., Wu, P., Cui, Y., & Hong, Y. (2018). A two-step framework for reconstructing remotely sensed land surface temperatures contaminated by cloud. *ISPRS Journal of Photogrammetry and Remote Sensing*, 141, 30–45.

Zhou, D., Xiao, J., Bonafoni, S., Berger, C., Deilami, K., Zhou, Y., … & Sobrino, J. A. (2018). Satellite remote sensing of surface urban heat islands: Progress, challenges, and perspectives. *Remote Sensing*, 11(1), 48.

Zhou, X., & Chen, H. (2018). Impact of urbanization-related land use land cover changes and urban morphology changes on the urban heat island phenomenon. *Science of the Total Environment*, 635, 1467–1476.

9 Satellite-Based Assessment of Urban Thermal Environments

Yuhei Yamamoto

INTRODUCTION

When evaluating urban thermal environments, Land Surface Temperature (LST) is primarily used as an indicator in satellite remote sensing (Henry et al., 1989; Weng, 2009; Stewart et al., 2021). LST is highly sensitive to surface materials, such as buildings, roads, and green spaces. As such, the direct impacts of urban surface materials and their spatial configuration on the thermal environment can be assessed by comparing LST with land use/land cover (LULC) and urban landscapes (e.g., Li et al., 2011; Connors et al., 2013; Bechtel et al., 2019). Satellite-derived LST differs from ground-observed surface air-temperature (T_a) (Voogt and Oke, 2003; Memon et al., 2009). T_a is greatly influenced by atmospheric turbulence and mixing near the surface, resulting in lower spatial variability than that of LST. Furthermore, T_a more strongly reflects anthropogenic heat emissions. Thus, T_a is useful for evaluating the impacts of meteorological factors and anthropogenic heat emissions (or energy consumption) (Takane et al., 2019). The urban heat island (UHI) effects represented by LST and T_a are distinguished as Surface UHI (SUHI) and Canopy UHI (CUHI), respectively (Tzavali et al., 2015; Memon et al., 2009; Du et al., 2021).

Satellite remote sensing offers a significant advantage in simultaneously monitoring LST across urban areas. This enables a quantitative assessment of the overall intensity and spatial variability of SUHI (Schwarz et al., 2011). The overall SUHI intensity (SUHII) is useful for comparing SUHI among various cities worldwide and for understanding their long-term trends (Peng et al., 2012; Li et al., 2020; Hsu and Manya, 2021; Li et al., 2023a). Such long-term analyzes aid in evaluating the impacts of urban growth as well as climate change and extreme heat waves. The spatial variability of SUHII helps in identifying "hotspots" within cities, allowing for a detailed analysis of its relationship with urban structures, including building density, road networks, and green spaces (Connors et al., 2013; Zhou et al., 2017). Such an analysis is crucial for urban planning and the formulation of thermal environment measures.

However, the use of satellite data requires the consideration of constraints owing to the observation geometric conditions and weather conditions.

DOI: 10.1201/9781003244561-12

Specifically, the proportion of roofs, walls, roads, and vegetation within a pixel of satellite imagery varies depending on the observation conditions (Voogt and Oke, 2003; Stewart et al., 2021). Daytime LST can further vary based on the duration of sunlight and the proportion of sunlit to shaded areas (Zakšek and Oštir, 2012; Yamamoto et al., 2022). Given these factors, discussions regarding the observed LST distribution should consider the thermal anisotropy of the urban surface and observation geometric conditions (Li et al., 2023b). Besides, LST retrieval by satellite thermal infrared (TIR) sensors is limited to clear-sky conditions. This introduces a clear-sky bias that differs from T_a observed under all-sky conditions (Du et al., 2021). These limitations can potentially lead to issues such as underestimation owing to cloud contamination and seasonal and climatic biases in data acquisition.

The aim of this chapter is to help readers choose appropriate satellite LST products and analytical methods to assess urban thermal environments. This chapter guide you through the advantages and limitations of each option. The following three subsections explain the methods of acquiring satellite LST data, quantification and analysis techniques of SUHI, and challenges and future prospects.

ACQUISITION AND PREPROCESSING OF SATELLITE LST DATA

Data Selection

The spatiotemporal resolution and observation period of the LST products vary according to the satellite's sensor, orbit height, and mission (Table 9.1). There is an inherent trade-off between spatial and temporal resolutions. LST products with high spatial resolutions have lower observation frequencies, whereas those with higher observation frequencies have lower spatial resolutions (Li et al., 2013). For instance, the TIR bands of the Landsat series sensors and the Advanced Spaceborne Thermal Emission and Reflection Radiometer (ASTER) offer high spatial resolutions of 60–120 m, but they observe only once every 16 days (Malakar et al., 2018; Hulley et al., 2015) (Figure 9.1(a)). In contrast, the Moderate Resolution Imaging Spectroradiometer (MODIS) and Visible Infrared Imaging Radiometer Suite (VIIRS) have coarser spatial resolutions of ~1 km, but they observe twice daily (Wan and Dozier, 1996; Wan, 2014) (Figure 9.1(b) and (c)). Geostationary satellite sensors, such as the Advanced Himawari Imager (AHI) and Advanced Baseline Imager (ABI), have spatial resolutions of ~2 km, but their observation frequencies are the highest, every 10–15 min (Xu et al., 2014; Yamamoto et al., 2018). While polar-orbiting satellites have relatively consistent overpass times (e.g., Landsat between 10–11 a.m.), the ECOsystem Spaceborne Thermal Radiometer Experiment on Space Station (ECOSTRESS) aboard the International Space Station (ISS) possesses a unique observation pattern. It revisits the same area at different times of the day at a frequency of 1–5 days (Hulley et al., 2021). Consequently, ECOSTRESS LST could be invaluable for investigating diurnal LST variations (Chang et al., 2021a, 2022; Yamamoto et al., 2022).

TABLE 9.1
Summary of Commonly Used Satellite LST Products

Sensor (Satellite)	Product ID	Time Span	Spatial Resolution	Temporal Resolution	Retrieval Algorithm
TM (Landsat 4–5)	TM	1982–1993	30 m (120 m)	16 days	SC
ETM+ (Landsat 7)	ETM+	1999–present	30 m (60 m)	16 days	SC
TIRS (Landsat 8)	TIRS	2013–present	30 m (100 m)	16 days	SC
ECOSTRESS (ISS)	ECO2LSTE	2018–present	70 m	–	TES
ASTER (Terra)	AST_08	2000–present	90 m	16 days	TES
SGLI (GCOM-C)	LST_SGLI	2018–present	250 m	2–3 days	SW
VIIRS (S-NPP/NOAA)	VLSTO	2012–present	750 m	Twice daily	SW
VIIRS (S-NPP/NOAA)	VNP21	2012–present	750 m–1 km	Twice daily	TES
AATSR (Envisat)	AATSR_LST_2	2002–2012	1 km	Twice daily	SW
MODIS (Terra)	MOD11	2000–present	1 km	Twice daily	SW
MODIS (Aqua)	MYD11	2002–present	1 km	Twice daily	SW
MODIS (Terra)	MOD21	2000–present	1 km	Twice daily	TES
MODIS (Aqua)	MYD21	2002–present	1 km	Twice daily	TES
SLSTR (Sentinel-3)	SL_2_LST	2016–present	1 km	Twice daily	SW
AVHRR (NOAA)	LST_ AVHRR	1998–present	1.1 km	Twice daily	SW
AHI (Himawari-8/9)	AHI LST	2016–present	2 km	10 min	SW
ABI (GOES-R)	ABI L2+ LST	2017–present	2 km	Hourly	SW
SEVIRI (MSG)	MLST	2009–present	3 km	15 min	SW

Source: Referenced from Li et al., 2023b.

Note: The values enclosed in parentheses under the "spatial resolution" item of the Landsat products correspond to the spatial resolutions of the TIR bands. The LST products are re-gridded at 30 m resolution. *TM: Thematic Mapper, ETM+: Enhanced Thematic Mapper Plus, TIRS: Thermal Infrared Sensor, ECOSTRESS: ECOsystem Spaceborne Thermal Radiometer Experiment on Space Station, ISS: International Space Station, ASTER: Advanced Spaceborne Thermal Emission and Reflection Radiometer, SGLI: Second-generation Global Imager, GCOM-C: Global Change Observation Mission – Climate, VIIRS: Visible Infrared Imaging Radiometer Suite, S-NPP: Suomi National Polar-orbiting Partnership, NOAA: National Oceanic and Atmospheric Administration, AATSR: Advanced Along-Track Scanning Radiometer, MODIS: Moderate Resolution Imaging Spectroradiometer, SLSTR: Sea and Land Surface Temperature Radiometer, AVHRR: Advanced Very High Resolution Radiometer, AHI: Advanced Himawari Imager, ABI: Advanced Baseline Imager, GOES-R: Geostationary Operational Environmental Satellite-R series, SEVIRI: Spinning Enhanced Visible and Infrared Imager, MSG: Meteosat Second Generation.*

The selection of LST products is based on the spatiotemporal scale of the SUHII. Landsat, ASTER, and ECOSTRESS products are suitable for investigations related to urban landscapes (e.g., building density, width, occupancy rate, and orientation) (Connors et al., 2013; Chang et al., 2021a). For studies at citywide or district scales, moderate-resolution products (≥1 km resolution), such as MODIS, VIIRS, and Sentinel-3, are also applicable (Bechtel et al., 2019; García, 2022). For evaluating long-term trends like the impacts of climate change or urban growth on

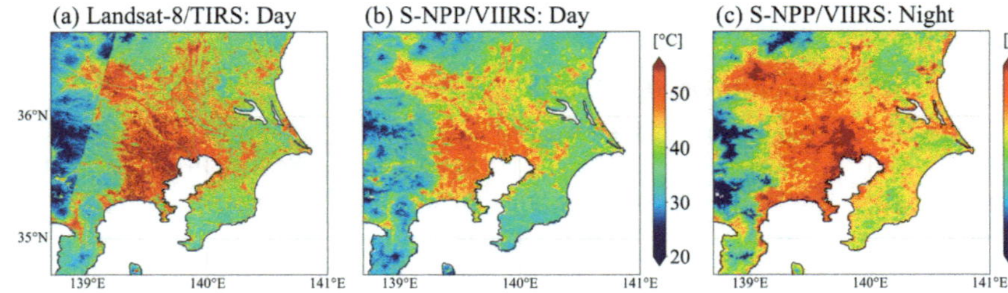

FIGURE 9.1 Composite LST images around Tokyo, Japan, for July 1–31, 2023. (a) Landsat/TIRS and (b) S-NPP/VIIRS at daytime, and (c) S-NPP/VIIRS at nighttime. For the TIRS LST, we selected two clear-sky images: one for the west side on July 2 (from the images of July 2 and 18) and the other for the east side on July 11 (from the ones of July 11 and 27). The VIIRS day/night LSTs are the median of 31 daily images. The TIRS has a high effective spatial resolution of ~100 m. However, when focusing on large urban areas, its narrow swath width (185 km) and infrequent observations (16 days) cause stripe-shaped discontinuities. VIIRS has a low spatial resolution of ~1 km, but it can monitor nighttime SUHIs, and its wider swath width of 3000 km allows for stable coverage of the entire urban area.

the thermal environment, Landsat and MODIS products, which provide over 20 years of LST data, are invaluable (Yao et al., 2019; Weng et al., 2019; Deng et al. 2023; Li et al., 2023a). When examining the impacts of smaller time-scale phenomena, such as the Föhn and sea-land breeze circulation, geostationary satellite products are the preferred choice (Chang et al., 2021b; Yamamoto et al., 2020). Additionally, diurnal LST derived from geostationary satellites provides unique information, such as daily maximum/minimum LST, their timing, and diurnal temperature range.

The quality of LST products should also be considered. LST retrieval assumes clear-sky conditions, so that data quality diminishes when clouds are contaminated in the pixel. Therefore, accurate cloud detection using multispectral bands is essential. For instance, MODIS and VIIRS possess various visible and infrared bands, ensuring some degree of accuracy in detecting clouds for both day- and nighttime LST products (Ackerman et al., 2008; Frey et al., 2008). In contrast, Landsat series sensors have only one or two infrared bands. Consequently, its cloud detection accuracy diminishes at night when the visible bands are ineffective. When evaluating nighttime thermal environments, one must either select products with guaranteed cloud detection accuracy or perform careful preprocessing.

Overview of LST Retrieval

Basic Theory

LST is retrieved from the observed radiance temperature in the TIR atmospheric window (8–12 μm). This retrieval involves atmospheric and land surface emissivity corrections (McMillin, 1975). The radiance, $R_i(\theta, \phi)$, observed at a certain

TIR band with a central wavelength λ μm for a given viewing zenith angle θ and viewing azimuth angle ϕ can be represented as (Li et al., 2013):

$$\begin{aligned} R_{\lambda(\theta,\varphi)} &= B_\lambda[T_{\lambda(\theta,\varphi)}] \\ &= \varepsilon_{\lambda(\theta,\varphi)} B_\lambda[LST_{(\theta,\varphi)}]\tau_{\lambda(\theta,\varphi)} + [1 - \varepsilon_{\lambda(\theta,\varphi)}] R_{at\downarrow\lambda(\theta,\varphi)}\tau_{\lambda(\theta,\varphi)} + R_{at\uparrow\lambda(\theta,\varphi)}, \end{aligned} \tag{9.1}$$

where B_λ is Planck's function, T_λ is the brightness temperature at the top of the atmosphere, ε_λ is the LSE, τ_λ is the atmospheric transmittance, and $R_{at\downarrow\lambda}$ and $R_{at\uparrow\lambda}$ are the downward and upward atmospheric thermal radiances, respectively. The first term on the right side represents the upward radiation emitted from the surface, and the second term represents the downward atmospheric thermal radiation reflected by the surface (Figure 9.2). This equation assumes that the surface–atmosphere is a single layer, and the scattering component of the thermal radiation from the surrounding pixel is neglected. It should be noted that the retrieved temperature (LST) is the "bulk" LST, which includes the LSTs of various surfaces within the pixel. Furthermore, LST reflects a certain aspect within a pixel, and its components vary with the viewing zenith and azimuth angles.

Given the knowledge of LSE and atmospheric effects (such as thermal radiance and transmittance), LST can be retrieved from the observed radiance temperature

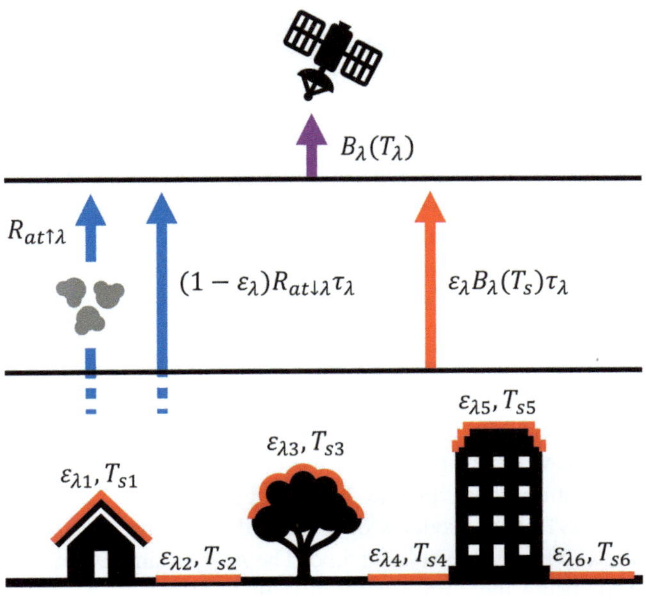

FIGURE 9.2 Schematic of LST retrieval of an urban pixel from satellite TIR sensors. The radiative components correspond to Eq. (9.1), but their dependence on the zenith and azimuth angles is ignored. T_{si} and $\varepsilon_{\lambda i}$ (i = 1–6) denote the LST and LSE of different surface materials, respectively; T_s and ε_λ denote the "bulk LST" and "bulk LSE", respectively, including surface materials within one pixel.

using Eq. (9.1). However, it is impossible to directly solve LST because LSE and atmospheric effects are usually unknown variables. Moreover, both LSE and atmospheric effects depend on λ, θ, and ϕ. Consequently, even when increasing the number of equations by incorporating measurements from different TIR bands, viewing zenith angles, or observation times, the number of unknown variables cannot be reduced (i.e., an ill-posed problem). To retrieve LST, one needs to reduce the number of unknown variables by estimating LSE or atmospheric effects from other sources, and/or making some assumptions or constraints. Thus, current satellite LST products generally adopt one of the following three approaches (Li et al., 2023b):

1. Single Channel (SC) algorithm: Assumes that both LSE and atmospheric effects are known a priori.
2. Split-Window (SW) algorithm: Assumes that only LSE is known a priori.
3. Temperature Emissivity Separation (TES) algorithm: Assumes only atmospheric effects are known a priori.

SC Algorithm

The SC algorithm typically uses brightness temperature (or radiance) from the TIR band, LSE, and atmospheric profiles as inputs to estimate LST through radiative transfer codes (Sobrino et al., 2004; Malakar et al., 2018). The SC algorithm is commonly applied to Landsat data because prior to Landsat 7 had one TIR band (see Table 9.1). It is difficult to obtain atmospheric profiles at satellite overpass times; hence, some methods employ empirical atmospheric corrections based on the relationship between water vapor content and air-temperature in the lower troposphere (Qin et al., 2001; Jiménez-Munoz and Sobrino, 2003). The advantages of the SC algorithm are that only one band is required, and the calculation is simple. It can be applied to historical satellite sensors with only one TIR band, thus allowing for long-term LST analysis. However, this results in increased uncertainties owing to errors in the radiative transfer codes, atmospheric profiles, and LSE.

Atmospheric profiles are commonly derived from numerical climate model data with grid sizes exceeding several kilometers or from radiosonde observation data around the target region. Such sources introduce errors in terms of temporal and spatial scales, viewing angle, and altitude. Several studies have used emissivity-corrected LST without atmospheric correction (Hung et al., 2006; Li et al., 2009; Li et al., 2011; Wang and Murayama, 2020). If focusing on only a few days within a single urban area, it would be possible to use emissivity-corrected LSTs, assuming a spatially uniform atmospheric profile. LSE is obtained from LSE products such as ASTER and ECOSTRESS products or estimated using the Normalized Difference Vegetation Index (NDVI) threshold method. The ASTER and ECOSTRESS products retrieve the LSE using the TES algorithm (Hulley et al., 2015, 2021). The NDVI threshold method empirically estimates LSE of a target pixel based on the emissivity of vegetated and non-vegetated surfaces, and the fractional vegetation cover as derived from NDVI (Valor and Caselles, 1996; Sobrino et al., 2008). The emissivity of vegetated/non-vegetated surfaces is based on LULC data and emissivity libraries (e.g., MODIS Emissivity Library: https://www.icess.ucsb.edu/modis/EMIS/html/em.

html; ECOSTRESS Spectral Library: https://speclib.jpl.nasa.gov/). Both the TES algorithm and NDVI threshold method have difficulty estimating LSEs corresponding to diverse surface materials in urban areas; thus, systematic errors arising from these approaches are unavoidable (Sobrino et al., 2012).

SW Algorithm

The SW algorithm utilizes the differences in atmospheric attenuation between two adjacent TIR bands (10–12.5 μm) for atmospheric correction (McMillin, 1975; Wan and Dozier, 1996). Owing to its simplicity and stable retrieval accuracy, the SW algorithm has been extensively applied to a multitude of satellite products (Table 9.1). The algorithm employs an empirical formula, primarily using the brightness temperatures (T_i and T_j) and LSEs (ε_i and ε_j) of two adjacent TIR bands centered at i and j μm. A typical formula can be written as:

$$LST = a_0 + a_1 T_i + a_2 (T_i - T_j), \tag{9.2}$$

$$a_k = f_k (g_i, g_j, \varepsilon_i, \varepsilon_j, WV, \theta) \ (k = 0, \ 1, \ \text{and} \ 2), \tag{9.3}$$

where the coefficients a_k (k = 0, 1, and 2) depend on the spectral response functions (g_i and g_j), LSE (ε_i and ε_j), total water vapor content (WV), and viewing zenith angle (θ). These coefficients are calibrated using radiative transfer models. However, Eq. (9.2) can provide insufficient atmospheric correction under warm/humid conditions (Sòria and Sobrino, 2007). To enhance atmospheric correction, several refined equations have been proposed that incorporate a quadratic term of $T_i – T_j$ or utilize more than two TIR bands (Li et al., 2013; Yamamoto et al., 2022). LSE of TIR bands can be obtained from LSE products or estimated using the NDVI threshold method, similar to the SC algorithm. Therefore, a limitation of the SW algorithm is its vulnerability to LSE errors.

TES Algorithm

The TES algorithm retrieves LST and LSE simultaneously by using atmospherically corrected brightness temperatures from multiple TIR bands (Gillespie et al., 1998; Hulley and Hook, 2011; Islam et al., 2017). Products such as ECOSTRESS, ASTER, MODIS (MOD21 and MYD21), and VIIRS (VNP21) have adopted TES algorithm (refer to Table 9.1). This algorithm requires at least three TIR bands. In addition to prior atmospheric correction, it further reduces unknown variables by using empirical relationships between the minimum emissivity and emissivity contrast among the TIR bands. The advantage of the TES algorithm is its high retrieval accuracy of LST and LSE, and the flexibility of LSE to respond to changes in surface conditions. However, the required processing is more complex than that of the SC and SW algorithms, and the applicable sensors are limited. In addition, the retrieval accuracy is reduced for water, snow, and vegetation surfaces with little change in spectral emissivity, and under warm/humid atmospheric conditions (Coll et al., 2007; Gillespie et al., 2011; Hulley and Hook, 2011).

Urban-Specific LST Errors and Thermal Anisotropy

Uncertainty in LST retrieval in urban areas is notably increased mainly due to the diversity and heterogeneity of surface materials (Sobrino et al., 2012). For instance, while the emissivity of forest surfaces at 11 μm is roughly 0.99, regardless of tree species, the emissivities of man-made materials such as paving asphalt, construction concrete, pine wood, plate window glass, and slate stone shingle are approximately 0.96, 0.95, 0.93, 0.86, and 0.90, respectively (ECO-STRESS Spectral Library: https://speclib.jpl.nasa.gov/). An LSE error of 0.01 can result in an LST error of approximately 1–2 K (Yang et al., 2015a). It is easy to imagine the high uncertainty in estimating representative LSE in urban areas, given the sporadic distribution of various surface materials at different heights. Additionally, urban geometries such as building height, building density, and topography further increase this uncertainty (Yang et al., 2015b). When retrieving LST using the SC algorithm without factoring in urban geometry, LST error can reach between 3–5 K (Li et al., 2013). As the spatial resolution of the pixels decreases, the LST uncertainty increases. Many products listed in Table 9.1 reported validation accuracies below 2 K, regardless of the retrieval algorithm used (Li et al., 2023b). However, the accuracies are for homogeneous and flat surfaces, thus not for urban surfaces.

The thermal anisotropy of the urban surfaces and observation angles significantly affected the retrieved LSTs. The complex spatial structure of urban areas can potentially induce directional variations in LST exceeding 10 K (Lagouarde et al., 2010; Morrison et al., 2023). Satellites observe such heterogeneous LST distributions at various viewing zenith and viewing azimuth angles. The nadir observations capture a larger portion of the road surface, whereas the off-nadir observations capture a larger portion of the wall surface. Moreover, during daylight, the proportion of sunlit to shaded areas varies depending on the relative azimuth angle between the sensor and sun. For instance, when geostationary satellites observe an urban area at mid-latitudes, they consistently target sunlit facades. As a result, the observed LST has reported a mean positive bias of >3 K when compared to LSTs from other polar-orbiting satellites resampled to a comparable spatial resolution (Zakšek and Oštir, 2012; Duan and Li, 2015; Yamamoto et al., 2022).

Scaling and Classification of LST

The retrieved LST is occasionally classified for a comprehensive analysis of SUHI. Such a classification can mitigate the uncertainties in LST retrieval, as discussed in the previous subsection. Before classification, LSTs are typically scaled via normalization, standardization, or relative deviation (Voogt and Oke, 2003; Xu and Chen, 2004; Weng et al., 2019; Lu et al., 2020; Ma et al., 2021). The relative-deviated LST is also termed the Urban Thermal Field Variance Index (UTFVI) (Guha et al., 2018; Najafzadeh et al., 2021; Zhang et al., 2023a). Such scaling can diminish the differences in atmospheric effects on different days (Figure 9.1(a)). It is also effective when using emissivity-corrected LST in which atmospheric correction is omitted (Hung et al., 2006; Li et al., 2009; Li et al., 2011; Wang et al., 2020).

The normalized LST ($NLST_i$), standardized LST ($SLST_i$), and $UTFVI_i$ for pixel i are represented by the following equations:

$$NLST_i = (LST_i - LST_{min})/(LST_{max} - LST_{min}), \quad (9.4)$$

$$SLST_i = (LST_i - LST_{mean})/\sigma, \quad (9.5)$$

$$UTFVI_i = (LST_i - LST_{mean})/LST_{mean}, \quad (9.6)$$

where LST_{max} and LST_{min}, LST_{mean}, and σ are the maximum, minimum, mean, and standard deviation of LSTs across the entire city, respectively. Note that if outliers exist in the image, $NLST_i$ is particularly vulnerable to them. Therefore, LST_{max} and LST_{min} are the values or medians of the upper and lower (e.g., 5th and 95th percentiles) samples, respectively, avoiding the highest and lowest LSTs.

Scaled LSTs are divided into five to seven temperature classes depending on their variability. For example, $NLST_i$ can be divided into five classes based on the mean ($NLST_{mean}$) and standard deviation (σ) of its spatial distribution: Very Low ($NLST_i \leq NLST_{mean}-1.5\sigma$), Low ($NLST_{mean}-1.5\sigma < NLST_i \leq NLST_{mean}-\sigma$), Medium ($NLST_{mean}-\sigma < NLST_i \leq NLST_{mean}+\sigma$), High ($NLST_{mean}+\sigma < NLST_i \leq NLST_{mean}+1.5\sigma$), and Very High ($NLST_{mean}+1.5\sigma \leq NLST_i$). The $UFTVI_i$ are divided into six environmental conditions: Excellent (< 0.000), Good (0.000–0.005), Normal (0.005–0.010), Bad (0.010–0.015), Worse (0.015–0.020), and Worst (> 0.020).

EVALUATION OF SUHI

The spatiotemporal patterns of SUHI measured using satellite LST can provide scientific evidence essential for sustainable urban development and heat mitigation strategies (Oke, 1982; Peng et al., 2018; Zhou et al., 2017; Chang et al., 2021b). Previous studies have demonstrated that SUHI is closely related to LULC, urban morphology, regional climate, and anthropogenic heat sources (Li et al., 2011; Zhao et al., 2014; Du et al., 2016; Cai et al., 2018). Consequently, measured SUHI has been utilized to assess the impacts of urban growth (Weng et al., 2019; Stewart et al., 2021; Deng et al., 2023; Li et al., 2023a), evaluate mitigation strategies through urban greening and ventilation (Yang et al., 2020; Yang et al., 2021), assess the effects of heat waves and climate change (Oleson et al., 2015; Cuerdo-Vilches et al., 2023; Cui et al., 2023), and examine interactions with energy consumption (Li et al., 2019).

SUHI is measured using the heat island's "overall height" and "elevation distribution". The overall height (i.e., the overall SUHII) enables efficient and straightforward comparisons of SUHI across different cities and the evaluation of potential heat risks over seasons and years. Indeed, this overall SUHII has been employed in research at urban scales (Memon et al., 2009; Najafzadeh et al., 2021; Chang et al., 2021b; García, 2022), national scales (Imhoff et al., 2010; Bounoua et al., 2018; Lai et al., 2018; Hu et al., 2022), and global scales (Schwarz et al., 2011;

Peng et al., 2012; Clinton and Gong, 2013; Li et al., 2020). Moreover, the overall SUHI can be useful for indicating indirect relationships with population dynamics, including mortality, morbidity, and energy consumption (Hung et al., 2006; Santamouris, 2015; Li et al., 2023a). On the other hand, the elevation distribution (i.e., the SUHII distribution) helps understand SUHI's causes by comparing it to LULC, landscape metrics, topography, and vegetation indices (Li et al., 2011; Connors et al., 2013; Peng et al., 2018; Chang et al., 2021a). The SUHII distribution is particularly suited for evaluations that focus on specific regions, rather than city-to-city comparisons.

Overall SUHII

The overall SUHII is commonly quantified by the difference in LST between urban and non-urban areas (Memon et al., 2009; Stewart, 2011). Owing to the diverse city sizes, shapes, and urban structures, multiple methods for defining urban and non-urban areas have been proposed (Clinton and Gong, 2013; Lai et al., 2018; Peng et al., 2012; Yao et al., 2019). There is no single method accepted by all for defining urban/non-urban areas that fits every urban type, making it a key challenge for consistent SUHI evaluation. Many methods rely on LULC data. However, when urban-suburban boundaries are unclear or land use frequently changes, methods independent of LULC are chosen. Naturally, different calculation methods yield different SUHIIs, and their variability is influenced by urban morphology, topography, time of day, and season (Schwarz et al., 2011). Therefore, it is essential to carefully consider the advantages and limitations of each method and select one based on urban characteristics and climatic conditions. When comparing the SUHII across multiple cities, using several methods might be beneficial. The following provides an overview of well-known SUHII calculation methods.

Simple SUHII with LULC

This SUHII is calculated as the difference between the mean LSTs in urban and non-urban pixels extracted based on LULC data:

$$SUHII = LST_u - LST_r, \tag{9.7}$$

where LST_u is the mean LST for the entire urban area or urban core, and LST_r is the mean LST for non-urban areas. Non-urban areas refer to suburban or agricultural areas based on LULC, or surrounding areas based on distance from the urban core. In coastal cities, water bodies that are less susceptible to urbanization may be selected as non-urban (Chen et al., 2006). The identification of urban/non-urban areas using LULC data is intuitive and highly versatile. As such, this method has been applied in comparative studies of SUHII covering numerous cities (Peng et al., 2012; Du et al., 2021; Hsu and Manya, 2021; Li et al., 2023a).

Prior studies have proposed various methods to define non-urban areas, broadly categorized into four types: Equal area, Equal distance, Sample point, and Administrative boundary (Zhang et al., 2023b).

1. Equal area: A non-urban area is defined as a ring surrounding an urban area with an equal number of pixels as the urban area (Peng et al., 2012). This definition method allows for inter-city comparisons considering differences in urban size. However, the ring zone might include high-elevation areas and diverse non-urban classes.
2. Equal distance: An area consistently urban from the past to the present is defined as a stable urban area (SUA), an area that has changed to urban is defined as an urbanized area (UA), and the surrounding area at a certain distance (e.g., 20–25 km) from the SUA is defined as a non-urban area (Yao et al., 2017). The surrounding areas exclude urban, industrial, and rural settlement pixels. This definition is useful for assessing the impact of urban expansion. However, the distance setting is subjective, and the obtained SUHII varies depending on the distance.
3. Sample point: Some typical urban core and non-urban pixels are selected (Zhao et al., 2014). In the selection, differences in elevation (e.g., ±100 m) and latitude (e.g., ±0.1°) between urban and non-urban pixels are considered. Although the selection process is subjective, it is particularly valuable in areas with complex urban structures. It also allows the selection of urban/non-urban pixels that best meet the objectives of the study without being restricted by certain criteria or algorithms.
4. Administrative boundary: Non-urban areas are defined as pixels within a city's administrative boundaries, excluding urban ones (Lai et al., 2018). This approach is suitable for regions with mixed urban and non-urban areas. The obtained SUHII can be compared consistently with data related to the population, economy, and infrastructure, which are compiled based on administrative boundaries. However, because these boundaries are man-made, they might neglect physical attributes such as topography. Therefore, the validity of the obtained SUHII itself cannot be guaranteed.

SUHII without LULC

Methods that are independent of LULC measure SUHI based on the mean LST across the entire urban area. Indices proposed in previous studies include the SUHI Ratio Index (SURI) (Xu and Chen, 2004; Weng et al., 2019), Hot Island Area (HIA) (Zhang et al., 2008), and Magnitude (Rajasekar and Weng 2009). Because these methods do not require LULC information, they are particularly useful in regions where LULC data are scarce, urban-suburban demarcations are ambiguous, or there are significant temporal land use variations. While the computation is straightforward, the results are significantly influenced by statistical parameters, such as sample size, mean, and standard deviation. Consequently, the indices might vary depending on the urban spatial scale (Zhang et al., 2023b). While these indices primarily provide insights into the relative temperature range, it is essential to incorporate data sources, such as LULC and NDVI, when exploring the physical mechanisms behind the derived SUHII.

SURI is calculated using temperature classes based on normalized LST (see subsection "Scaling and Classification of LST" for details on obtaining the temperature classes) as follows:

$$SURI = \frac{1}{100m} \sum_{i=1}^{n} w_i P_i, \tag{9.8}$$

where m is the number of temperature classes (e.g., m = 5 if there are five classes), n is the number of temperature classes higher than the "medium temperature" class (e.g., n = 2 for five classes), w_i is the value of each temperature class (e.g., if there are five classes, $w_1 = 4$ for i = 1, $w_2 = 5$ for i = 2), and P_i is the percentage (%) of the temperature class relative to the total area. Essentially, higher-temperature classes are assigned greater weights, making them more influential in the SURI calculation. The resulting values range between 0 and 1.

The calculation methods for HIA and Magnitude are more straightforward than SURI. The HIA quantifies SUHII as the percentage of high-temperature areas relative to the entire urban expanse. High-temperature areas are defined as regions where the LST is greater than the citywide mean LST + σ. Given that HIA accounts for multiple hotspots, it is likely to be beneficial for cities where hotspots are sporadically distributed. The magnitude is calculated by subtracting the mean LST of the entire urban area from its maximum LST.

SUHII based on Linear Relationship with Impervious Surface Area (ISA)

This SUHII is calculated from the linear relationship between ISA and LST, with the assumption that LST is highly dependent on ISA (Zhang et al., 2010; Shi et al., 2023):

$$SUHII = LST_{100} - LST_0, \tag{9.9}$$

where LST_{100} and LST_0 are the LSTs with ISA of 100% and 0%, respectively. These LSTs are estimated using a linear regression model with ISA as the explanatory variable (i.e., $LST \approx a \times ISA + b$). This method requires only LST and ISA. Given that the ISA serves as an indicator of urbanization, reflecting population and building densities, the SUHII can be easily related to urban heat mitigation strategies (Angel et al., 2011; Bounoua et al., 2018). However, relying solely on the ISA might introduce uncertainties in the SUHII estimation. The linear relationship between LST and ISA works well for cities located in biomes dominated by forests and grasslands. However, this does not hold true for cities in arid/semi-arid regions (Imhoff et al. 2010; Shi et al., 2023). Hence, this method may not be suitable for cities in arid/semi-arid regions (Li et al. 2018).

SUHII based on Gaussian Surface Model

The SUHII calculation assumes that the LST distribution within an urban area follows a Gaussian surface (i.e., temperatures within the urban area decrease uniformly from the center toward the outside) (Streutker et al., 2002; 2003). By fitting a Gaussian surface model to the spatial distribution of LST in urban and surrounding areas, the height and extent of SUHI can be quantified. The Gaussian surface model characterizing the spatial distribution of LST is given by the following equation:

$$LST\,(x,\,y) = LST_0 + a_1x + a_2y + SUHII\,(x,\,y), \tag{9.10}$$

$$SUHII\,(x,\,y) = a_0 \exp\left[-\frac{\{(x - x_0)\cos\phi + (y - y_0)\sin\phi\}^2}{(0.5a_x)^2} - \frac{\{(y - y_0)\cos\phi - (x - x_0)\sin\phi\}^2}{(0.5a_y)^2} \right], \tag{9.11}$$

where LST(x, y) is the LST at pixel location (x, y), LST_0 is the mean LST of the surrounding non-urban area, and a_1 and a_2 are the spatial gradients of the LST distribution in the surrounding non-urban area in the x- and y-directions, respectively. The first three terms on the right-hand side of Eq. (9.10) assumes that the LST distribution in a non-urban area is planar. The Gaussian function in Eq. (9.11) describes the LST distribution within the urban area, providing the magnitude (a_0), spatial extent (${}_{ax}$ and a_y), orientation (Φ), and central location (x_0 and y_0) of the SUHI. a_0 corresponds to the overall SUHII.

First, the LSTs of the non-urban pixels are fitted to the plane (the first three terms on the right side of Eq. (9.10)), which determines LST_0, a_1 and a_2. Here, given the uncertainty in the LST, one sometimes adds a buffer of approximately 2 K to the non-urban temperature component (i.e., $LST_0 + a_1x + a_2y + 2$) (Streutker et al., 2003; Hung et al., 2006). Second, the differences between the original LSTs and the temperature components of the first three terms in Eq. (9.10) are fitted to a Gaussian surface (the exponential term on the right-hand side of Eq. (9.11)), which determines a_0, ${}_{ax}$, a_y, Φ, x_0, and y_0. The footprint area (FA) of SUHI is calculated using the derived ${}_{ax}$ and a_y in the following equation:

$$FA = \frac{\pi a_x a_y}{4}. \tag{9.12}$$

The Gaussian surface model offers an intuitive visualization of the three-dimensional structure of SUHI. Compared to the simple SUHII with LULC, it has been reported that the definition of boundaries between urban and non-urban has a smaller impact on the SUHII (Hu et al., 2022). However, the model may be difficult to apply to cities with complex urban geometry and topography (e.g., coastal cities). The validity of the derived SUHII and FA should be assessed using evaluation metrics such as the coefficient of determination and RMSE in the fitting process.

SUHII Distribution

The spatial variability of SUHII is represented by LST, scaled or classified LST, or the difference from the mean LST of non-urban pixels. Previous studies have primarily linked SUHII distribution to LULC (Chang et al., 2021a; Cui et al., 2023), landscape patterns (Connors et al., 2013; Lu et al., 2020; Ma et al., 2021), and spectral indices such as NDVI and Normalized Difference Built-up Index (NDBI)

(Weng et al., 2004; Meng et al., 2009; Li et al., 2011; Peng et al., 2018). Considering LULC and landscapes, the Local Climate Zone (LCZ: Stewart and Oke, 2012), which accounts for building height and density, and landscape metrics (McGarigal and Marks, 1995), which address urban feature arrangement, shape, distribution, and connectivity, are likely appropriate. Within the typical land use class "built-up land", the SUHII varies based on building height and density (Bechtel et al., 2019). Comparisons with LCZ and landscape metrics typically utilize LST from polar-orbiting satellites with a high spatial resolution (Zhou et al., 2017). Because LST products with high spatial resolution have a low observation frequency, the focus on temporal variations in SUHII distribution is directed towards seasonal and interannual changes (Du et al., 2016; Najafzadeh et al., 2021; Deng et al., 2023). Identifying seasonal patterns of hot and cool spots in urban areas is crucial for urban planning and enhancing energy efficiency. Moreover, monitoring the interannual variations in the SUHII assists in evaluating the effects of urban growth or cooling strategies on urban sustainability.

Quantifying Urban Growth

Based on an analysis of long-term SUHII trends, various approaches have been employed to classify or quantify urban growth. Deng et al. (2023) classified temporal variations of hot/cold spots within the Guangdong-Hong Kong-Macao Greater Bay Area from 2000 to 2020. Based on the interannual trends of LST, pixels were classified into Intensifying hot/cold spots, Persistent hot/cold spots, Oscillating hot/cold spots, Sporadic hot/cold spots, and Diminishing hot/cold spots. Additionally, centroid movement analysis to evaluate interannual shifts in SUHI centroids (Wang et al., 2022), spatial coupling analysis to assess the coupling coefficient of growth rate between SUHI and built-up land (Yang et al., 2019), and the GeoDetector model to evaluate the spatial alignment between SUHI and built-up land patterns (Wang et al., 2010) have been utilized.

Weng et al. (2019) applied degree-of-sprawl, degree-of-freedom, and degree-of-goodness to evaluate the spatiotemporal variation of SUHI in Babol, Iran, from 1985 to 2017 over 32 years. The degree-of-sprawl is based on Shannon's entropy method. A high degree-of-sprawl indicates an extensive distribution of hotspots, suggesting dispersed urban expansion. The degree-of-freedom is based on Pearson's chi-square test. Elevated degree-of-freedom shows substantial discrepancies between the observed and expected SUHII, indicating higher heterogeneity in urban growth patterns and processes. The degree-of-goodness metric, derived from the degree-of-sprawl and degree-of-freedom, gauges the quality of urban expansion. A positive value for degree-of-goodness indicates that the observed SUHII closely aligns with the expected SUHII, reflecting a more compact urban form, in contrast to sprawl (Bhatta et al., 2010). The expected SUHII values are statistically inferred from past growth trends. Hence, these metrics can be applied in developing countries, even without strategic urban development policies. However, the inherent uncertainty in the expected SUHII warrants caution, implying that the expected SUHII should be interpreted only as reference values (Dadras et al., 2015).

Diurnal Scale Analysis using Next-Generation Satellites

Recent advancements have enabled the acquisition of diurnal LST patterns within urban areas using next-generation geostationary satellites and ECOSTRESS aboard the ISS (Yamamoto and Ishikawa, 2018; Chang et al., 2021a, 2022). Third-generation geostationary satellites, such as Himawari-8, Geostationary Operational Environmental Satellite R-Series (GOES-R), Meteosat Third Generation (MTG), and Geostationary Korea Multi-Purpose Satellite-2A (GEO-KOMPSAT-2A), can provide LSTs with a high frequency of 10–15 min, albeit at a coarse spatial resolution of ~2 km. On the other hand, ECOSTRESS can provide diurnal LSTs with a resolution of approximately 70 m through data collection spanning 1–4 months (Chang et al., 2021; Yamamoto et al., 2022). However, the collected LST reflects not only diurnal variations but also the meteorological conditions of each acquisition day. Daily meteorological differences may be complemented by LST from geostationary satellites.

Diurnal LST offers invaluable insights, such as daily maximum/minimum LST, their timings, and diurnal temperature range, which are not accessible by polar-orbiting satellites (Duan et al., 2012; Lai et al., 2018; Yamamoto et al., 2023). The daily maximum/minimum LST can provide insight into the extreme hot/cold environments experienced during heat and cold waves. Diurnal peak times indicate the spatial distribution of the timing when hotspots appear (Lai et al., 2018). Temperature variations, such as daytime warming, nighttime cooling, and diurnal temperature range, correlate with the thermal inertia of urban surfaces. For instance, urban centers primarily composed of high thermal inertia materials such as concrete become nighttime hotspots, while residential areas mainly constructed from low inertia materials such as wood become daytime hotspots. Geostationary satellites can capture diurnal temperature variations consistent with the thermal inertia of urban land use (Yamamoto et al., 2018). Interestingly, a relationship between daytime warming and building density has also been observed (Yamamoto et al., 2020).

To extract the characteristics of diurnal LST variations, two primary approaches have been adopted: (1) the Diurnal Temperature Cycle (DTC) model (Weng et al., 2014; Lai et al., 2018) and (2) Principal Component Analysis (PCA) (Yamamoto et al., 2020). The DTC model is a semi-empirical model consisting of functions based on the thermal diffusion equation during the daytime and Newton's law of cooling during the nighttime (Figure 9.3). By fitting the DTC model to the observed diurnal LSTs, parameters such as daily maximum/minimum LST, peak timings, and diurnal temperature range can be obtained. Various functions of the DTC model have been proposed, each with its own fitting bias (Duan et al., 2012). The PCA-based approach allows for the extraction of primary modes from diurnal LST waveforms observed across urban areas. Typically, the first mode is associated with the amplitude and the second with the phase (Ignatov and Gutman, 1998; Aires et al., 2004). However, PCA is sensitive to outliers and data gaps caused by cloud contamination. When characterizing diurnal LST variations in humid climates, the DTC model may be a preferable (Yamamoto et al., 2023).

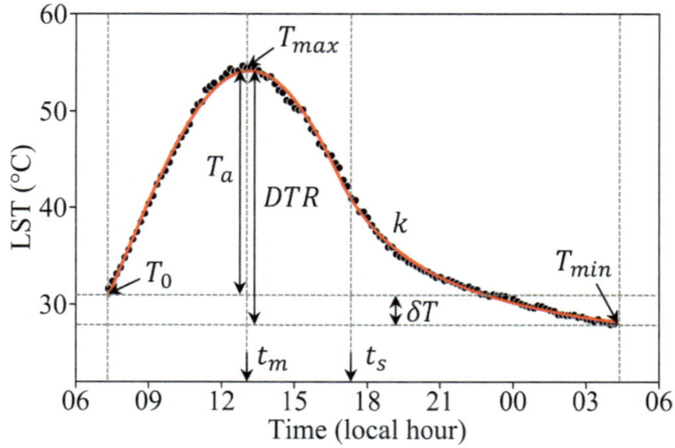

FIGURE 9.3 Parameters obtained by fitting a DTC model (red line) to diurnal variation in Himawari-8 LSTs (dot) of a pixel on a clear-sky day. T_0 is the residual temperature around sunrise; T_a is the temperature amplitude; δT is the temperature difference between T_0 and $T\ (t \rightarrow \infty)$; T_{max} is the daily maximum temperature ($T_{max} = T_0 + T_a$); T_{min} is the daily minimum temperature ($T_{min} = T_0 + \delta T$); DTR is the diurnal temperature range ($DTR = T_a - \delta T$); t_m is the time of the T_{max} (solar time); t_s is the start time of free attenuation (solar time); k is the attenuation constant (hour). The Himawari-8 LSTs were retrieved on 4 August 2018 at a pixel in Osaka, Japan (34.7°N, 135.6°E).

CHALLENGES AND FUTURE PERSPECTIVES

Because LST retrieval from satellites assumes clear-sky conditions, the resulting SUHII may lack consistency due to clear-sky bias. To address this, the development of all-sky LST data is expected. Current approaches for estimating all-sky LST include interpolating cloudy-sky LSTs utilizing passive microwave satellite data (Yoo et al., 2020) and assimilating clear-sky LSTs into all-sky LSTs simulated by a numerical model (Cho et al., 2022; Jia et al., 2022; Zhang et al., 2023c). These approaches currently focus on moderate-resolution (~1 km) LST. However, for SUHI research, higher spatial resolutions (~100 m), such as those of the Landsat sensors and ECOSTRESS, are preferred (Chang et al., 2022). Hence, future efforts are anticipated to further downscale all-sky LST (Desai et al., 2021).

Satellite-derived LST represents a certain aspect of urban temperatures and does not convey the complete surface temperature within a pixel, termed as "incomplete surface temperature". As mentioned previously, urban thermal anisotropy can potentially induce directional variations in LST exceeding 10 K (Lagouarde et al., 2010; Morrison et al., 2023). Discussing these discrepancies with complete surface temperatures is currently challenging (Voogt and Oke, 2003; Stewart et al., 2021). In addition, thermal anisotropy complicates the interpretation and comparison of SUHI obtained from different satellites. In the future, angular normalization of LST will be required for broader applications (Li et al., 2023b).

With the progression of climate change, both the frequency and intensity of heat waves are increasing (Imada et al., 2019; Zhang et al., 2022). The interaction between heat waves and SUHI raises concerns regarding unprecedented thermal environments (Cuerdo-Vilches et al., 2023; Cui et al., 2023). The spatiotemporal understanding of SUHII during heat wave events using satellite-derived LST (Dousset et al., 2011; García, 2022) will become increasingly crucial. This knowledge can help identify heat wave risk regions and timing, facilitating the development of mitigation and adaptation strategies. Additionally, it is important to assess SUHI in terms of human thermal comfort by incorporating meteorological elements such as T_a, humidity, and wind speed (Lin et al., 2011; Oleson et al., 2015). These meteorological elements typically rely on numerical weather/climate models, and LST is the only observational data that covers a wide area. Hence, the satellite LST exhibits the highest sensitivity to extreme temperatures. The need for LST-driven estimates of the thermal environment and heat stress indices is expected to increase in the future (Jiang et al., 2015; Najafzadeh et al., 2021).

CONCLUSIONS

This chapter provided a detailed discussion of various satellite LST products and SUHI evaluation methods. At present, there are no satellite LST data or SUHI evaluation methods that are universally applicable to various spatiotemporal scales and diverse urban structures. Thus, making an appropriate selection among various satellite LST data and SUHI evaluation methods is essential for understanding urban thermal environments and evaluating mitigation strategies. When using satellite LST data, we should consider aspects such as the spatiotemporal resolution, time span, and swath width depending on the sensor, satellite orbit height, and mission, as well as the error characteristics caused by retrieval algorithms. When comparing multiple satellite LST data, additional attention should be paid to thermal anisotropy. SUHI evaluation methods should be selected based on urban/non-urban configuration, background climate and topographic characteristics, the focused time scale and number of cities, and the availability of LULC data.

Future research is expected to address the clear-sky bias and thermal anisotropy in satellite-derived LST, advance heat wave risk assessment methods, and conduct assessments at diurnal scales. Correcting the clear-sky bias and thermal anisotropy will provide consistent urban thermal insights, aligning with ground-observed and model-derived T_a (i.e., CUHI). Such consistency with other meteorological factors will further enhance the reliability of the thermal environment and heat stress indices in heat wave risk assessments. Additionally, diurnal-scale LSTs and thermal indices will provide insights into the most extreme thermal conditions experienced by humans. These advancements and initiatives will be crucial for future considerations on the sustainability of cities.

REFERENCES

Ackerman, S. A., Holz, R. E., Frey, R., Eloranta, E. W., Maddux, B. C., McGill, M., 2008. Cloud detection with MODIS. Part II: Validation. *J. Atmos. Ocean. Technol.* 25, 1073–1086. 10.1175/2007JTECHA1053.1

Aires, F., Prigent, C., Rossow, W. B., 2004. Temporal interpolation of global surface skin temperature diurnal cycle over land under clear and cloudy conditions. *J. Geophys. Res. Atmos.* 109, 1–18. 10.1029/2003JD003527

Angel, S., Parent, J., Civco, D. L., Blei, A., Potere, D., 2011. The dimensions of global urban expansion: Estimates and projections for all countries, 2000–2050. *Prog. Plann.* 75, 53–107. 10.1016/j.progress.2011.04.001

Bechtel, B., Demuzere, M., Mills, G., Zhan, W., Sismanidis, P., Small, C., Voogt, J., 2019. SUHI analysis using Local Climate Zones—A comparison of 50 cities. *Urban Clim.* 28, 100451. 10.1016/j.uclim.2019.01.005

Bhatta, B., Saraswati, S., Bandyopadhyay, D., 2010. Quantifying the degree-of-freedom, degree-of-sprawl, and degree-of-goodness of urban growth from remote sensing data. *Appl. Geogr.* 30, 96–111. 10.1016/j.apgeog.2009.08.001

Bounoua, L., Nigro, J., Zhang, P., Thome, K., Lachir, A., 2018. Mapping urbanization in the United States from 2001 to 2011. *Appl. Geogr.* 90, 123–133. https://doi.org/10.1016/j.apgeog.2017.12.002

Cai, Z., Han, G., Chen, M., 2018. Do water bodies play an important role in the relationship between urban form and land surface temperature? *Sustain. Cities Soc.* 39, 487–498. 10.1016/j.scs.2018.02.033

Chang, Y., Xiao, J., Li, X., Middel, A., Zhang, Y., Gu, Z., Wu, Y., He, S., 2021a. Exploring diurnal thermal variations in urban local climate zones with ECOSTRESS land surface temperature data. *Remote Sens. Environ.* 263, 112544. 10.1016/j.rse.2021.112544

Chang, Y., Xiao, J., Li, Xuxiang, Frolking, S., Zhou, D., Schneider, A., Weng, Q., Yu, P., Wang, X., Li, Xing, Liu, S., Wu, Y., 2021b. Exploring diurnal cycles of surface urban heat island intensity in Boston with land surface temperature data derived from GOES-R geostationary satellites. *Sci. Total Environ.* 763, 144224. 10.1016/j.scitotenv.2020.144224

Chang, Y., Xiao, J., Li, X., Zhou, D., Wu, Y., 2022. Combining GOES-R and ECOSTRESS land surface temperature data to investigate diurnal variations of surface urban heat island. *Sci. Total Environ.* 823, 153652. 10.1016/j.scitotenv.2022.153652

Chen, X. L., Zhao, H. M., Li, P. X., Yin, Z. Y., 2006. Remote sensing image-based analysis of the relationship between urban heat island and land use/cover changes. *Remote Sens. Environ.* 104, 133–146. 10.1016/j.rse.2005.11.016

Cho, D., Bae, D., Yoo, C., Im, J., Lee, Y., Lee, S., 2022. All-sky 1 km MODIS land surface temperature reconstruction considering cloud effects based on machine learning. *Remote Sens.* 14. 10.3390/rs14081815

Clinton, N., Gong, P., 2013. MODIS detected surface urban heat islands and sinks: Global locations and controls. *Remote Sens. Environ.* 134, 294–304. 10.1016/j.rse.2013.03.008

Coll, C., Caselles, V., Valor, E., Niclòs, R., Sánchez, J. M., Galve, J. M., Mira, M., 2007. Temperature and emissivity separation from ASTER data for low spectral contrast surfaces. *Remote Sens. Environ.* 110, 162–175. 10.1016/j.rse.2007.02.008

Connors, J. P., Galletti, C. S., Chow, W. T. L., 2013. Landscape configuration and urban heat island effects: Assessing the relationship between landscape characteristics and land surface temperature in Phoenix, Arizona. *Landsc. Ecol.* 28, 271–283. 10.1007/s10980-012-9833-1

Cuerdo-Vilches, T., Díaz, J., López-Bueno, J. A., Luna, M. Y., Navas, M. A., Mirón, I. J., Linares, C., 2023. Impact of urban heat islands on morbidity and mortality in heat waves: Observational time series analysis of Spain's five cities. *Sci. Total Environ.* 890, 164412. 10.1016/j.scitotenv.2023.164412

Cui, F., Hamdi, R., Kuang, W., Yang, T., He, H., Termonia, P., De Maeyer, P., 2023. Interactions between the summer urban heat islands and heat waves in Beijing during 2000–2018. *Atmos. Res.* 291, 106813. 10.1016/j.atmosres.2023.106813

Dadras, M., Shafri, H. Z. M., Ahmad, N., Pradhan, B., Safarpour, S., 2015. Spatio-temporal analysis of urban growth from remote sensing data in Bandar Abbas city, Iran. *Egypt. J. Remote Sens. Sp. Sci.* 18, 35–52. 10.1016/j.ejrs.2015.03.005

Deng, X., Gao, F., Liao, S., Liu, Y., Chen, W., 2023. Spatiotemporal evolution patterns of urban heat island and its relationship with urbanization in Guangdong-Hong Kong-Macao greater bay area of China from 2000 to 2020. *Ecol. Indic.* 146, 109817. 10.1016/j.ecolind.2022.109817

Desai, A. R., Khan, A. M., Zheng, T., Paleri, S., Butterworth, B., Lee, T. R., Fisher, J. B., Hulley, G., Kleynhans, T., Gerace, A., Townsend, P. A., Stoy, P., Metzger, S., 2021. Multi-sensor approach for high space and time resolution land surface temperature. *Earth Sp. Sci.* 8, 1–18. 10.1029/2021EA001842

Dousset, B., Gourmelon, F., Laaidi, K., Zeghnoun, A., Giraudet, E., Bretin, P., Mauri, E., Vandentorren, S., 2011. Satellite monitoring of summer heat waves in the Paris metropolitan area. *Int. J. Climatol.* 31, 313–323. 10.1002/joc.2222

Du, H., Wang, D., Wang, Y., Zhao, X., Qin, F., Jiang, H., Cai, Y., 2016. Influences of land cover types, meteorological conditions, anthropogenic heat and urban area on surface urban heat island in the Yangtze River Delta Urban Agglomeration. *Sci. Total Environ.* 571, 461–470. 10.1016/j.scitotenv.2016.07.012

Du, H., Zhan, W., Liu, Z., Li, J., Li, L., Lai, J., Miao, S., Huang, F., Wang, Chenguang, Wang, Chunli, Fu, H., Jiang, L., Hong, F., Jiang, S., 2021. Simultaneous investigation of surface and canopy urban heat islands over global cities. *ISPRS J. Photogramm. Remote Sens.* 181, 67–83. 10.1016/j.isprsjprs.2021.09.003

Duan, S. B., Li, Z. L., 2015. Intercomparison of operational land surface temperature products derived from MSG-SEVIRI and Terra/Aqua-MODIS data. *IEEE J. Sel. Top. Appl. Earth Obs. Remote Sens.* 8, 4163–4170. 10.1109/JSTARS.2015.2441096

Duan, S. B., Li, Z. L., Wang, N., Wu, H., Tang, B. H., 2012. Evaluation of six land-surface diurnal temperature cycle models using clear-sky in situ and satellite data. *Remote Sens. Environ.* 124, 15–25. 10.1016/j.rse.2012.04.016

Frey, R. A., Ackerman, S. A., Liu, Y., Strabala, K. I., Zhang, H., Key, J. R., Wang, X., 2008. Cloud detection with MODIS. Part I: Improvements in the MODIS cloud mask for Collection 5. *J. Atmos. Ocean. Technol.* 25, 1057–1072. 10.1175/2008JTECHA1052.1

García, D. H., 2022. Analysis of urban heat island and heat waves using sentinel-3 images: A study of andalusian cities in Spain. *Earth Syst. Environ.* 6, 199–219. 10.1007/s41748-021-00268-9

Gillespie, A., Rokugawa, S., Matsunaga, T., Steven Cothern, J., Hook, S., Kahle, A. B., 1998. A temperature and emissivity separation algorithm for advanced spaceborne thermal emission and reflection radiometer (ASTER) images. *IEEE Trans. Geosci. Remote Sens.* 36, 1113–1126. 10.1109/36.700995

Gillespie, A. R., Abbott, E. A., Gilson, L., Hulley, G., Jiménez-Muñoz, J. C., Sobrino, J. A., 2011. Residual errors in ASTER temperature and emissivity standard products AST08 and AST05. *Remote Sens. Environ.* 115, 3681–3694. 10.1016/j.rse.2011.09.007

Guha, S., Govil, H., Dey, A., Gill, N., 2018. Analytical study of land surface temperature with NDVI and NDBI using Landsat 8 OLI and TIRS data in Florence and Naples city, Italy. *Eur. J. Remote Sens.* 51, 667–678. 10.1080/22797254.2018.1474494

Henry, J. A., Dicks, S. E., Wetterqvist, O. F., Roguski, S. J., 1989. Comparison of satellite, ground-based, and modeling techniques for analyzing the urban heat island. *Photogramm. Eng. Remote Sens.* 55, 69–76.

Hsu, A., Manya, D., 2021. Disproportionate exposure to urban heat island intensity across major US cities. *Nat. Commun.* 1–11. 10.1038/s41467-021-22799-5

Hu, J., Yang, Y., Zhou, Y., Zhang, T., Ma, Z., Meng, X., 2022. Spatial patterns and temporal variations of footprint and intensity of surface urban heat island in 141 China cities. *Sustain. Cities Soc.* 77, 103585. 10.1016/j.scs.2021.103585

Hulley, G. C., Hook, S. J., 2011. Generating consistent land surface temperature and emissivity products between ASTER and MODIS data for earth science research. *IEEE Trans. Geosci. Remote Sens.* 49, 1304–1315. 10.1109/TGRS.2010.2063034

Hulley, G. C., Hook, S. J., Abbott, E., Malakar, N., Islam, T., Abrams, M., 2015. The ASTER Global Emissivity Dataset (ASTER GED): Mapping Earth's emissivity at 100 meter spatial scale. *Geophys. Res. Lett.* 42, 7966–7976. 10.1002/2015GL065564

Hulley, G. C., Gottsche, F. M., Rivera, G., Hook, S. J., Freepartner, R. J., Martin, M. A., Cawse-Nicholson, K., Johnson, W. R., 2021. Validation and quality assessment of the ECOSTRESS level-2 land surface temperature and emissivity product. *IEEE Trans. Geosci. Remote Sens.* 1–23. 10.1109/TGRS.2021.3079879

Hung, T., Uchihama, D., Ochi, S., Yasuoka, Y., 2006. Assessment with satellite data of the urban heat island effects in Asian mega cities. *Int. J. Appl. Earth Obs. Geoinf.* 8, 34–48. 10.1016/j.jag.2005.05.003

Ignatov, A., Gutman, G., 1998. Diurnal cycles of land surface temperatures. *Adv. Sp. Res.* 22, 641–644. 10.1016/S0273-1177(97)01125-3

Imada, Y., Watanabe, M., Kawase, H., Shiogama, H., Arai, M., 2019. The July 2018 high temperature event in Japan could not have happened without human-induced global warming. *Sola* 15, 8–12. 10.2151/sola.15a-002

Imhoff, M. L., Zhang, P., Wolfe, R. E., Bounoua, L., 2010. Remote sensing of the urban heat island effect across biomes in the continental USA. *Remote Sens. Environ.* 114, 504–513. 10.1016/j.rse.2009.10.008

Islam, T., Hulley, G. C., Malakar, N. K., Radocinski, R. G., Guillevic, P. C., Hook, S. J., 2017. A physics-based algorithm for the simultaneous retrieval of land surface temperature and emissivity from VIIRS thermal infrared data. *IEEE Trans. Geosci. Remote Sens.* 55, 563–576. 10.1109/TGRS.2016.2611566

Jia, A., Liang, S., Wang, D., 2022. Generating a 2-km, all-sky, hourly land surface temperature product from Advanced Baseline Imager data. *Remote Sens. Environ.* 278, 113105. 10.1016/j.rse.2022.113105

Jiang, Y., Fu, P., Weng, Q., 2015. Downscaling GOES land surface temperature for assessing heat wave health risks. *IEEE Geosci. Remote Sens. Lett.* 12, 1605–1609. 10.1109/LGRS.2015.2414897

Jiménez-Munoz, J. C., Sobrino, J. A., 2003. A generalized single-channel method for retrieving land surface temperature from remote sensing data. *J. Geophys. Res. Atmos.* 108. 10.1029/2003jd003480

Lagouarde, J. P., Hénon, A., Kurz, B., Moreau, P., Irvine, M., Voogt, J., Mestayer, P., 2010. Modelling daytime thermal infrared directional anisotropy over Toulouse city centre. *Remote Sens. Environ.* 114, 87–105. 10.1016/j.rse.2009.08.012

Lai, J., Zhan, W., Huang, F., Voogt, J., Bechtel, B., Allen, M., Peng, S., Hong, F., Liu, Y., Du, P., 2018. Identification of typical diurnal patterns for clear-sky climatology of surface urban heat islands. *Remote Sens. Environ.* 217, 203–220. 10.1016/j.rse.2018.08.021

Li, H., Zhou, Y., Li, X., Meng, L., Wang, X., Wu, S., Sodoudi, S., 2018. A new method to quantify surface urban heat island intensity. *Sci. Total Environ.* 624, 262–272. 10.1016/j.scitotenv.2017.11.360

Li, J., Song, C., Cao, L., Zhu, F., Meng, X., Wu, J., 2011. Impacts of landscape structure on surface urban heat islands: A case study of Shanghai, China. *Remote Sens. Environ.* 115, 3249–3263. 10.1016/j.rse.2011.07.008

Li, J., Wang, Xiang-rong, Wang, Xin-jun, Ma, W., Zhang, H., 2009. Remote sensing evaluation of urban heat island and its spatial pattern of the Shanghai metropolitan area, China. *Ecol. Complex.* 6, 413–420. 10.1016/j.ecocom.2009.02.002

Li, L., Zhan, W., Hu, L., Chakraborty, T.C., Wang, Z., Fu, P., Wang, D., Liao, W., Huang, F., Fu, H., Li, J., Liu, Z., Du, H., Wang, S., 2023. Divergent urbanization-induced impacts on global surface urban heat is land trends since 1980s. *Remote Sens. Environ.* 295, 113650. https://doi.org/10.1016/j.rse.2023.113650

Li, X., Zhou, Y., Yu, S., Jia, G., Li, H., Li, W., 2019. Urban heat island impacts on building energy consumption: A review of approaches and findings. *Energy* 174, 407–419. 10.1016/j.energy.2019.02.183

Li, Z., Tang, B., Wu, H., Ren, H., Yan, G., Wan, Z., Trigo, I. F., Sobrino, J. a., 2013. Satellite-derived land surface temperature: Current status and perspectives. *Remote Sens. Environ.* 131, 14–37. 10.1016/j.rse.2012.12.008

Li, Z., Wu, H., Duan, S., Zhao, W., Ren, H., Liu, X., Leng, P., Tang, R., Ye, X., Zhu, J., Sun, Y., Si, M., Liu, M., Li, J., Zhang, X., Shang, G., Tang, B., Yan, G., Zhou, C., 2023. Satellite Remote Sensing of Global Land Surface Temperature: Definition, Methods, Products, and Applications. *Rev. Geophys.* 61, 1–77. https://doi.org/10.1029/2022rg000777

Li, X., Gong, P., Zhou, Y., Wang, J., Bai, Y., Chen, B., Hu, T., Xiao, Y., Xu, B., Yang, J., Liu, X., Cai, W., Huang, H., Wu, T., Wang, X., Lin, P., Li, Xun, Chen, J., He, C., Li, Xia, Yu, L., Clinton, N., Zhu, Z., 2020. Mapping global urban boundaries from the global artificial impervious area (GAIA) data. *Environ. Res. Lett.* 15. 10.1088/1748-9326/ab9be3

Lin, T. P., De Dear, R., Hwang, R. L., 2011. Effect of thermal adaptation on seasonal outdoor thermal comfort. *Int. J. Climatol.* 31, 302–312. 10.1002/joc.2120

Lu, L., Weng, Q., Guo, H., Li, Q., Hui, W., 2020. Spatiotemporal variation of surface urban heat islands in relation to land cover composition and configuration: A multi-scale case study of Xi'an, China. *Remote Sens.* 12, 2713. 10.3390/rs70403670

Ma, Y., Zhang, S., Yang, K., Li, M., 2021. Influence of spatiotemporal pattern changes of impervious surface of urban megaregion on thermal environment: A case study of the Guangdong – Hong Kong – Macao Greater Bay Area of China. *Ecol. Indic.* 121, 107106. 10.1016/j.ecolind.2020.107106

Malakar, N. K., Hulley, G. C., Hook, S. J., Laraby, K., Cook, M., Schott, J. R., 2018. An operational land surface temperature product for Landsat thermal data: Methodology and validation. *IEEE Trans. Geosci. Remote Sens.* 56, 5717–5735. 10.1109/TGRS.2018.2824828

McGarigal, K., Marks, B. J., 1995. FRAGSTATS: spatial pattern analysis program for quantifying landscape structure, General Technical Report – US Department of Agriculture, Forest Service.

McMillin, L. M., 1975. Estimation of sea surface temperatures from two infrared window measurements with different absorption. *J. Geophys. Res.* 80, 5113–5117. 10.1029/JC080i036p05113

Memon, R. A., Leung, D. Y. C., Liu, C. H., 2009. An investigation of urban heat island intensity (UHII) as an indicator of urban heating. *Atmos. Res.* 94, 491–500. 10.1016/j.atmosres.2009.07.006

Meng, D., Li, X., Zhao, W., Gong, H., 2009. Quantitative exploration of the mechanisms behind the urban thermal environment in Beijing. *Prog. Nat. Sci.* 19, 1757–1763. 10.1016/j.pnsc.2009.07.005

Morrison, W., Grimmond, S., Kotthaus, S., 2023. Simulating satellite urban land surface temperatures: sensitivity to sensor view angle and assumed landscape complexity. *Remote Sens. Environ.* 293, 113579. 10.1016/j.rse.2023.113579

Najafzadeh, F., Mohammadzadeh, A., Ghorbanian, A., Jamali, S., 2021. Spatial and temporal analysis of surface urban heat island and thermal comfort using landsat satellite images between 1989 and 2019: A case study in tehran. *Remote Sens.* 13. 10.3390/rs13214469

Oke, T. R., 1982. The energetic basis of the urban heat island. *Q. J. R. Meteorol. Soc.* 108, 1–24. 10.1002/qj.49710845502

Oleson, K. W., Monaghan, A., Wilhelmi, O., Barlage, M., Brunsell, N., Feddema, J., Hu, L., Steinhoff, D. F., 2015. Interactions between urbanization, heat stress, and climate change. *Clim. Change* 129, 525–541. 10.1007/s10584-013-0936-8

Peng, J., Jia, J., Liu, Y., Li, H., Wu, J., 2018. Seasonal contrast of the dominant factors for spatial distribution of land surface temperature in urban areas. *Remote Sens. Environ.* 215, 255–267. 10.1016/j.rse.2018.06.010

Peng, S., Piao, S., Ciais, P., Friedlingstein, P., Ottle, C., Bréon, F. M., Nan, H., Zhou, L., Myneni, R. B., 2012. Response to comment on "Surface urban heat island across 419 global big cities." *Environ. Sci. Technol.* 46, 6889–6890. 10.1021/es301811b

Qin, Z., Karnieli, A., Berliner, P., 2001. A mono-window algorithm for retrieving land surface temperature from Landsat TM data and its application to the Israel-Egypt border region. *Int. J. Remote Sens.* 22, 3719–3746. 10.1080/0143116001 0006971

Rajasekar, U., Weng, Q., 2009. Urban heat is land monitoring and analysis using a non-parametric model: A case study of Indianapolis. *ISPRS J. Photogramm. Remote Sens.* 64, 86–96. https://doi.org/10.1016/j.isprsjprs.2008.05.002

Santamouris, M., 2015. Analyzing the heat island magnitude and characteristics in one hundred Asian and Australian cities and regions. *Sci. Total Environ.* 512–513, 582–598. 10.1016/j.scitotenv.2015.01.060

Schwarz, N., Lautenbach, S., Seppelt, R., 2011. Exploring indicators for quantifying surface urban heat islands of European cities with MODIS land surface temperatures. *Remote Sens. Environ.* 115, 3175–3186. 10.1016/j.rse.2011.07.003

Shi, Z., Li, X., Hu, T., Yuan, B., Yin, P., Jiang, D., 2023. Modeling the intensity of surface urban heat island based on the impervious surface area. *Urban Clim.* 49, 101529. 10.1016/j.uclim.2023.101529

Sobrino, J. A., Jiménez-Munoz, J. C., El-Kharraz, J., Gómez, M., Romaguera, M., Sòria, G., 2004. Single-channel and two-channel methods for land surface temperature retrieval from DAIS data and its application to the Barrax site. *Int. J. Remote Sens.* 25, 215–230. 10.1080/0143116031000115210

Sobrino, J. A., Jiménez-Muñoz, J. C., Sòria, G., Romaguera, M., Guanter, L., Moreno, J., Plaza, A., Martínez, P., 2008. Land surface emissivity retrieval from different VNIR and TIR sensors. *IEEE Trans. Geosci. Remote Sens.* 46, 316–327. 10.1109/TGRS.2 007.904834

Sobrino, J. A., Oltra-Carrió, R., Jiménez-Muñoz, J. C., Julien, Y., Sòria, G., Franch, B., Mattar, C., 2012. Emissivity mapping over urban areas using a classification-based approach: Application to the Dual-use European Security IR Experiment (DESIREX). *Int. J. Appl. Earth Obs. Geoinf.* 18, 141–147. 10.1016/j.jag.2012.01.022

Sòria, G., Sobrino, J. A., 2007. ENVISAT/AATSR derived land surface temperature over a heterogeneous region. *Remote Sens. Environ.* 111, 409–422. 10.1016/j.rse.2007.03.017

Stewart, I. D., 2011. A systematic review and scientific critique of methodology in modern urban heat island literature. *Int. J. Climatol.* 31, 200–217. 10.1002/joc.2141

Stewart, I. D., Krayenhoff, E. S., Voogt, J. A., Lachapelle, J. A., Allen, M. A., Broadbent, A. M., 2021. Time evolution of the surface urban heat island. *Earth's Futur.* 9. 10.1029/ 2021EF002178

Stewart, I. D., Oke, T. R., 2012. Local climate zones for urban temperature studies. *Bull. Am. Meteorol. Soc.* 93, 1879–1900. 10.1175/BAMS-D-11-00019.1

Streutker, D. R., 2002. A remote sensing study of the urban heat island of Houston, Texas. *Int. J. Remote Sens.* 23, 2595–2608. 10.1080/01431160110115023

Streutker, D. R., 2003. Satellite-measured growth of the urban heat island of Houston, Texas. *Remote Sens. Environ.* 85, 282–289. 10.1016/S0034-4257(03)00007-5

Takane, Y., Kikegawa, Y., Hara, M., Grimmond, C. S. B., 2019. Urban warming and future air-conditioning use in an Asian megacity: Importance of positive feedback. *npj Clim. Atmos. Sci.* 2, 1–11. 10.1038/s41612-019-0096-2

Tzavali, A., Paravantis, J. P., Mihalakakou, G., Fotiadi, A., Stigka, E., 2015. Urban heat island intensity: A literature review. *Fresenius Environ. Bull.* 24, 4537–4554.

Valor, E., Caselles, V., 1996. Mapping land surface emissivity from NDVI: Application to European, African, and South American areas. *Remote Sens. Environ.* 57, 167–184. 10.1016/0034-4257(96)00039-9

Voogt, J.A., Oke, T.R., 2003. Thermal remote sensing of urban climates. *Remote Sens. Environ.* 86, 370–384. https://doi.org/10.1016/S0034-4257(03)00079-8

Wan, Z., 2014. New refinements and validation of the collection-6 MODIS land-surface temperature/emissivity product. *Remote Sens. Environ.* 140, 36–45. 10.1016/j.rse.2013.08.027

Wan, Z., Dozier, J., 1996. A generalized split-window algorithm for retrieving land-surface temperature from space. *IEEE Trans. Geosci. Remote Sens.* 34, 892–905. 10.1109/36.508406

Wang, J. F., Li, X. H., Christakos, G., Liao, Y. L., Zhang, T., Gu, X., Zheng, X. Y., 2010. Geographical detectors-based health risk assessment and its application in the neural tube defects study of the Heshun Region, China. *Int. J. Geogr. Inf. Sci.* 24, 107–127. 10.1080/13658810802443457

Wang, R., Murayama, Y., 2020. Geo-simulation of land use/cover scenarios and impacts on land surface temperature in Sapporo, Japan. *Sustain. Cities Soc.* 63, 102432. 10.1016/j.scs.2020.102432

Wang, Y., Yao, Y., Chen, S., Ni, Z., Xia, B., 2022. Spatiotemporal evolution of urban development and surface urban heat island in Guangdong-Hong Kong-Macau greater bay area of China from 2013 to 2019. *Resour. Conserv. Recycl.* 179, 106063. 10.1016/j.resconrec.2021.106063

Weng, Q., 2009. Thermal infrared remote sensing for urban climate and environmental studies: Methods, applications, and trends. *ISPRS J. Photogramm. Remote Sens.* 64, 335–344. https://doi.org/10.1016/j.isprsjprs.2009.03.007

Weng, Q., Firozjaei, M. K., Sedighi, A., Kiavarz, M., 2019. Statistical analysis of surface urban heat island intensity variations: A case study of Babol city, Iran. *GIScience Remote Sens.* 56, 576–604. 10.1080/15481603.2018.1548080

Weng, Q., Fu, P., 2014. Modeling diurnal land temperature cycles over Los Angeles using downscaled GOES imagery. *ISPRS J. Photogramm. Remote Sens.* 97, 78–88. 10.1016/j.isprsjprs.2014.08.009

Weng, Q., Lu, D., Schubring, J., 2004. Estimation of land surface temperature – vegetation abundance relationship for urban heat island studies. *Remote Sens. Environ.* 89, 467–483. 10.1016/j.rse.2003.11.005

Xu, H. Q., Chen, B. Q., 2004. Remote sensing of the urban heat island and its changes in Xiamen City of SE China. *J. Environ. Sci.* 16, 276–281.

Xu, H., Yu, Y., Tarpley, D., Gottsche, F. M., Olesen, F. S., 2014. Evaluation of GOES-R land surface temperature algorithm using SEVIRI satellite retrievals with in situ measurements. *IEEE Trans. Geosci. Remote Sens.* 52, 3812–3822. 10.1109/TGRS.2013.2276426

Yamamoto, Y., Ichii, K., Ryu, Y., Kang, M., Murayama, S., 2022. Uncertainty quantification in land surface temperature retrieved from Himawari-8/AHI data by operational algorithms. *ISPRS J. Photogramm. Remote Sens.* 191, 171–187. 10.1016/j.isprsjprs.2022.07.008

Yamamoto, Y., Ichii, K., Ryu, Y., Kang, M., Murayama, S., Kim, S., Cleverly, J.R., 2023. Detection of vegetation drying signals using diurnal variation of land surface temperature: Application to the 2018 East Asia heatwave. *Remote Sens. Environ.* 291, 113572. https://doi.org/10.1016/j.rse.2023.113572

Yamamoto, Y., Ishikawa, H., 2018. Spatiotemporal variability characteristics of clear-sky land surface temperature in urban areas of Japan observed by Himawari-8. *Sola* 14, 179–184. 10.2151/sola.2018-032

Yamamoto, Y., Ishikawa, H., 2020. Influence of urban spatial configuration and sea breeze on land surface temperature on summer clear-sky days. *Urban Clim.* 31, 100578. 10.1016/j.uclim.2019.100578

Yamamoto, Y., Ishikawa, H., Oku, Y., Hu, Z., 2018. An algorithm for land surface temperature retrieval using three thermal infrared bands of Himawari-8. *J. Meteorol. Soc. Japan* 96B, 59–76. 10.2151/jmsj.2018-005

Yang, J., Wang, Y., Xue, B., Li, Y., Xiao, X., Xia, J. (Cecilia), He, B., 2021. Contribution of urban ventilation to the thermal environment and urban energy demand: Different climate background perspectives. *Sci. Total Environ.* 795, 148791. 10.1016/j.scitotenv.2021.148791

Yang, J., Wong, M. S., Menenti, M., Nichol, J., 2015a. Study of the geometry effect on land surface temperature retrieval in urban environment. *ISPRS J. Photogramm. Remote Sens.* 109, 77–87. 10.1016/j.isprsjprs.2015.09.001

Yang, J., Wong, M. S., Menenti, M., Nichol, J., 2015b. Modeling the effective emissivity of the urban canopy using sky view factor. *ISPRS J. Photogramm. Remote Sens.* 105, 211–219. 10.1016/j.isprsjprs.2015.04.006

Yang, Q., Huang, X., Tang, Q., 2020. Irrigation cooling effect on land surface temperature across China based on satellite observations. *Sci. Total Environ.* 705, 135984. 10.1016/j.scitotenv.2019.135984

Yang, Z., Chen, Y., Qian, Q., Wu, Z., Zheng, Z., Huang, Q., 2019. The coupling relationship between construction land expansion and high-temperature area expansion in China's three major urban agglomerations. *Int. J. Remote Sens.* 40, 6680–6699. 10.1080/01431161.2019.1590877

Yao, R., Wang, L., Huang, X., Gong, W., Xia, X., 2019. Greening in rural areas increases the surface urban heat island intensity. *Geophys. Res. Lett.* 46, 2204–2212. 10.1029/2018GL081816

Yao, R., Wang, L., Huang, X., Niu, Z., Liu, F., Wang, Q., 2017. Temporal trends of surface urban heat islands and associated determinants in major Chinese cities. *Sci. Total Environ.* 609, 742–754. 10.1016/j.scitotenv.2017.07.217

Yoo, C., Im, J., Cho, D., Yokoya, N., Xia, J., Bechtel, B., 2020. Estimation of all-weather 1 km MODIS land surface temperature for humid summer days. *Remote Sens.* 12, 1–23. 10.3390/RS12091398

Zakšek, K., Oštir, K., 2012. Downscaling land surface temperature for urban heat island diurnal cycle analysis. *Remote Sens. Environ.* 117, 114–124. 10.1016/j.rse.2011.05.027

Zhang, J., Wang, Y., 2008. Study of the relationships between the spatial extent of surface urban heat islands and urban characteristic factors based on Landsat ETM + data. *Sensors* 8, 7453–7468. 10.3390/s8117453

Zhang, M., Tan, S., Zhang, C., Han, S., Zou, S., Chen, E., 2023. Assessing the impact of fractional vegetation cover on urban thermal environment: A case study of Hangzhou, China. *Sustain. Cities Soc.* 96, 104663. 10.1016/j.scs.2023.104663

Zhang, P., Imhoff, M. L., Wolfe, R. E., Bounoua, L., 2010. Characterizing urban heat islands of global settlements using MODIS and nighttime lights products. *Can. J. Remote Sens.* 36, 185–196. 10.5589/m10-039

Zhang, Q., She, D., Zhang, L., Wang, G., Chen, J., Hao, Z., 2022. High sensitivity of compound Drought and heatwave events to global warming in the future. *Earth's Futur.* 10, 1–22. 10.1029/2022EF002833

Zhang, W., Li, Y., Zheng, C., Zhu, Y., 2023. Surface urban heat island effect and its driving factors for all the cities in China: Based on a new batch processing method. *Ecol. Indic.* 146, 109818. https://doi.org/10.1016/j.ecolind.2022.109818

Zhang, Y., Yang, Y., Pan, X., Ding, Y., Hu, J., Dai, Y., 2023. Multiinformation Fusion Network for Mapping Gapless All-Sky Land Surface Temperature Using Thermal Infrared and Reanalysis Data. *IEEE Trans. Geosci. Remote Sens.* 61, 1–15. https://doi.org/10.1109/TGRS.2023.3269622

Zhao, L., Lee, X., Smith, R. B., Oleson, K., 2014. Strong contributions of local background climate to urban heat islands. *Nature* 511, 216–219. 10.1038/nature13462

Zhou, W., Wang, J., Cadenasso, M.L., 2017. Effects of the spatial configuration of trees on urban heat mitigation: A comparative study. *Remote Sens. Environ.* 195, 1–12. https://doi.org/10.1016/j.rse.2017.03.043

10 Urban Air Pollution Mitigation for Sustainable Cities

Observation, Modeling, and Control Strategies

Hai Guo, Yu Wang, Yangzong Zeren, and Xiaoyu Jiao

INTRODUCTION: BACKGROUND AND DRIVING FORCES

O_3 and Its Primary Emitted Precursors

Tropospheric O_3, a major constituent of photochemical smog, significantly affects atmospheric chemistry, human health and climate change (Monks et al., 2015; NARSTO, 2000; T. Wang et al., 2017). O_3 is formed through a series of photochemical reactions among volatile organic compounds (VOCs) and nitrogen oxides (NO_x) in the atmosphere, while O_3 pollution is a complex phenomenon involving precursor emissions, photochemical formation and dynamic transport on different scales (Guo et al., 2017; Jacob, 1999). With the rapid urbanization and industrialization, China has experienced severe O_3 pollution over the past two decades (Ding et al., 2004; Li et al., 2020; Lu et al., 2020; T. Wang et al., 2017; Y. Wang et al., 2017; Xu et al., 2008; Zeren, Guo, et al., 2022; Zhang et al., 2007; Zhang et al., 2014). The maximum hourly O_3 mixing ratio often exceeds the national air quality standard (100 ppbv), mainly attributed to the elevated emissions of local precursors and regional/super-regional transport of O_3 and its precursors from other parts of China (Guo, Jiang, et al., 2009; Guo, Ling, Cheung, Jiang, et al., 2013; Wang et al., 2003). In addition, the non-linear relationship between O_3 and its precursors makes developing appropriate control measures for O_3 problematic in megacities (Sillman, 1999).

As currently the most important urban air pollutant in most Chinese cities, O_3 formation and its control policies have been well documented, particularly in the Pearl River Delta (PRD) region of China, including VOC speciation and abundance (Guo et al., 2011; Guo et al., 2007; Guo et al., 2006; Lee et al., 2002; Liu, Shao, Lu, et al., 2008; Ou et al., 2015), in-situ photochemical O_3 production (Lyu et al., 2016; Y. Wang et al., 2018; Zhang et al., 2007), process analysis of regional O_3 formation

DOI: 10.1201/9781003244561-13

(Cheng, Guo, Saunders, et al., 2010; Guo, Jiang, et al., 2009; Jiang et al., 2010; Wang et al., 2003; Y. Wang et al., 2019; Zeren et al., 2019; Zheng et al., 2010), long-term variations of O_3 and its control suggestions (X. Liu et al., 2019; Lyu, Zeng, et al., 2017; T. Wang et al., 2019; Wang et al., 2009; Y. Wang et al., 2017; Yao et al., 2019; Zeren, Guo, et al., 2022).

Due to the effects of complex topography of Hong Kong, O_3 pollution is modulated by not only mountain-valley breeze circulations but also sea-land breeze circulations (Wang et al., 2015; Zeren et al., 2019). To understand the characteristics of O_3 at different elevations and to quantify the influences of different chemical and physical processes on elevated O_3 formation. A comprehensive study was conducted at Mt. Tai Mo Shan (TMS) in Hong Kong based on sampling campaign and chemical transport model simulation (Wang et al., 2015). The combination of field measurement and model simulation is also appropriate to other studies related to physical transport processes and chemical production of air pollutants. In specific, sea-land breeze circulation has been regarded as the one of key factors enhancing O_3 pollution in coastal cities through returning the aged air masses from marine area (H. Wang et al., 2018). However, most of the previous studies on O_3 pollution in Hong Kong were either solely based on the observations (Shao et al., 2009; Wang et al., 2008; Zheng et al., 2018) or solely based on the simulations of the synoptic processes (Fung et al., 2005; Li et al., 2012; Lo et al., 2006; Lu et al., 2009). Thus, the integrated processes (including physical and chemical processes) resulting in the build-up of O_3 in Hong Kong remain unclear. To fill the gaps, an intensive observation in the coast and 3-dimension model simulation were conducted in Hong Kong and its neighboring coastal cities (Zeren et al., 2019). As such, the chemical and dynamic processes dominating O_3 pollution were figured out.

Moreover, Hong Kong government devoted great efforts and developed manifold control programs to alleviate O_3 pollution in the past decade, investigations on the long-term variations of air pollutants can be used to assess the effectiveness of air pollution control measures (Lyu, Zeng, et al., 2017). Besides, several studies focusing on the long-term variations of O_3 and its precursors in Hong Kong provided the knowledge for long-term variations of relationship between O_3 and its precursors (Y. Wang et al., 2017), radical chemistry, air pollutants levels emitted from different sources (X. Liu et al., 2019), and influences of meteorological conditions (Zeren, Guo, et al., 2022). Apart from the long-term observations of O_3 and its precursors, the photochemical box models were also applied to study the variation of atmospheric chemistry. In addition, the long-term impact of meteorological conditions could be quantified through a statistical method combing the Kolmogorov-Zurbenko (KZ)-filter with multiple linear regression. In this chapter, the methodologies and major findings on most of the above-mentioned studies are introduced in the following sections.

Airborne Particulate and Aerosols

Aerosols are multi-facet agents that affect air quality, weather, and climate through many mechanisms, as reviewed extensively in a series of IPCC reports. Aerosols are colloids of fine solid particles or liquid droplets suspended in the atmosphere.

Through the effects of aerosol–radiation interactions (ARI), aerosol–cloud interactions (ACI), or both, aerosols can dramatically affect Earth's climate by perturbing the Earth's radiation budget and water cycle processes (Z. Li et al., 2016; Liao et al., 2015; Rosenfeld et al., 2008). As some observational analysis have indicated, cloud and precipitation properties are remarkably affected by elevated aerosols, which suppress light rainfall but intensify heavy rainfall and lightning in the coastal regions of Southeast China (Wang et al., 2011). Aerosol pollution in the planetary boundary layer (PBL) adversely affects human health (Pope & Dockery, 2006). Chemical reactions can occur on the surface of non-gaseous particles or within the body of liquid droplets, and these processes are key components of the biogeochemistry of our planet (Holloway & Wayne, 2010).

Airborne particulate pollution, also known as "haze", is caused by ultrafine particles (UFPs, with an aerodynamic diameter less than 100 nm), such as high concentration of aerosol particles in 2.5 μm ($PM_{2.5}$) and 10 μm (PM_{10}) concentrations, and has become a major concern of scientific communities and local governments in terms of air quality impairment, public health, and global climate (Guo et al., 2003; Huang et al., 2014; Lu et al., 2022; Lyu et al., 2016). Since air pollution is related to regional contributions, the information on the chemical components of aerosol can help us identify the sources and even quantify source contributions. Moreover, analysis of aerosol spatial and temporal variations can provide information on the effectiveness of the formation and implementation of air pollution measures. A great understanding of aerosols' chemical and physical properties can also result in good performance of numerical model predictions of aerosol characteristics.

Organic aerosol (OA) constitutes a considerable fraction (20–90% globally) of airborne fine particulate matter (PM), comprising a broad spectrum of compounds with varying volatilities and properties, followed by ammonium sulfate or ammonium nitrate, and black carbon (Jimenez et al., 2009; Zhang et al., 2011). While some OA directly emitted from diverse primary anthropogenic and natural sources, such as traffic, coal combustion, biomass burning, and cooking-termed as primary OA (POA)- secondary organic aerosols (SOA) are formed through chemical evolutions of organic compounds account for a large fraction of the OA burden and are found to be more important in the formation of severe haze episodes (Li et al., 2016; Sun et al., 2011). It is believed that SOA is a collection of hundreds to thousands of organic chemicals featuring relatively low volatilities.

To identify sources of OA and quantify their contributions, receptor models based on known tracers of specific sources are widely adopted (Huang et al., 2014; Jaeckels et al., 2007). SOA tracer-based approach is a simplified method widely used to estimate the amount, precursors, and sources of SOA compared to more advanced instruments, such as aerosol mass spectrometry and thermal desorption aerosol gas chromatography, which are either fragment-based or highly dependent upon the skills and knowledge of users (Ding et al., 2012; Williams et al., 2014). The filter-based technique uses fine PM collected and accumulated on filters, which are chemically characterized following solvent extraction (SE) (X. Ding et al., 2013) or thermal desorption (TD) (Q. Wang et al., 2017). Data obtained from filters are generally of low time resolution (12–24 h) (X. Ding et al., 2013).

As an alternative, Aerosol Mass Spectrometry (AMS) is extensively used in source apportionment of OA, based on the fast detection of OA at the resolution of minutes (Lyu et al., 2020; Yao et al., 2022; Yao et al., 2021). However, the AMS-based technique usually fails to identify specific sources of OAs and determine the precursors and formation pathways of SOA, due to the lack of compound-based separation during analysis and the chemical similarity of many source markers (Lee et al., 2013; Sun et al., 2011). Thermal desorption Aerosol Gas-chromatograph (TAG), an in-situ chemical marker-oriented instrument, has been developed for the time-resolved analysis of organic molecular makers and is increasingly applied in OA studies (Lyu et al., 2020; Q. Wang et al., 2020; Yao et al., 2022). TAG families including impactor TAG, 2D-TAG, SV-TAG, and TAG-AMS developed by U.S. groups, and Q-TAG, Dual-trap TAG, and 2D-Q-TAG developed by a China group, have contributed to comprehensive measurements of semi-volatile, intermediate-volatility and condensed organic compounds, study on gas/particle partitioning, identification of sources of airborne organics, and research on SOA formation mechanisms (Hunter et al., 2017; Williams et al., 2010; Zhao, Kreisberg, Worton, Isaacman, Gentner, et al., 2013; Zhao, Kreisberg, Worton, Isaacman, Weber, et al., 2013).

The efficient aerosol nucleation, combined with aerosol growth, is closely associated with severe haze episodes in China. It has been well recognized that the high anthropogenic emissions and rapid secondary aerosol formation are the key factors leading to the frequent occurrence of severe haze episodes, characterized by efficient new particle formation (NPF) and growth (Guo et al., 2014). The formation and growth of atmospheric aerosols have received increasing attention as important sources of aerosol particles. Aerosol nucleation has been measured under diverse environmental conditions, and globally NPF accounts for about 50% of the aerosol number production in the troposphere (Zhang, 2010; Zhang et al., 2012). NPF is a widely recognized source of secondary nanoparticles in the atmosphere and over half of the atmospheric cloud condensation nuclei at various locations, thus influencing cloud properties and Earth's energy balance (Kerminen et al., 2018; Kulmala et al., 2004). So far, the generally recognized NPF undergoes a two-step process: nucleation of gaseous vapors precursors, which form ~1 nm molecular clusters prior to the formation of a new phase during the transformation from vapor to liquid to solid, and the growth of clusters to the detectable sizes or even larger particles (Kulmala et al., 2013; Kulmala et al., 2005). The growth of a cluster can be represented as a reversible, stepwise kinetic process. During NPF, the newly formed particles are also lost through coagulation and further growth into larger size particles (Kulmala et al., 2001). After reaching a critical size (the critical cluster or nucleus), further growth of the cluster becomes spontaneous. Both the primarily emitted and secondarily formed particles can grow up to the size of 50–100 nm, acting as cloud condensation nuclei (CCN) in the atmosphere. At each step, the formation and decomposition of a cluster can be described by fundamental kinetic rate theories. Condensable vapors, such as sulfuric acid and organic condensable vapors generated from the oxidation of sulfur dioxide (SO_2) and volatile organic compounds (VOCs), respectively, play critical roles in both nucleation and particle growth (Sihto et al., 2006; Tröstl et al., 2016).

METHODOLOGY

In general, the combination of observation and simulation methods is commonly used to deeply explore the characteristics, causes and possible solutions of urban air pollutants. Figure 10.1 summarizes measurement and model techniques for air pollution studies. In terms of observation, the representative observation sites should be first selected, and then sampling equipment can be installed and set up. On the other hand, both observation-based models and emission-based models can be applied for numerical simulations. Moreover, source apportionment models and backward trajectory models can be used to assist the data analysis.

Sampling Site Selection

Sampling sites can be divided into long-term fixed monitoring stations and short-term intensive sampling sites. Taking Hong Kong as an example, the air quality monitoring network in Hong Kong operated by the Environmental Protection Department (HKEPD) comprises 18 fixed monitoring stations (Figure 10.2 (6)), where online trace gases data were available. Three roadside sites and 15 general urban sites in different districts covering the entire territory of Hong Kong are included, which could provide a full picture of routine urban air quality measurements. Particularly, Tap Mun and Tung Chung (TC) can be regarded as the upwind and downwind sites, respectively, when the northeast Asian monsoon dominated. More descriptions refer to the website of the HKEPD (https://cd.epic.epd.gov.hk/EPICDI/air/station/?lang=en).

Apart from fixed sites, sampling campaigns can provide more in-depth analysis of urban air pollution formation mechanisms with the help of multiple measurements, particularly for air pollution events, such as O_3 episodes in Hong Kong and

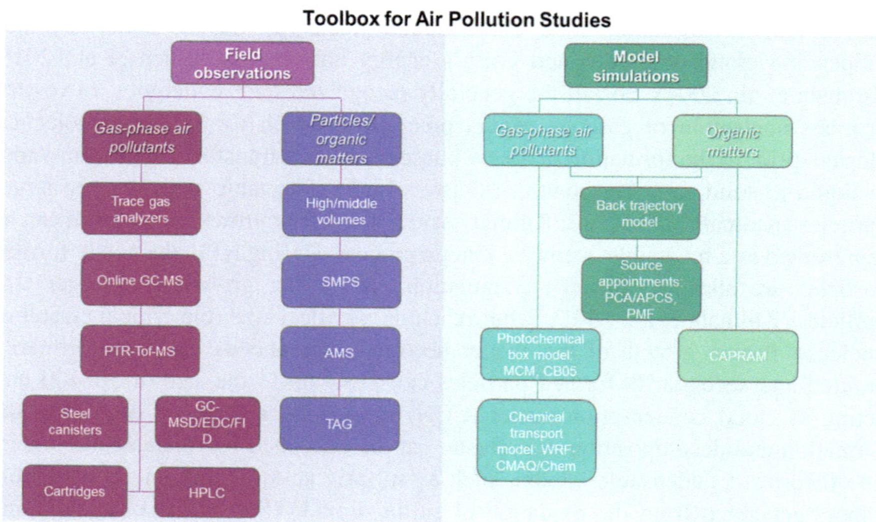

FIGURE 10.1 Measurement and model techniques for air pollution studies.

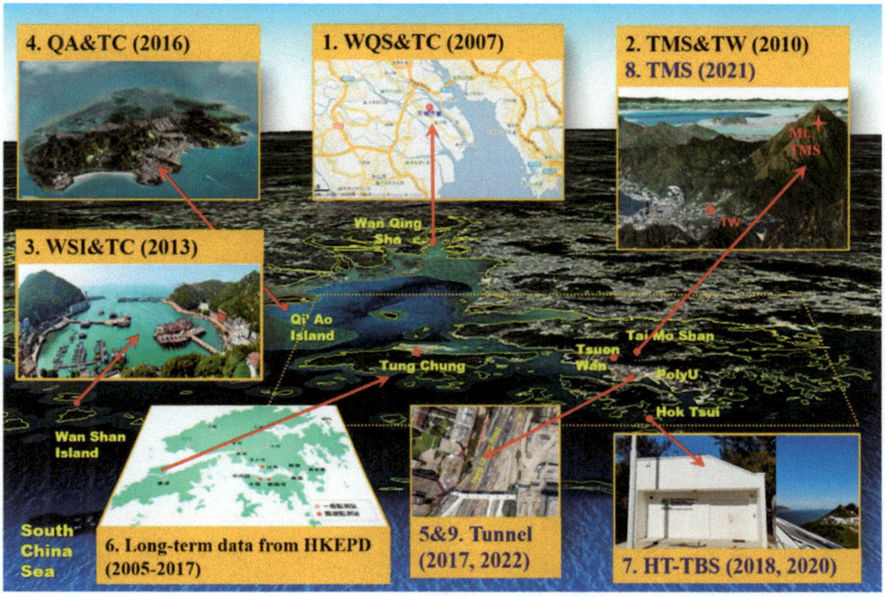

FIGURE 10.2 Intensive sampling locations in Hong Kong and its surrounding areas (most of the sampling campaigns were conducted by air quality research groups of the Hong Kong Polytechnic University between 2007 and 2022).

its surrounding areas. Figure 10.2 displays the intensive sampling locations, namely Wanqingsha (WQS), TC, TMS, Tsuen Wan (TW), Wanshan Island (WSI), Qi'ao Island (QA), Hung Hom Cross Harbour Tunnel (Tunnel) and Hok Tsui (HT) sites, which were conducted by air quality research groups of the Hong Kong Polytechnic University between 2007 and 2022. The special issues, such as the impacts of inland regional transport (1. WQS & TC), mountain-valley breezes (2. TMS & TW), the interaction between marine and continental air (3. WSI & TC), sea-land breezes (4. QA & TC), coastal regional transport (7. HT) on O_3 formation in Hong Kong were investigated. Moreover, the long-term variations of O_3-precursor relationships were deciphered (6. TC and 8. TMS). At last, The OA formation mechanisms were illustrated based on the roadside samplings (5&9. Tunnel).

Furthermore, to better understand the primary driving factors of O_3 pollution in China, concurrent sampling campaigns were conducted in ten major cities in 2018 (Figure 10.3), which were funded by the National Key R&D Program of China, under the Ministry of Science and Technology. The scientific findings and control strategies are provided in sections 10.3–10.5.

Measurement Techniques

O_3, CO, SO_2 and NO_x

Trace gases including O_3, NO_x, CO, and SO_2 are simultaneously monitored perennially. Briefly, the target pollutants are measured with commercial

FIGURE 10.3 Map showing the locations of the ten sampling cities in China.

analyzers. Taking TC site as an example, O_3 was measured using a commercial UV photometric instrument (Advanced Pollution Instrumentation (API), model 400 A) with a detection limit of 0.6 ppbv. CO was measured with a gas filter correlation CO analyzer (Thermo Electron Corp. (TECO), model 48 C) with a detection limit of 0.04 ppmv. SO_2 was measured using a pulsed fluorescence analyzer (TECO, model 43 A) with a detection limit of 1.0 ppbv. NO-NO_2-NO_x was detected using a commercial chemiluminescence with an internal molybdenum converter (API, model 200 A) and a detection limit of 0.4 ppbv. All the time resolutions for these gas analyzers were 1 h. To ensure a high degree of accuracy and precision, the QA/QC procedures for gaseous pollutants were identical to those in the US air quality monitoring program (http://epic.epd.gov.hk/ca/uid/airdata). The accuracy of the monitoring network was assessed by performance audits, while the precision, a measure of the repeatability, of the measurements was checked in accordance with HKEPD's quality manuals. For the gaseous pollutants, accuracy and precision within the limits of ±15 and ±20% were adopted, respectively (HKEPD, 2015; Y. Wang et al., 2017).

VOCs

Online VOCs

Compared with trace gases, real-time VOC data is only measured at few fixed monitoring stations. VOCs at TC site was online measured by the HKEPD. An online gas chromatography - flame ionization detector (GC-FID, Synspec GC 955, Series 600/800) was used to collect VOC speciation data continuously with a time resolution of 30 minutes. The VOC analyzer consists of two separate systems for detection of C_2–C_5 and C_6–C_{10} hydrocarbons, respectively. Detailed description about the real-time VOC analyzer can be found in (Lyu, Guo, et al., 2016). There were twenty-eight C_3–C_{10} VOC species identified and quantified with this method. In terms of the QA/QC for VOC analysis, built-in computerized programs of quality control systems such as auto-linearization and auto-calibration were used. Weekly calibrations were conducted by using NPL standard gas (National Physical Laboratory, Teddington, Middlesex, UK). In general, the detection limits of the target VOCs ranged from 2 to 56 pptv. The accuracy of each species measured by online GC-FID was determined by the percentage difference between measured mixing ratio and actual mixing ratio based on weekly span checks and monthly calibrations. The precision was based on the 95% probability limits for the integrated precision check results. The accuracy of the measurements was about 1–7%, depending on the species, and the measurement precision was about 1–10% (Y. Wang et al., 2017). In addition, the quality of the real-time data was assured by regular comparison with whole-air canister samples collected and analyzed by University of California at Irvine. More details can be found from previous studies in Hong Kong (Lyu, Guo, et al., 2016; Ou et al., 2015; Y. Wang et al., 2017; 2016; Xue et al., 2014).

Offline VOCs

Figure 10.4 displays the schematic diagram of the analysis process of offline sampling samples by GC with mass spectrometer detection (MSD) method. Taking the study of the interaction between marine and continental air as an example, concurrent VOC samples (*i.e.,* non-methane hydrocarbons (NMHCs) and carbonyls) were collected on 21 selected days (including both non-O_3 episodes and O_3 episodes) at both TC and WSI sites. The whole-air samples of NMHCs were collected using 2-L electro-polished stainless-steel canisters. The canisters were cleaned, conditioned and evacuated before being used for sampling. A metal bellows pump was used to fill up the canisters with sample air over one-hour integration (with a flow restrictor) to a pressure of 40 psi.

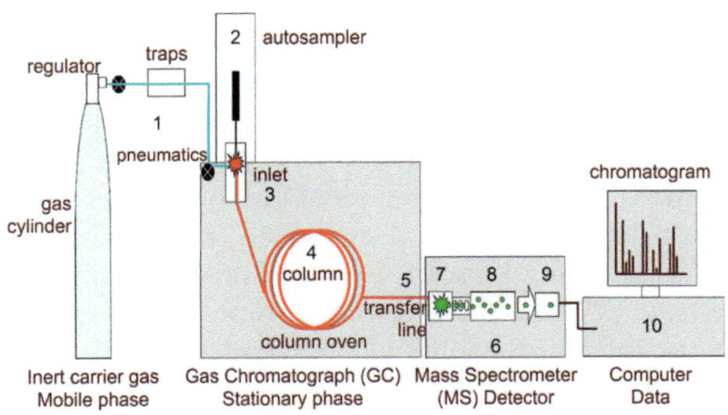

FIGURE 10.4 A simplified diagram of a gas chromatograph–mass spectrometer showing (1) carrier gas, (2) autosampler, (3) inlet, (4) analytical column, (5) interface, (6) vacuum, (7) ion source, (8) mass analyzer, (9) ion detector and (10) PC. Credit: Anthias Consulting.

Seven one-hour VOC samples (every two hours during 7:00 – 19:00 inclusive) were collected simultaneously at each site. Intensive VOC sampling was also carried out at WSI in selected seven days (*i.e.*, 3, 4, 9, and 22–25 October) with 11 one-hour samples (every two hours during 1:00 – 22:00 inclusive). Totally, 311 valid VOC samples (144 at TC and 167 at WSI) were collected in addition to about 5% field blanks and 5% parallel samples for quality assurance purpose. The speciation and abundance of 59 C_2-C_{11} NMHCs in the canisters were determined by a Model 7100 preconcentrator (Entech Instruments Inc., California, USA) coupled with an Agilent 5973 N (GC-MSD/FID, Agilent Technologies, USA). The detection limit of NMHCs was 3 pptv with a measurement precision of 2-5%, and a measurement accuracy of 5%. Detailed information of the analysis system and quality control and quality assurance for VOC samples can be found elsewhere (Simpson et al., 2010).

OVOCs

Furthermore, carbonyl samples were collected using silica gel filled cartridges impregnated with acidified 2,4-dinitrophenylhydrazine (DNPH). Air samples were drawn through the cartridge at a flow rate of 0.4–0.6 L min^{-1} for 2 hours; the flow rate through the cartridges was monitored with a rotameter which was calibrated before and after each sampling. An O_3 scrubber was connected to the inlet of the DNPH–silica gel cartridge to prevent interference from O_3. In total, 227 carbonyl samples (124 at TC and 103 at WSI) were collected with 5 and 6 samples per non-O_3 and O_3 episode day (every two hours during 7:00 – 18:00 inclusive), respectively. All cartridges were stored in a refrigerator at 4 °C after sampling. The sampled carbonyl cartridges were eluted slowly with <5 ml acetonitrile in the direction opposite to sampling flow into a 5-ml brown volumetric flask, followed by adding acetonitrile to a constant volume of 5 ml. A 20-µl aliquot was injected into the high-performance liquid chromatography (HPLC) system through an auto-sampler. Typically, C_1–C_9 carbonyl compounds were measured efficiently with a detection limit of ~0.2 ppbv. The procedures including the pretreatment, the configuration of the analysis system, and the methods of the quality control and quality assurance for carbonyl samples can be found in detail elsewhere (Guo, Jiang, et al., 2009; H. Wang et al., 2018).

SOA Tracers

The method of SOA analysis was in line with that introduced by Ding et al. (2012). Briefly, the procedures for each sample include solvent extraction, derivation, GC-MSD analysis, and the identification and quantification of SOA tracers, which can be found in detail elsewhere (Lyu, Guo, et al., 2017). The identification process for SOA tracers was based on comparing their mass spectra with previous studies, and their retention time in GC with other known compounds as the references. Some of the targeted tracers were quantified by authentic standards. However, due to lack of standards, other α-pinene SOA tracers, -such as isoprene SOA tracers, *β*-Caryophyllenic acid and 2,3-dihydroxy-4-oxopentanoic acid- were quantified using pinic acid (PA), erythritol, octadecanoic acid and azelaic acid, respectively.

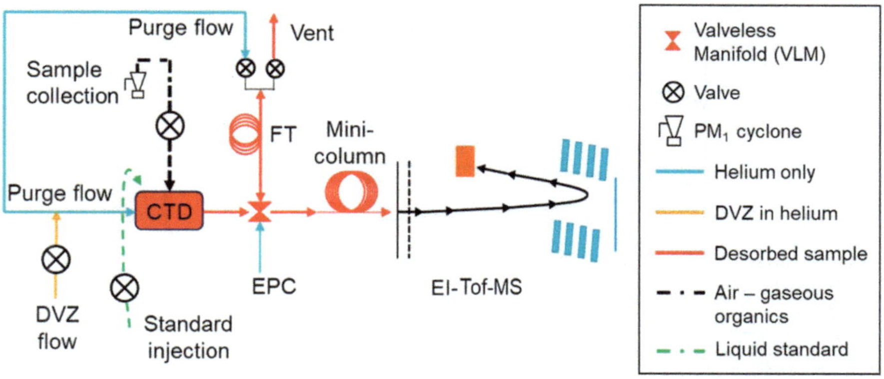

FIGURE 10.5 Schematic of TAG.

VOCs Molecular Markers

A Thermal desorption Aerosol Gas-chromatograph-Time of Flight-Mass Spectrometer (TAG-Tof-MS, hereinafter referred to as TAG) was deployed to identify and quantify the OA markers in airborne PM_1 (PM with aerodynamic diameters less than 1 μm). Figure 10.5 shows a schematic of TAG. Detailed information on the instrument and operation procedure are elaborated on previous TAG studies (Kreisberg et al., 2009; Lyu et al., 2020). In the TAG system, ambient atmospheric $PM_{2.5}$ or PM_1 (depending on the particle cyclone employed) is collected by means of humidification and inertial impaction. The sample is then thermally desorbed, separated by a GC, and identified and quantified at the molecular level using electron impact ionization quadrupole MS.

NR-PM_1 Compositions

The non-refractory submicron particulate matters (NR-PM_1) including organics, sulfate, nitrate, ammonium, and chloride, were measured using a High Resolution-Time of Flight-AMS (HR-Tof-AMS). Figures 10.6 and 10.7 show a schematic of TAG and a timing diagram for use of the HR-ToF-AMS in PTOF mode and MS mode, respectively. Detailed information on the instrument is described in previous AMS studies and methods of processing collected NR-PM_1 mass concentrations can be found in detail elsewhere (DeCarlo et al., 2006; Yao et al., 2021). In brief, an aerodynamic lens is used in AMS to sample and focus ambient particles into a narrow beam. The time zero of particle flight is defined by a rotating mechanical chopper, and the end of particle flight is defined as the time of mass spectrometric detection. The particles are transmitted to be vaporized by impaction on a resistively heated surface (~ 600 °C) with non-refractory species detected and measured at this vaporizer temperature and ionized by electron ionization (70 eV) (Canagaratna et al., 2007).

Particle Size Distribution

Atmospheric particle number concentrations within the size range of 5.5–350.4 nm were monitored by a Scanning Mobility Particle Sizer (SMPS) coupled with a

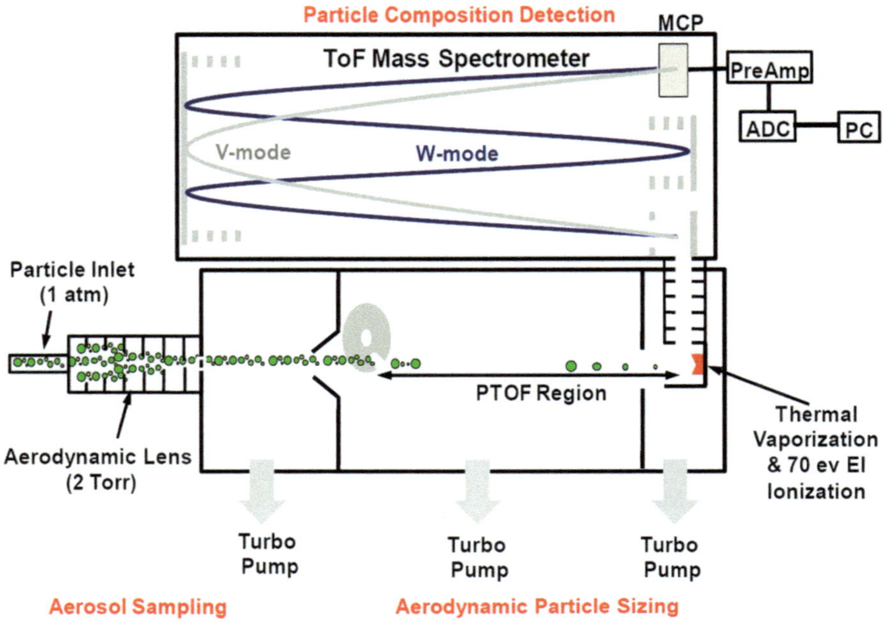

FIGURE 10.6 Schematic of the HR-ToF-AMS showing its two ion optical modes.

butanol-based Condensation Particle Counter (CPC). Details about the protocols of operating instruments and the calculations can be found in previous work (Kulmala et al., 2012; Lyu et al., 2018; Wang, Guo, & Chan, 2014). Particles were counted in 44 size bins every 4 min, according to their mobility through a differential mobility analyzer mounted in the SMPS. Particles were categorized into three modes: nucleation (5.5–24.7 nm), Aitken (24.7–100 nm), and accumulation modes (100–350 nm) according to particle size. The nucleation and Aitken modes are referred to as ultrafine particles (5.5–100 nm).

Modeling and Calculation Techniques

Photochemical Box Model (PBM)

General Description

The photochemical formation of O_3, reaction rates, radicals and intermediate products at a given site can be simulated using a photochemical box model. The box model can incorporate either condensed mechanism (*e.g.*, carbon bond mechanism (CB05)) or nearly explicit mechanism (*i.e.*, Master Chemical Mechanism (MCM)). The CB05 mechanism is a condensed mechanism with high computational efficiency and reliable simulation. In the CB05 module, VOCs are grouped according to carbon bond type and the reactions of individual VOC were condensed using a lumped structure technique (Y. Wang et al., 2017; Yarwood et al., 2005). In contrast, the MCM adopts a near-explicit mechanism, involving 5,900 chemical species and around 16,500 reactions

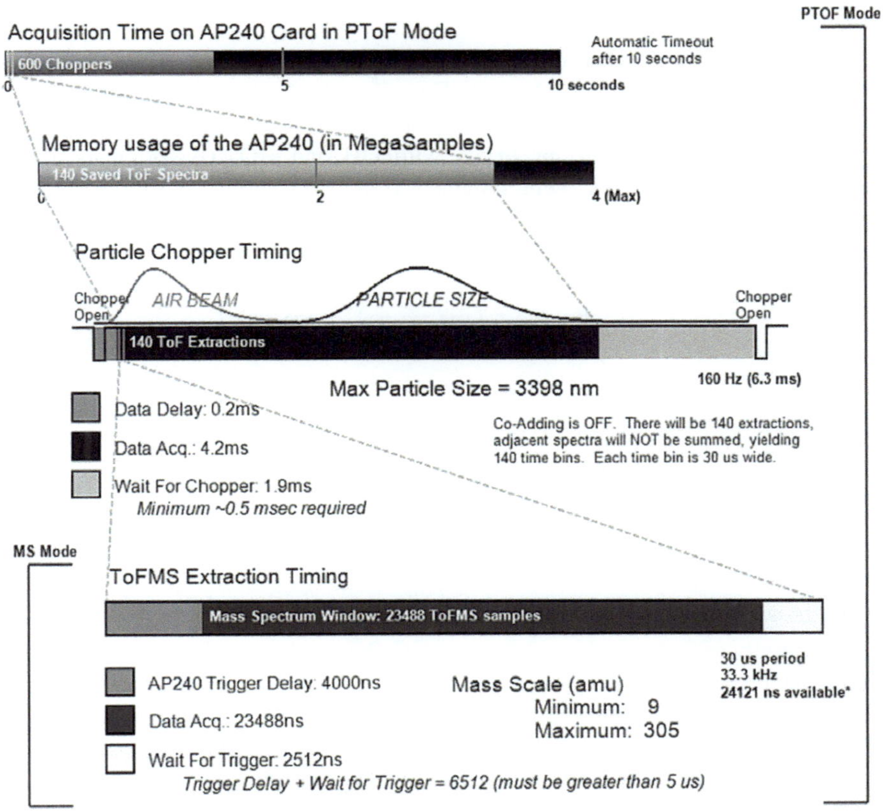

FIGURE 10.7 Timing diagram for use of the HR-ToF-AMS in PTOF mode and MS mode.

(Jenkin et al., 2003; Jenkin et al., 2015; Saunders et al., 2003). The observed values of air pollutants (O_3, NO_x, CO, SO_2, and VOCs) and meteorological parameters (temperature and relative humidity) were used to constrain the model.

Apart from the chemical reactions, the model considered other modules such as photolysis rate, dry deposition and the boundary layer height. The photolysis rates of different chemicals in the model were parameterized using the photon flux determined from the Tropospheric Ultraviolet and Visible Radiation (TUVv5) model (Lam et al., 2013), while the dry deposition module included the dry deposition of air pollutants, which were parameterized as an average deposition rate within the mixed layer height (MLH). The MLH was assumed to vary from 300 m at night to 1400 m during the daytime (Lam et al., 2013). It should be noted that the PBM-MCM model did not consider physical processes, *i.e.*, vertical and horizontal transport. More detailed descriptions of the PBM-MCM are provided in previous studies (Guo, Ling, Cheung, Jiang, et al., 2013; Ling et al., 2014; X. Liu et al., 2019; Lyu et al., 2019; Lyu et al., 2015; Lyu, Guo, et al., 2016; Y. Wang et al., 2018).

Simulation of Photochemical Pathways and Radicals

The net O_3 production ($P_{O_3\text{-}NO}$) was the difference between O_3 gross production ($G_{O_3\text{-}NO}$) and O_3 destruction ($D_{O_3\text{-}NO}$) (Equation 10.1). $G_{O_3\text{-}NO}$ was calculated by the oxidation of NO by HO_2 and RO_2 (Equation 10.2), while $D_{O_3\text{-}NO}$ was calculated by O_3 photolysis and reactions of O_3 with OH and HO_2 (Equation 10.3). All radicals and intermediates were obtained from the outputs of the PBM. The constants (k) represent the rate coefficients of corresponding reactions, respectively. The subscript '*i*' in Equation 10.2 refers to individual RO_2.

$$P_{O_3-NO} = G_{O_3-NO} - D_{O_3-NO} \tag{10.1}$$

$$G_{O_3-NO} = k_{HO_2+NO}[HO_2][NO] + \sum k_{RO_{2i}+NO}[RO_{2i}][NO] \tag{10.2}$$

$$\begin{aligned} D_{O_3-NO} = {} & k_{HO_2+O_3}[HO_2][O_3] + k_{OH+O_3}[OH][O_3] + k_{O(^1D)+H_2O}[O(^1D)][H_2O] \\ & + k_{OH+NO_2}[OH][NO_2] + k_{alkenes+O_3}[alkenes][O_3] \end{aligned} \tag{10.3}$$

Estimate of Relative Incremental Reactivity (RIR)

The sensitivity of O_3 formation to precursors can be usually assessed using the relative incremental reactivity (RIR), which was calculated from the output parameters of the PBM (Cardelino & Chameides, 1995). RIR reflects the relative variation in O_3 production rate for a change of mixing ratios in precursors (Equation 10.4). A larger positive RIR of a given precursor indicates a higher probability that O_3 production will be more significantly reduced by cutting down emissions of this precursor. The calculation of RIR values refers to the following equation:

$$RIR(X) = \frac{[P^S_{O_3-NO}(X) - P^S_{O_3-NO}(X - \Delta X)]/P^S_{O_3-NO}(X)}{\frac{\Delta S(X)}{S(X)}} \tag{10.4}$$

where X is a specific precursor (*i.e.*, VOCs, NO_x, or CO); S(X) is the observed mixing ratio of precursor X (ppbv); ΔS(X) is the hypothetical change of mixing ratio of X; $P^s_{O_3-NO}(X)$ is the net O_3 production in a base run with original mixing ratios, while $P^s_{O_3-NO}(X - \Delta X)$ is the net O_3 production in a second run with a hypothetical change (ΔS(X)) (10% in this study). Ozone titration by NO was considered in both runs. The O_3 production $P^S_{O_3-NO}$ was calculated according to Equations 10.1–10.3. The superscript "S" represents the specific sampling site.

Chemical Transport Model

Weather Research and Forecasting (WRF) coupled with Community Multiscale Air Quality model (CMAQ) and Chemistry model (Chem) (i.e., WRF-CMAQ, WRF-Chem) are usually applied to simulate the atmospheric dynamics and chemical processes that led to the occurrence of high O_3 days. WRF-Chem has been widely used and confirmed to be robust in modeling mesoscale O_3 photochemistry in previous studies (Grell et al., 2005; Jiang et al., 2010; Kumar et al., 2012; Tie et al., 2010). The model was mainly developed by the National Oceanic and Atmospheric Administration and continuously updated with the involvement of numerous collaborators. Detailed descriptions of the model can be found in Grell et al. (2005). Briefly, the WRF and Chem were used to simulate the meteorological and chemical processes, respectively. They shared the same domain settings and physical parameters, such as irradiance, planetary boundary layer, and cloud cover. The meteorological data simulated by WRF were used to drive the Chem.

The feedbacks between chemical and meteorological processes are also modulated by WRF. In contrast, WRF model and CMAQ model are not online, which means the CMAQ model would be run after completion of WRF simulation. However, WRF-CMAQ has more efficient computation comparing with WRF-Chem. In addition, the WRF-CMAQ model has multifunction on tracing the contributions of air pollutants produced from the specific source regions, and quantifying the contributions from different processes, such as horizontal/vertical advection and diffusion, gas-phase chemistry, cloud processes, and dry deposition.

The updated MODIS underlying surface data is applied in WRF model to replace the default land cover data due to its outdated information. The Chem model shares the same domain setting with WRF model, while in the CMAQ (v5.0.2) model, one grid cell would be subtracted from each side of WRF domains. The WRF model is driven by Final Operational Global Analysis data from the U.S. National Center of Environmental Prediction (NCEP). For the simulation cases in China, the MIX Asian emission inventory would be applied for anthropogenic emissions of Asian countries excluding China (Li et al., 2017). The Multi-resolution Emission Inventory for China (MEIC v.1.3) is used for anthropogenic emissions within China (Zhang et al., 2009). The biogenic emissions are estimated using the Model of Emissions of Gases and Aerosols from Nature (MEGAN v2.04) (Guenther et al., 2006). In addition, the ship emission inventory is obtained from Shipping Emission Inventory Model (SEIM) in East Asia (H. Liu et al., 2019). Simulations are usually independently conducted for individual O_3 episode event in each area, and 48 hours preceding O_3 episode days are run as the spin-up.

In addition, the integrated process rate (IPR) module could be implemented within the boundary layer to calculate the O_3 variation rates contributed by different processes including horizontal/vertical advection and diffusion, gas-phase chemistry, cloud processes, and dry deposition (Byun & Schere, 2006). Furthermore, the Integrated Source Apportionment Method (ISAM) coupled in the CMAQ model could be applied to track the contributions of O_3 produced from the emissions of coastal cities and the precursors over marine areas (Kwok et al., 2015).

Source Apportionment Model

Positive Matrix Factorization (PMF) model (USEPA version 5.0) is commonly applied to the measurement data for receptor-based source appointment. The PMF model is a multivariate factor analysis tool that decomposes a matrix of speciated sample data into two matrices—factor contribution and factor profile which can be interpreted by an analyst as to what sources are represented based on observations at the receptor site (USEPA: https://cfpub.epa.gov/si/si_public_record_report.cfm?Lab=NERL&direntryid=308292). The details of PMF applied to data for source profiles and the contributions of individual species have been introduced somewhere else (Guo et al., 2011; Ling et al., 2011; Lyu, Guo, et al., 2016; Lyu, Zeng, et al., 2017; Ou et al., 2015). Briefly, the sample/species are input into the PMF model and the uncertainties for each sample/species are determined as the sum of 5–15% of concentration and two times the method detection limit (MDL) of the species (USEPA website). Different number of factors and uncertainties need to be tested, and an optimum solution is determined based on a good fit to the observed data and the most meaningful results by comparing with previous studies.

Backward Particle Release Model

Backward particle release simulations can be carried out using HYSPLIT model (Stein et al., 2015) at given sites during the sampling period (Draxler & Rolph, 2003). The backward particle release simulation, which considers the dispersion processes in the atmosphere, is capable of identifying the history of air masses (Ding, Fu, et al., 2013; Ding, Wang, et al., 2013; Guo, Jiang, et al., 2009). More details can be found in previous studies (Guo, Jiang, et al., 2009; H. Wang et al., 2018).

Quantification of Meteorological Impact

To quantify the impact of meteorology on the ground-level O_3, a statistical calculation combining Kolmogorov-Zurbenko (KZ) filter and stepwise multiple linear regression (MLR) was implemented in this study. KZ filter is a moving average method to preserve the low-pass signal processed by repeating iterations (Equation 10.5) (Rao & Zurbenko, 1994):

$$Y_i = \frac{1}{m} \sum\nolimits_{j=-k}^{k} A_{i+j} \tag{10.5}$$

where k represents the number of variables included on each side of target values, and the moving window length is calculated as m = 2k + 1. A_{i+j} and Y_i denote the time series of input and output, respectively. i is the time interval, and j stands for the number of windows. The output of ith pass turns to be the input of $(i+1)^{\text{th}}$ pass. KZ (m, n) indicates the filter calculated through n iterations with a window length of m, aiming to remove the fluctuations higher than period of $m \times n^{1/2}$ (days). Previous studies applied filters of KZ (15, 5) and KZ (365, 3) to separate the observed O_3 (O(t), Equation 10.6) into three time-scale components (Ma et al., 2016; Rao et al., 1997; Yang et al., 2019).

$$O(t) = W(t) + S(t) + L(t) \tag{10.6}$$

$$L(t) + S(t) = KZ\,(15,\;\; 5) = B(t) \tag{10.7}$$

$$W(t) = O(t) - KZ(15, \ 5) \quad (10.8)$$

$$\mathrm{L(t) = KZ(365, \ 3)} \quad (10.9)$$

$$S(t) = KZ(15, \ 5) - KZ(365, \ 3) \quad (10.10)$$

where O(t) is the time series of observed daily maximum 8-hour average (DM8A) O_3 concentration. L(t) (Equation 10.9) and S(t) (Equation 10.10) refer to the long-term and seasonal components, respectively. W(t) (Equation 10.8) reflects the short-term variation calculated by mesoscale- and/or synoptic-scale factors. The sum of L (t) and S(t) is defined as the baseline (B(t)) (Equation 10.7) for convenience of the following calculations. The stepwise MLR was established between DM8A O_3 and the meteorological components in different time scales.

$$A_W(t) = \sum b_{W_i} \times M_{W_i} + c_W \quad (10.11)$$

$$A_B(t) = \sum b_{B_i} \times M_{B_i} + c_B \quad (10.12)$$

$$\varepsilon_W(t) = W(t) - A_W(t), \quad \varepsilon_B(t) = B(t) - A_B(t) \quad (10.13)$$

$$\varepsilon(t) = \varepsilon_W(t) + \varepsilon_B(t) \quad (10.14)$$

$$A(t) = \varepsilon(t) + \sum b_{W_i} \times \bar{M}_{W_i} + \sum b_{B_i} \times \bar{M}_{B_i} + c_W + c_B \quad (10.15)$$

where $A_W(t)$ (Equation 10.11) and $A_B(t)$ (Equation 10.12) calculated by the multiple linear regressions are the short-term and baseline O_3 concentrations, respectively. M_{W_i} and M_{B_i} stand for the meteorological variables in the two-time scales. The variables incorporate daily maximum Temp, daily minimum RH, daytime (07:00–19:00 LST) average of SR, U & V wind components, and PBLH, which are acknowledged as the important factors affecting the ambient O_3 concentrations (He et al., 2017; Lee et al., 2014; Liu & Wang, 2020; Sánchez-Ccoyllo et al., 2006). b_{W_i} and b_{B_i} represent the regression coefficients, while c_W and c_B indicate the interceptions of regression formulas. The residuals between observations and calculations can be expressed by $\varepsilon(t)$ (Equation 10.14). In addition, a base condition related to the mean values of meteorological data on the same date throughout entire period ($\bar{M}$) was calculated to preserve the annual variation pattern. The O_3 concentration after ruling out the impact of meteorological variability can be determined by $A(t)$ (Equation 10.15), which is the sum of $\varepsilon(t)$ and base conditions. Moreover, the impact of each meteorological parameter was calculated by considering the individual meteorological variable in M_{W_i} and M_{B_i}. This calculation method has been proved valid in a previous work studying the O_3 variation in the PRD region (Yang et al., 2019).

NPF-Related Calculations

To investigate NPF processes, the formation rate of 5.5 nm particles ($J_{5.5}$), condensation sink (CS), and the production of condensable vapors were calculated. $J_{5.5}$ was calculated using Equation 10.16 (Guo et al., 2012; Sihto et al., 2006). $J_{5.5}$ was calculated due to the fact that 5.5 nm was the smallest size that could be detected by our SMPS. It is important to note that 5.5 nm was not the initial size of clusters formed by condensable vapors, $J_{5.5}$ actually reflected the overall rate of cluster formation and subsequent growth to 5.5 nm, with the consideration of sinks of particles due to coagulation and further growth into larger-size particles.

$$J_{5.5} = \frac{dN_{5.5-10}}{dt} + CoagS_{d_p=7\ nm}N_{5.5-10} + \frac{1}{(10\ -\ 5.5)\ nm}GR_{5.5-10}N_{5.5-10} \tag{10.16}$$

where $N_{5.5-10}$ refers to the total particle number in the size range of 5.5–10 nm. The coagulation sink of particles in this size range is approximated by the coagulation loss of 7 nm particles, under the assumption that 7 nm is approximately the geometric mean diameter (GMD) of particles between 5.5 and 10 nm. $GR_{5.5-10}$ denotes the growth rate of 5.5–10 nm size range particles, which was calculated through the linear regression between $GMD_{5.5-10}$ and time. Apart from $GR_{5.5-10}$, growth rates for particles in modes of nucleation ($GR_{5.5-24.7}$) and Aitken ($GR_{24.7-101.4}$) were also calculated in the same way.

Growth rate (GR) and geometric mean diameter (GMD) for nucleation mode particles (5.5–24.7 nm) were calculated using the following equations (Guo et al., 2012):

$$\mathrm{GR} = \frac{d_{\mathrm{GMD}}}{d_t} \tag{10.17}$$

$$\mathrm{GMD} = \mathrm{e}^{(\sum n_i \ln^{d_i})/N} \tag{10.18}$$

$$\frac{d_N}{d_{log^{D_p}}} = \frac{\Delta N}{log_{D_p^2} — log_{D_p^1}} \tag{10.19}$$

where n_i is the particle number concentration in the i th bin with an upper diameter of d_i, N represents the total number concentration (cm^{-3}). ΔN is the particle number concentration in the size bin with upper and lower limit diameters of $D_p{}^2$ and $D_p{}^1$, respectively.

Condensation sink (CS), which describes the loss rate of vapor molecules and newly formed particles on the pre-existing aerosol particles, was calculated as follows (Kulmala et al., 2005):

$$\mathrm{CS} = 2\pi D \int D_p \beta_m (D_\mathrm{p}) n (D_\mathrm{p}) dD_p = 2\pi D \sum_\mathrm{i} \beta_\mathrm{mi} D_\mathrm{pi} N_\mathrm{i} \tag{10.20}$$

where D is the diffusion coefficient of the condensing vapor, β_m is the transitional regime correction factor, D_p denotes the particle diameter, n and N represent the particle numbers. β_{mi}, D_{pi}, and N_i are the specific values for a given size bin (i). The growth factor calculated was used to calibrate the dry particle size measured by SMPS (Laakso et al., 2004).

To investigate the relationship of sulfuric acid vapor (Q_{sa}, $cm^{-3}\ s^{-1}$) with NPF, a predictive proxy based on solar radiation, SO_2 concentration, CS, and relative humidity was used to estimate sulfuric acid concentration (Mikkonen et al., 2011) (Equation 10.21):

$$Q_{SA} = 8.21 \times 10^{-3} \cdot k \cdot [SO_2]^{0.62} \cdot [SR] \cdot (CS \cdot RH)^{-0.13} \quad (10.21)$$

where k is a constant value of 1.035, $[SO_2]$ is the measured concentration in ppb, SR is the solar radiation in $W \cdot m^{-2}$, RH is the relative humidity in % and CS is the condensation sink in s^{-1}.

OBSERVATIONS OF URBAN AIR POLLUTANTS

Intensive Samplings of Urban Air Pollutants

Characters of O_3 and Its Precursors

Figure 10.8 summarizes the spatial distributions of O_3 and its precursors (*i.e.*, NO_x and VOCs) in Hong Kong and adjoining areas based on intensive samplings in autumns during 2007 to 2016. The samplings were conducted in autumn because O_3 episodes in southern China occurred frequently in autumn and O_3 and its precursors usually showed low values in summer, which is different from other locations. The differences in autumn from those in summer were probably because marine air significantly diluted air pollution in summer, while continental air masses remarkably burdened air pollution in autumn (Wang et al., 2009). The levels of all pollutants (*i.e.*, O_3, NO_2, CO, SO_2, VOCs and OVOCs) were significantly higher on O_3 episode days, except for the comparable or even lower NO due to its titration to O_3. The higher concentrations of both primary and secondary pollutants on episode days than those on non-episode days were likely due to more intense photochemical reactions, more local pollutant accumulation as well as the regional transport of more highly polluted air masses.

According to the figure, it can be seen that higher VOC levels and O_3 were found in inland areas (WQS) due to the higher emissions of VOCs and more intensive inland O_3 formations (Cheng, Guo, Saunders, et al., 2010; Cheng, Guo, Wang, et al., 2010; Guo, Jiang, et al., 2009; Jiang et al., 2010). The urban (TW) and new town (TC) sites in Hong Kong had higher levels of NO_x and O_3 were partially titrated by NO (Guo, Ling, Cheung, Jiang, et al., 2013; Y. Wang et al., 2017). In areas with high NO levels, the NO titration ($O_3 + NO \rightarrow NO_2 + O_2$) is a main process consuming O_3. In comparison, higher O_3 levels were measured at coastal (WS) and mountainous (TMS) sites, mainly due to the weaker NO titration and higher O_3 production rate with stronger oxidative capacity of the atmosphere (Guo, Ling, Cheung, Jiang, et al., 2013; Ling et al., 2014; Y. Wang et al., 2018).

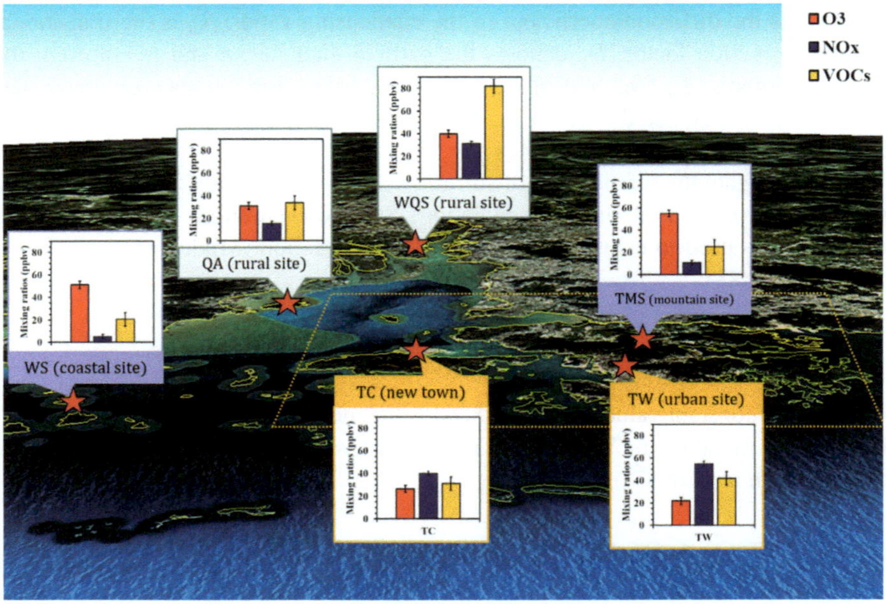

FIGURE 10.8 Characters of O_3 and its precursors in Hong Kong and adjoining areas based on intensive samplings during 2007 to 2016.

In addition, O_3 were much higher at WS and TMS than those at TC and TW, though the levels of O_3 precursors at WS and TMS were lower. Insight into VOC ratios found that ethene/ethane and toluene/benzene at WS and TMS were significantly lower than those at TC and TW, likely indicating that the air masses at WS and TMS were more aged (Guo et al., 2007). Therefore, the higher O_3 at WS and TMS might be partially attributable to the aging of air masses (*e.g.*, during the transport of continental air).

Characters of VOCs

Figure 10.9 displays the spatial distributions of VOCs concentration and their compositions in ten major cities of China. In general, TVOCs concentrations in northern cities, such as Jinan, Beijing, and Shijiazhuang, were higher than those in southern cities. The highest TVOCs level was found in Lanzhou, with the mixing ratio of 75.9±8.9 ppbv, while the lowest TVOCs was in Shanghai (26.1±2.6 ppbv). As a highly industrialized city, the mixed industrial processes in Lanzhou contributed the most to VOCs (Jia et al., 2016). Shanghai had the lowest levels of TVOCs, which was mainly due to the unstable weather conditions with continuous precipitation caused by a few tropical storms over the East China Sea (CMA, 2019).

As for VOCs groups, alkanes and OVOCs contributed more than half of the TVOCs, with the percentage between 52% and 69%. Alkanes contributed the most in six cities, namely, Shijiazhuang (46%), Lanzhou (42%), Wuhan (40%), Shanghai (38%), Zhengzhou (35%) and Chengdu (31%), while OVOCs contributed the most in the rest 4 cities (Beijing (48%), Guiyang (40%), Jinan (36%), and

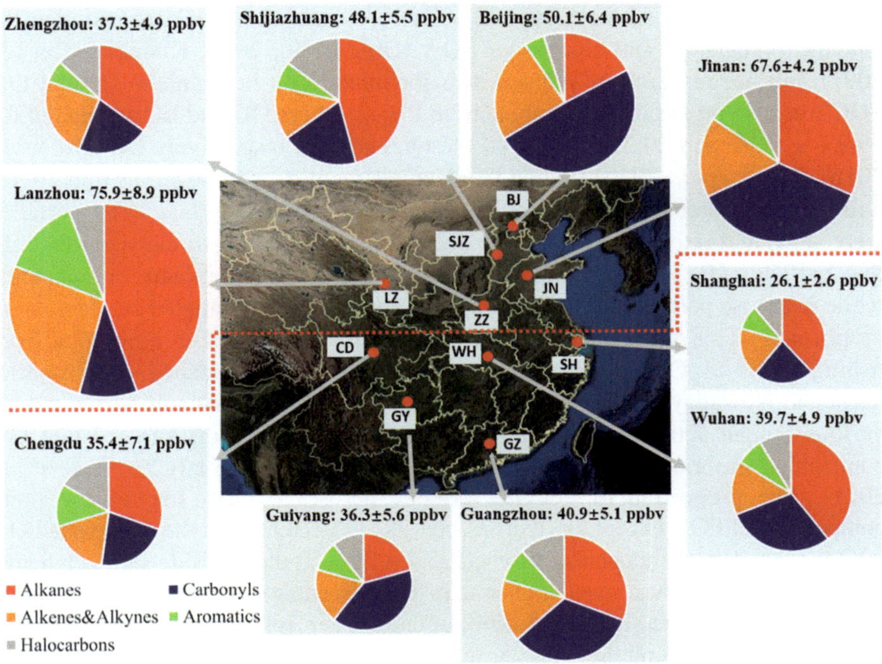

FIGURE 10.9 VOCs concentrations and their compositions in ten major cities of China.

Guangzhou (33%)), implying varied chemical compositions in different cities though all the cities had severe O_3 pollution problem.

Chemical Compositions of Aerosols

Numerous results and findings on aerosol composition, sources, and processes have been presented. Polycyclic aromatic hydrocarbons (PAHs) are ubiquitous constituents of particulate matter and are well-known to be carcinogenic and mutagenic. They are mostly derived from incomplete combustion and pyrolysis of fossil fuels, partly from natural and anthropogenic sources, and are present in the atmosphere in gaseous and particulate form. Based on the analysis of $PM_{2.5}$ and PM_{10} samples collected at residential/industrial/commercial sites in the wintertime of 2000 and summertime of 2001, Guo et al. (2003) found that the dominant PAHs in $PM_{2.5}$ and PM_{10} included benzo[b]fluoranthene, pyrene, fluoranthene, indeno[1,2,3-cd]pyrene and chrysene, accounting for 50–82% of total PAHs, and most of the PAHs were in the $PM_{2.5}$ fraction. As for mountain sites, primary organic carbon (POC) emitted from biomass burning accounted for a considerable proportion of OC (11.6 ± 3.2%), and secondary organic carbon (SOC, carbon fraction of SOA) derived from isoprene, monoterpenes, sesquiterpenes, and aromatics constituted 40.6 ± 5.7%, 30.4 ± 5.5%, 5.6 ± 1.7% and 21.3 ± 8.2% of the total estimated SOC at TMS in Hong Kong, respectively (Lyu, Guo, et al., 2017). Furthermore, the findings that biogenic SOA correlated well with the indicators of anthropogenic emissions (Goldstein et al., 2009;

Hoyle et al., 2011), so the SOC was selected as the SOA tracer to estimate the amount, precursors, and sources of SOA (Ding et al., 2012; Kleindienst et al., 2007). Further research in Wuhan, a major inland city in China, suggested that SOC was also an important component in $PM_{2.5}$, where OC and EC accounted for 79.9% ± 0.3% and 20.2% ± 0.3% of the total carbon, respectively (X. Lyu et al., 2016). Zhang *et al.* (2020) studied the detailed chemical compositions in particles with different size ranges emitted from ships in severe shortage. The particles were separated into nine-size bins by the impactor and the particle size range. In fine particles ($D_p < 1.1$ μm), the OC and EC were the dominant components, while in coarse particles ($D_p > 3.3$ μm), OC and EC only accounted for very small proportions. This size-dependent toxicity, particularly noted in emissions from fishing boats in China, underscores the health risks associated with smaller particles. Nowadays, with the increasing population, the aerosols generated by various human activities have dramatically increased. Results showed that OA contributed the most to NR-PM_1 with a proportion of 57.7 ± 0.2% (Yao et al., 2021). The proportion of cooking organic aerosol (COA), one of the main components of POA except for hydrocarbon-like OA (HOA) and the oxygenated OA (OOA, secondary components), reached 25.4 ± 0.3% at the roadside site with heavy traffic fleet, which was even higher than the percentage of HOA, implying that cooking activities are a significant source of organic aerosols in Hong Kong, even on a busy road and exempliying the substantial impact of urban lifestyle on aerosol composition. As the important component of OA, 86.5% of the organics in OA was contributed by the OOA. Two kinds of OOA, less-oxidized OOA (labeled as LO-OOA1 in this study) and the more-oxidized OOA (MO-OOA) are derived from the anthropogenic organic compounds, based on the multiple linear regression with molecular markers of OA, highlighting the important roles of anthropogenic emissions (Yao et al., 2022). Molecular markers are some individual compounds contained in OA, which are highly source-specific or indicative of secondary transformation of certain precursors. Thus, these organic molecular markers are highly valuable in serving as source tracers for the apportionment of source contributions and formation processes of OA and help formulate effective control measures (Jaeckels et al., 2007). Using in-situ measurements method (*i.e.*, TAG, AMS) with high time resolution, Lyu *et al.* (2020) measured the air compositions at a relatively remote coastal site at the southeast tip of Hong Kong to investigate the air photochemical pollution at the site. Results showed that continental air contained both POA markers (*i.e.*, levoglucosan, fatty acids, and oleic acid) and SOA markers (*i.e.*, SOA markers derived from isoprene and monoterpenes). The marine air and continental-marine mixed air had more abundant hydroxyl dicarboxylic acids (OHDCA), with anthropogenic unsaturated organics as potential precursors.

Much effort focused on the inorganic components in aerosol particles were also conducted. Due to the change of $PM_{2.5}$ component, i.e. sulfate (SO_4^{2-}), nitrate (NO_3^-) and ammonium (NH_4^+), total acidity ($[H^+]_{total}$) and in-situ acidity ($[H^+]_{in\text{-}situ}$) of $PM_{2.5}$ significantly decreased, further causing significant variation of concentration of organic matter in winter in PRD areas (Fu et al., 2015). Secondary inorganic ions were the most predominant species in $PM_{2.5}$ in the form of NH_4NO_3 and $(NH_4)_2SO_4$ and stationary and mobile sources had equivalent importance in Wuhan according to comparable levels of SO_4^{2-} and NO_3^- (Lyu et al., 2016). Potassium was the most abundant element in $PM_{2.5}$, implying biomass burning in and around Wuhan during the sampling campaign.

Particle Size Distribution

To date, numerous measurements have been conducted in different locations to study particle number (PN) concentrations, particle size distributions (PSD) and new particle formation (NPF) processes, including clean areas, *i.e.*, free troposphere (Gordon et al., 2016), boreal forests (Carracedo et al., 2022; Jokinen et al., 2022; Leino et al., 2019), coastal rural areas (Cheung et al., 2015; Lyu et al., 2018), mountain sites (Guo et al., 2012; Lyu, Guo, et al., 2017), Antarctic/Arctic areas (Brean et al., 2021; Quelever et al., 2022); and polluted urban areas, i.e. Hong Kong (Lyu et al., 2018; Wang, Guo, Cheung, et al., 2014), Shanghai (Lu et al., 2022), Mexico City (Zhou et al., 2023), East St. Louis (Qian et al., 2007), and Brisbane (Cheung et al., 2011; Guo et al., 2008). According to the spatial discrepancy of air pollutants emitted by different sources and meteorological parameters in different land use (i.e., roadside, urban, and rural) and cities, there are differences between UFP and NPF. For example, as shown in Figure 10.10, significantly higher particle number concentrations in all modes were observed at the suburban site than at the background site ($p < 0.05$) during the sampling periods, possibly due to stronger primary emissions/regional transport and more intensive NPF (Lyu et al., 2018). The GMDs of UFPs and acidic ultrafine particles (AUFPs) in the urban areas between Hong Kong and Shanghai were similar ($p > 0.05$), implying similar emission sources and/or chemical formation mechanisms of UFPs and AUFPs in these two cities (Lu et al., 2022). Generally, intensive NPF events were always observed in pristine environments where the condensation sinks were weak with a low concentration of pre-existing particles in the atmosphere based on the analysis of spatiotemporal variations in PN concentrations (Guo et al., 2012b; Lyu et al., 2018; Modini et al., 2009; Wang, Guo, Cheung, et al., 2014).

A variety of formation mechanisms have been suggested for atmospheric aerosols (Kulmala et al., 2000; Lehtipalo et al., 2018; L. Liu et al., 2021; M. Wang et al., 2020). Typically, the formation of nucleation mode particle requires two processes: the nucleation of stable atmospheric clusters (1–3 nm) and the growth of these clusters into observable sizes (> 3 nm). Altogether, the driving force for nucleation and the growth of freshly formed particles is the sufficiently high concentration of condensational vapors, such as sulfuric acid and possibly other organic compounds with a low saturation vapor pressure formed by photochemical oxidation reaction. Sulfuric acid has been identified as a key compound in atmospheric nucleation in many studies, demonstrating that NPF requires the presence of gas-phase sulfuric acid in concentrations exceeding 10^5 molecules cm^{-3} (Kulmala et al., 2006; Sipilä et al., 2010; Yin et al., 2021). Furthermore, in terms of the positive correlation between Q_{sa} and the proportion of AUFPs in UFPs ($R^2 = 0.71$), the results suggested that sulfuric acid vapors contributed to the transformation from UFPs to AUFPs through heterogeneous reaction of sulfuric acid vapor with non-acidic UFPs and/or condensation of sulfuric acid vapor on non-acidic UFPs at a high Q_{sa} level (Lu et al., 2022). Also, it has been

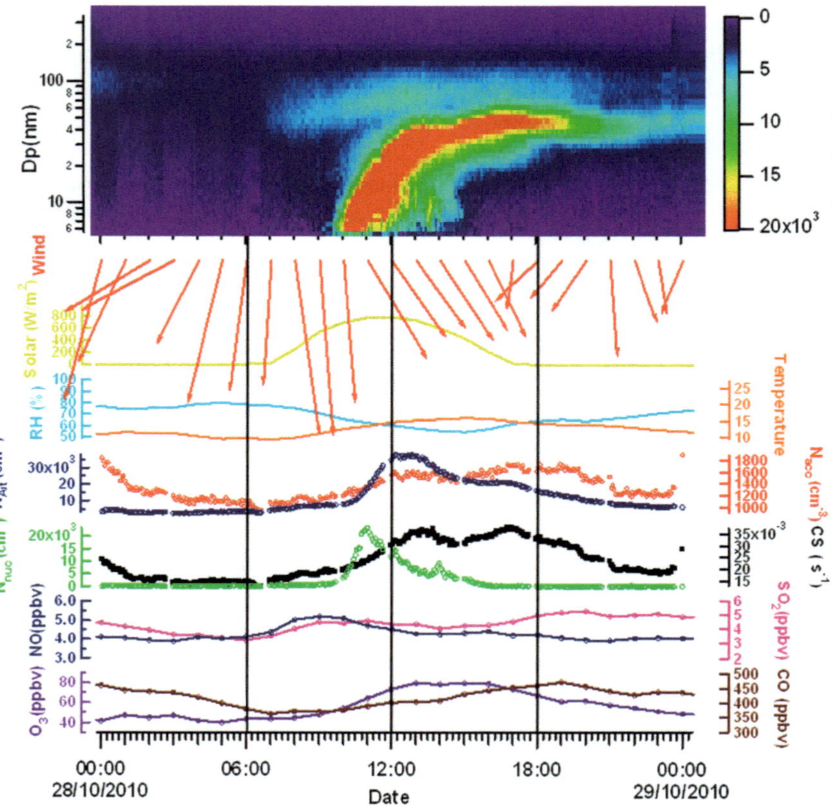

FIGURE 10.10 Contour plots of the average diurnal variations of particle size distribution, the meteorological conditions and the concentrations of airborne pollutants.

shown that some organic acids from the oxidation of terpenes (especially monoterpene) (Kerminen et al., 2010), amines (Brean et al., 2021), ammonia (Jorga et al., 2023) and iodine species (O'Dowd et al., 2002; Yu et al., 2019) may also be involved in aerosol nucleation. For particle growth, both sulfuric acid and VOCs especially many BVOCs, *i.e.*, isoprene and monoterpenes, and some AVOCs such as toluene could make contributions are believed to be critical contributors, while the relative importance of them needs to be identified in different cases (Guo et al., 2012b; Lyu et al., 2018; Lyu, Guo, et al., 2017). On the other hand, particle growth is generally size dependent ruled by the Kelvin effect (smaller particles have a barrier, requiring a higher saturation ratio of condensable vapors to drive particle growth) (Zhang et al., 2012). Therefore, the roles of sulfuric acid and organic condensable vapors in particle growth may be also size-dependent.

Method Developing on AUFPs

In order to guarantee the accuracy of the results, continuous improvement of the sampling method and the analysis is also an important aspect of studies. As shown in Figure 10.11, Wang *et al.* (2014) developed a diffusion sampler (DS)

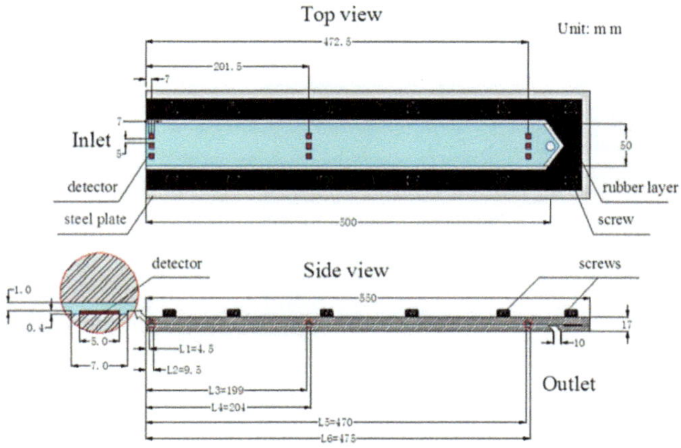

FIGURE 10.11 Schematic diagram of the DS.

with iron nanofilm detectors combining an atomic force microscope (AFM) to effectively measure the number concentration and size distribution of airborne AUFPs in indoor and outdoor environments and the sampling results showed very good agreement in terms of total particle number concentration and size distribution obtained by a commercially available online measurement system, *i.e.*, SMPS incorporated with a CPC (Wang, Guo, & Chan, 2014). Lu et al. (2022) also deployed the above method to illustrate the important roles of interaction between sulfuric acid vapor and non-acidic UFPs in AUFPs formation, which further demonstrates the utility of the method. Purposeful identification, selection, and collection of particles are of great significance in environmental research. Although microscopy is the common technique used in previous studies of particle identification, studies proved that the technique was complex and time-consuming. Lu et al. (2022) attempted three methods, including air jet, nanobubble, and ultrasonic methods, that could selectively abandon the non-targeted particles and only retain the targeted particles on the surface of a substrate. The results illustrated that ultrasonic treatment with ethanol was fully efficient for differential removal of non-acidic particles on the surface since the treatment was able to entirely remove the non-acidic particles from the surface compared to air jet treatment and nanobubble treatment with the removal efficiencies of 5.1 ± 3.4 and 89.3 ± 4.1%, respectively (Lu et al., 2021).

Figure 10.12 shows the schematic of the air jet treatment and nanobubble treatment. We now possess much better knowledge of chemical constitutions and detailed processes as well as how they affect SOA formation, properties and possible environmental impacts. Despite this, there are still significant gaps in our knowledge which places limitations on our ability to quantify and predict SOA in the ambient atmosphere. In light of the progress made and problems identified, several emerging issues pose challenges that need to be addressed via a combination of laboratory, field, and modeling studies. In general, it is pointed out that many of the issues (i.e., SOA budget, SOA formation mechanism, measurements of aerosol composition, and modeling of SOA) require faster and

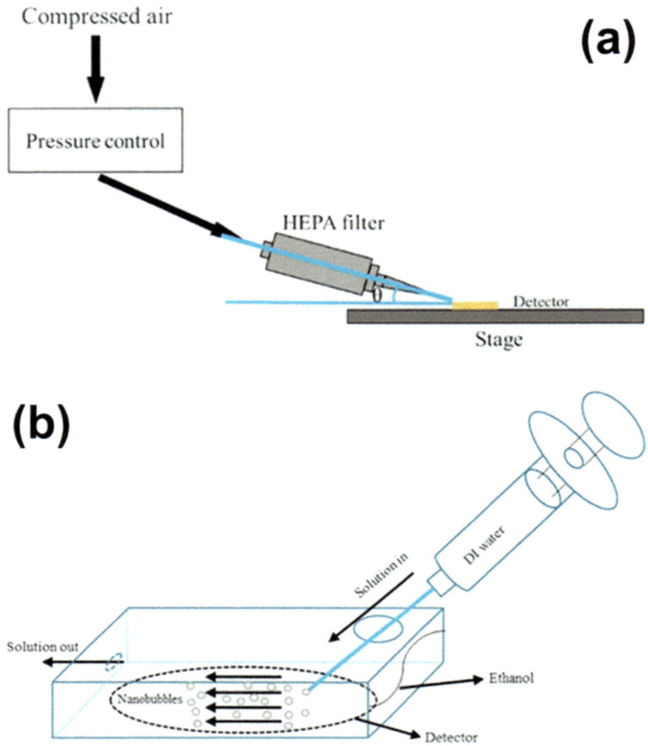

FIGURE 10.12 Schematic of (a) the air jet treatment and (b) the nanobubble process.

more accurate analytical tools which may open avenues for exploring a wider range of phenomena.

Long-term Variations of Urban Air Pollutants

Investigation on the long-term variations of air pollutants enables us to understand the seasonal and inter-annual characteristics of air pollution, the influence of meteorological parameters on the formation and accumulation of air pollutants, and the relationships between primary and secondary air pollutants in different years. In addition, the published control measures focusing on air pollution can be assessed following the variations of air pollutants before and after implementation of control measures, which would be helpful for the formulating the advanced control strategies in future.

The comprehensive datasets including O_3, CO, SO_2, NO-NO_2-NO_x, VOCs, and meteorological parameters were continuously recorded in Hong Kong since 2005. Y. Wang et al. (2017) found that NO_x and CO decreased while TVOCs remained unchanged during 2005–2014, suggesting the effective reduction of some emissions in Hong Kong. However, ambient O_3 levels increased at rate of 0.56 ppbv/year ($p < 0.01$) in these years. Specifically, the locally produced O_3 increased in spring while decreased in autumn, and remained unchanged in reast seasons. However, the increase of regional transport of O_3 in autumn offset the decrease in locally produced O_3. This was also confirmed by the results from Xue et al. (2014), who discovered that the efficient control of local aromatics resulted in the reduction of autumn O_3, while the enhanced regional background O_3 negated the local efforts during 2002–2013. It is noteworthy that the situation after 2013 contrasted with that before 2013. X. Liu et al. (2019) reported that the observed and locally produced O_3 decreased ($p < 0.05$) from the VOC sampling days in autumn of 2013 to those in 2016 at a rate of -5.31 ± 0.07 and -5.52 ± 0.05 ppbv/year, respectively. However, a leveling-off ($p = 0.32$) was simulated for the regionally transported O_3 during this period, indicating the alleviation of regional transport of O_3, especially from mainland of China, where the emission reductions (particularly for NO_x) started from 2013, the year when China's Clean Air Action Plan was launched, which might more or less play a role in ceasing the increase in regional O_3.

Moreover, to comprehensive analyze the long-term and seasonal variations of O_3 levels before and after the post-2013 period, the variations of O_3 and its precursors are presented in Figure 10.13. The overall increase of O_3 during 2005–2017 was found, including a rising stage (2005–2012, period I) and a steady stage (2013–2017, period II). The elevated O_3 in period I was attributable to the stimulation of meteorological variability, increase of regional transport, and enhancement of local photochemical production. In contrast, meteorological suppression, decrease of regional transport, and insignificant change of local photochemical production resulted in stable O_3 pattern in period II. Seasonally, leveled-off O_3 was observed in autumn during 2005–2017 due to the same reasons as those for long-term variations in period II, which even led to a significant decrease of autumn O_3 in period II as the mitigation of local production. Although the meteorological variability enhanced the O_3 concentrations in summer and winter during 2005–2017, the rise of O_3 in these two seasons have terminated since 2013. However, it was surprisingly found that the continuous NO_x reduction led to the increase in spring O_3 during 2005–2017, particularly the sharp increase in period II, which overwhelmed the decrease of autumn O_3 in the same period. As a consequence, an overall increase of O_3 was observed throughout 2005–2017. In summary, the spring increase prevailed over the hard-earned autumn decrease in O_3 pollution between 2005 and 2017, leading to the increase in the overall O_3 in the past 13 years (Zeren, Guo, et al., 2022).

To improve the air quality in Hong Kong, the local government implemented a series of control strategies, including a solvent program implemented during 2007–2011, and a diesel commercial vehicle (DCV) program since 2007. Lyu, Zeng, et al. (2017) has evaluated the efficiency of those programs based the measurement data and model simulation. The VOCs emitted from both solvent usage (-204.7 ± 39.7 pptv/year, $p < 0.05$) and diesel exhaust (-304.5 ± 17.7 pptv/year, $p < 0.01$) decreased significantly from 2005 to 2013, while VOCs emitted from gasoline and liquefied petroleum gas (LPG) vehicles showed an increasing trend with rate of (1086 ± 34 pptv/year, $p < 0.01$). In terms of the impacts on O_3 production, both the solvent and DCV programs lessened the O_3 rate of increase, whereas O_3 was enhanced by the gasoline and LPG vehicle emissions. Further assessment on the efficiency of DCV control programs was explored by Zeng et al.

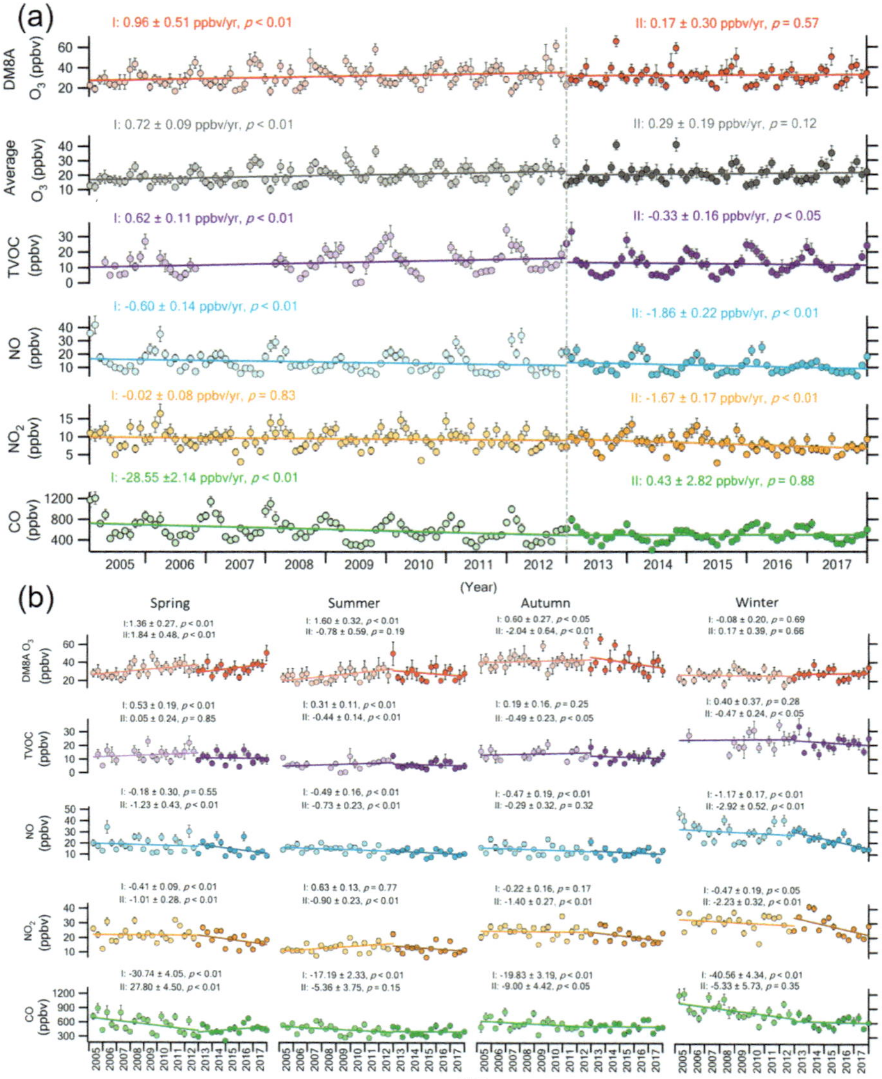

FIGURE 10.13 (a) Long-term and (b) seasonal variations of DM8A O_3 and O_3 precursors during 2005–2017 (data points are integrated into monthly averages and 95% confidence intervals are presented as error bars).

(2022) at a road site of Hong Kong. It was found that NO and NO_2 levels decreased during the study period but more dramatically after the implementation of DCV Phase III (2014) than pre-DCV Phase III. Source apportionment analysis confirmed that the ambient NO and NO_2 in Hong Kong attributed to the regulated DCV emissions in Phase III reduced at rates of 5.1 – 14.4 ppbv/yr in roadside environment and 1.6 – 3.1 ppbv/yr in suburban area. Despite overall NO_x reduction, increased NO_2/NO_x ratios were recorded during the study period possibly due to the application of diesel particulate filter (DPF) in DCVs. However, after introducing DCV Phase III, observed O_3 values experienced more dramatic increasing trends in most areas of Hong Kong than pre-DCV Phase III. Model simulations revealed that O_3 production rate kept increasing and turned to be less sensitive to NO_x from 2014 to 2019. On the roadside, net O_3 production rate was more than doubled during 2014–2019 owing to NO_x reduction. Moreover, the levels of oxidants (OH, HO_2 and RO_2) were 1.5–5 times those before 2014.

NUMERICAL SIMULATIONS ON URBAN AIR POLLUTION

SIMULATION OF ATMOSPHERIC DYNAMICS

The mountain-valley breeze circulation is one of important mesoscale circulations influencing urban air pollution. An intensive field measurement was simultaneously conducted at a mountain site (Mt. TMS) and an urban site at the foot of the mountain TMS (TW) in Hong Kong. Wang et al. (2015) discovered the consecutive high O_3 days from 08:00 28 Oct. to 23:00 03 Nov. 2010 at Mt. TMS with the highest mixing ratio of 110.4 ppbv, whereas no similar polluted event at the urban site. A comprehensive air quality model (WRF-CMAQ) was used to understand this event. Horizontally, photochemical processes dominated O_3 concentrations in the southwestern PRD and PRE, while in eastern and northern PRD the gas-phase chemistry was O_3-consumed. Vertically, higher O_3 values were found at ~600–1100 m height at both sites (Figure 10.14(a)). The difference in O_3 values between 1000 m height and ground-level suggested the existence of a vertical gradient over Hong Kong. Apart from the less NO titration at this mountainous location, the high O_3 values observed at Mt. TMS were attributable to the facts that

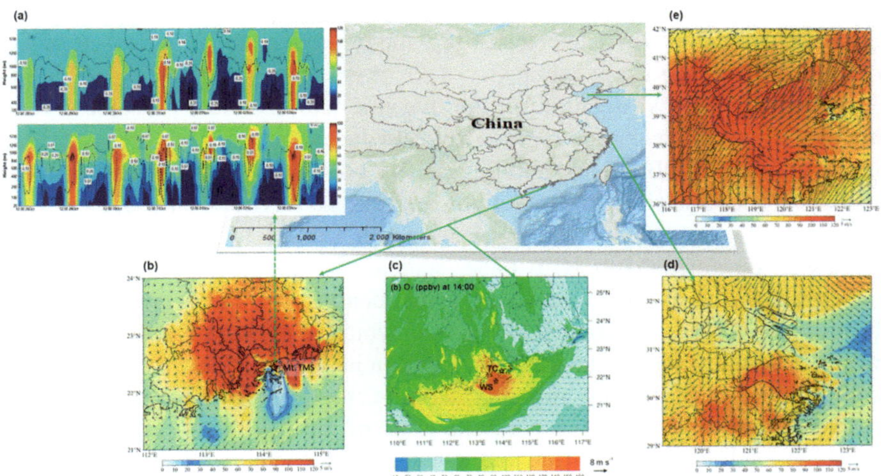

FIGURE 10.14 Distributions of O_3 and wind fields in (a) the vertical side of the Mt. Tai Mo Shan, (b) Pearl River Estuary, (c) South China Sea, (d) Hangzhou Bay, and (e) Bohai Bay.

O_3 mixing ratio in the middle of PBL was higher than that at ground-level, and vertical transport was an important factor in determining O_3 distribution at Mt. TMS.

In addition, sea-land breeze circulation is widely identified as the common mesoscale circulation affecting the air quality in coastal cities. Zeren et al. (2019) observed that Pearl River Estuary (PRE) and downwind areas experienced the highest levels of O_3, though the hot spots of O_3 were also identified in central and western PRD. It should be noted that the central and western PRD were characterized by high emissions of air pollutants, therefore potentially accounting for the elevated levels of O_3 therein. In contrast, the build-up of O_3 over PRE was more likely attributable to external rather than the very sparse local emissions. Thus, WRF-Chem model was applied to investigate the dynamic and chemical processes for the high O_3 pollution over PRE region. Driven by the temperature difference, the air was lifted above the land areas and subsided over PRE, forming a vertical circulation. the downdrafts may suppress the dispersion of air pollutants over PRE. Furthermore, divergent winds were developed on water surface following the downdrafts, which interacted with the synoptic winds and led to the decrease of wind speeds over PRE. Thus, the air pollutants transported to this area had longer residence time, when intensive photochemical reactions might occur (Figure 10.14(b)). Moreover, due to the sparse emissions of NO_x over PRE and the relatively short lifetimes (quick depletion in transport) made PRE an area with far less saturated NO_x than the surrounding land areas. As such, this can be regarded as an important factor accounting for the high O_3 production rates over PRE, which were even higher than those in surrounding cities, due to the weak loss of OH by reacting with NO_2. The effect was commonly seen in regions remote from but still influenced by anthropogenic emissions. Another example is the intensive O_3 production over the South China Sea (SCS) (Figure 10.14(c)), which was believed to be fueled by the continental air pollutants with relatively low NO_x. In fact, the observation data also supported the simulation results, according to H. Wang et al. (2018) who reported much higher concentrations of O_3 over SCS than in coastal areas.

Moreover, the investigation on the dynamic and chemical processes for the high O_3 events in the coastal regions were conducted on other bay areas of China, i.e., Bohai Bay (BHB), Hangzhou Bay (HZB), which are the important regions of transition between city agglomerations and marine environments (Zeren, Zhou, et al., 2022). Analogous to the physical environment of PRE, the onshore winds were well developed over HZB under stagnant weather, due to the temperature difference between land and sea. Consequently, the divergent winds were formed above the surface of bay areas, which intercepted with background winds, leading to the accumulation of substantial air pollutants transported from coastal cities over the bay area. Furthermore, the less dry deposition on water surface further enhanced the concentrations of air pollutants. The trapped air pollutants subsequently carried out intense photochemimainly due to smaller contributionsmainly due to smaller contributionscal reactions. As such, the net O_3 production rates over HZB were even comparable to those in adjacent cities which had much more primary emissions (Figure 10.14(d)). These characteristics were also feasible over BHB but the net O_3 production rates were lower

(Figure 10.14(e)), mainly due to smaller contributions of O_3 precursors in coastal cities. In addition, the high background O_3 level in North China and the regional transport of O_3 from the farther upwind regions of BHB superimposed the O_3 concentrations.

Simulation of Atmospheric Chemistry

Figure 10.15 displays the average diurnal profiles of the simulated O_3 production and destruction rates in each city. Among the five cities, Lanzhou had the highest net O_3 production rate (8.9 ± 1.7 ppbv h^{-1}), while Shanghai had the lowest value (2.8 ± 0.7 ppbv h^{-1}) ($p<0.05$). The net O_3 production rate in Beijing (6.4 ± 1.3 ppbv h^{-1}), Wuhan (5.8 ± 1.2 ppbv h^{-1}), and Chengdu (4.2 ± 0.9 ppbv h^{-1}) was at a moderate level ($p = 0.08$ and $p = 0.11$ related to Beijing and Wuhan, respectively). HO_2 + NO (51.9–65.2%) and RO_2 + NO (34.6–47.3%) were the dominant pathways of O_3 production, in accordance with previous studies in China (Lyu et al., 2019; Lyu et al., 2015; H. Wang et al., 2018; Y. Wang et al., 2018). The higher contributions of RO_2 +NO in Beijing (41.5 ± 8.3%) and Lanzhou (47.3 ± 8.5%) than other cities (34.6–39.9%) might be attributable to the higher RO_2 concentrations enhanced by the higher local VOCs. In contrast, the prominent destruction pathway of O_3 was OH + NO_2 (45.0–72.4%) in all the five cities. However, the second important destruction pathway of O_3 was different in the cities. It was the O_3 reaction with alkenes in Beijing (16.1%) and Lanzhou (29.5%), but photolysis of O_3 in Chengdu (11.8%), Shanghai (18.6%) and Wuhan (19.2%). Overall, the simulated net O_3 production rates among the five cities were consistent with those in previous studies conducted in warm seasons in China. More details can be found in X. Liu et al. (2021).

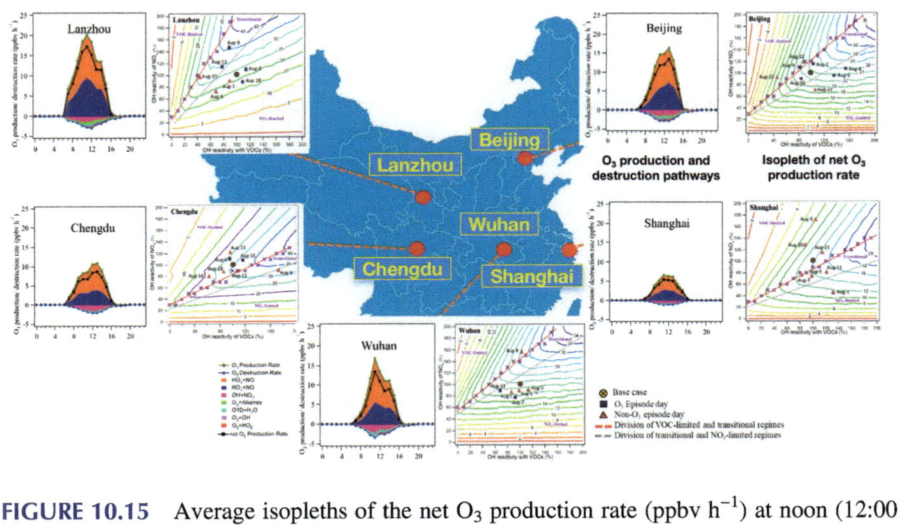

FIGURE 10.15 Average isopleths of the net O_3 production rate (ppbv h^{-1}) at noon (12:00 LT) based on the variations of OH reactivity of VOCs and NO_x on all VOCs sampling days in the five cities (X. Liu et al., 2021).

Source Apportionments of Urban Air Pollutants

VOC Source Identification

In the five cities, NO_x and 21 VOC species including ten alkanes, five alkenes/alkynes, and six aromatics in 199 out of the 207 VOC samples were processed by PMF model for source identification and quantification. Note: eight samples were not used as there were no corresponding NO_x values for these samples. The selection of the 22 species was based on the criteria: 1) their abundances and 2) tracers of NO_x and VOC sources. To reduce the uncertainty of the source apportionment results, compared to separate source apportionments for each city, the samples collected in all the five cities as a whole were used for model input. Figure 10.16 shows the six sources resolved based on the distributions of the NO_x and VOC tracers. The first source, having moderate to high levels of propane and *i*/*n*-butanes, represented LPG usage (Guo, Ling, Cheung, Wang, et al., 2013; Jorquera & Rappenglück, 2004; Lyu, Zeng, et al., 2017; Song et al., 2018;

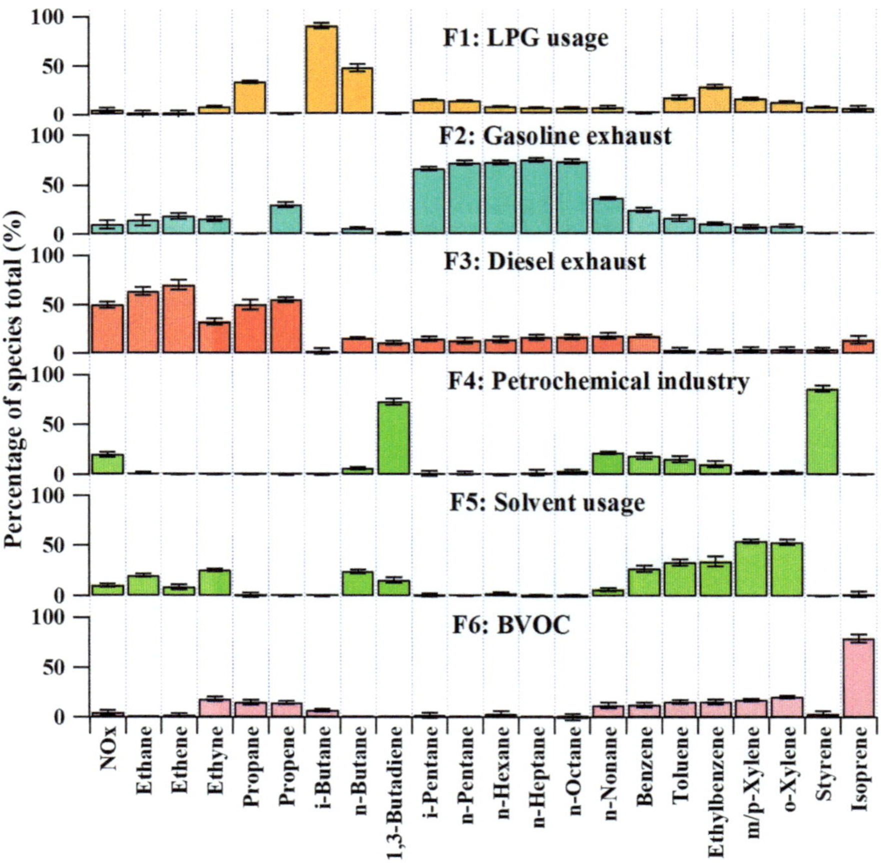

FIGURE 10.16 Average profile of the six sources in the five cities extracted from PMF (X. Liu et al., 2021).

F. Wu et al., 2016; R. Wu et al., 2016). The considerable loadings of *i*/*n*-pentanes and *n*-hexane and moderate amount of C_2 hydrocarbons suggested the second source as gasoline exhaust (Ho et al., 2009; Ling & Guo, 2014; Liu, Shao, Fu, et al., 2008; Lyu et al., 2019). The third source containing moderate levels of C_6–C_9 hydrocarbons, benzene and high percentages of NO_x and C_2-C_3 hydrocarbons especially alkenes, which was defined as diesel exhaust (Liu, Shao, Fu, et al., 2008; Liu, Shao, Lu, et al., 2008; Lyu et al., 2019; Sahoo et al., 2011). Considering the great amount of 1,3-butadiene and styrene, and moderate loadings of NO_x and aromatics, the fourth source was assigned to industrial emissions (Jones, 2008; Knighton et al., 2012; Liu, Shao, Fu, et al., 2008; Liu, Shao, Lu, et al., 2008). The fifth source was more related to solvent usage due to the high levels of C_7–C_9 aromatics (Ling & Guo, 2014; Shao et al., 2016; Yuan et al., 2010). Last, the source of biogenic emission was identified because of the dominance of isoprene in factor 6 (Guenther et al., 2006).

Source Apportionment on OA

Source apportionment (SA) plays a critical role in making control policies of aerosol pollution. Aerosol sources include all types of combustion activities (motor vehicles, marine traffic, power plants, biomass burning) and certain industrial processes (metal smelters, chemical plants, paper and pulp industries, refineries) which may also include various emissions from anthropogenic construction (cement industry) dust, road dust, cooking, etc (Engelbrecht et al., 2014). SA involves one or more techniques that are used to quantify how individual sources contribute to aerosol concentrations, which is a crucial task in air pollution management, control, and policy options (Chelani, 2013). SA techniques that rely on a statistical analysis of observations at monitor sites are receptor models, including PMF (Yao et al., 2021), principal component analysis (PCA) (Guo, Ding, et al., 2009), back-trajectory analysis (Guo, Ding, et al., 2009), and nonparametric wind regression (Chelani, 2013).

PCA is often used as an exploratory tool to identify the major sources of air pollutant emissions and to select statistically independent source tracers (Guo et al., 2003). Deploying the diagnostic ratios between PAHs and PCA analysis, potential sources of PAHs in aerosols were identified, implying that PAHs in aerosol predominantly from gasoline and diesel engines, were the main contributors of particle-associated PAHs, and stationary combustion sources may also contribute to the particulate PAHs in Hong Kong in wintertime and summertime (Guo et al., 2003). Applying the PCA with absolute principal component scores (APCS) technique identified and quantified the fact that, the major sources of $PM_{2.5}$ in Hong Kong were vehicular emissions/road erosion, secondary sulfate, residual oil combustion, soil suspension, and sea salt regardless of sampling sites and sampling periods at the rural, urban and roadside sites in Hong Kong (Guo, Ding, et al., 2009). Afterwards, PMF was widely applied for source apportionment using PM data. A lot of studies show that different areas have different air pollution sources, which leads to different proportions of $PM_{2.5}$ composition consisting of sulfate, nitrate, ammonium, organics, carbon, and heavy metals (Cheung et al., 2015; X. Lyu et al., 2016). Results through combined organic tracers and PMF analysis illustrated that vehicular emissions, regional transport, and photochemical formation were the three most important contributors to the observed PNC at the Hong Kong University of Science and Technology (HKUST) Air Quality Research Supersite (Cheung et al., 2015).

While the PMF is one of the most popular methods for $PM_{2.5}$ source apportionment due to its convenience and practicable performance, SA results are often rough estimations since such models generally do not fully leverage features in highly time-resolved data (Vedantham et al., 2014). To deal with this, source apportionment results should be checked by the spatial and temporal patterns of local $PM_{2.5}$ characteristics, regional emissions, meteorology, atmospheric transmission, etc. The integration of PMF with AMS and TAG in PM providing high time resolution data of aerosol offers a more nuanced approach to source apportionment. PMF-AMS results showed that SOA could have much higher mass fractions (5 times) than did POA (mainly COA and HOA) in non-refractory components in PM at a suburban site in Hong Kong (Li et al., 2015). PMF of the AMS particle-phase mass spectrum revealed that four OA components in non-refractory aerosols were identified, including a HOA factor, a COA factor and two oxygenated OOA factors of different oxidation levels: LO-OOA and MO-OOA with the highest contribution to NR-PM_1 at the roadside site (Yao et al., 2021). Many works utilizing organic molecular markers particularly those measured by TAG in conjunction with the PMF model also was conducted to address the challenge of source apportionment organic aerosols (Lyu et al., 2020; Q. Wang et al., 2020), and PM (Yao et al., 2022). Regional biomass burning (BB) was identified as the largest contributor of OA in continental air, with contributions fluctuating from 8% to 74%. Instead, anthropogenic SOA accounted for the highest fraction of OA in marine (37 ± 4%) and mixed air (31 ± 3%), overriding the contributions from BB according to different content of SOA molecular markers (Lyu et al., 2020). Diagnostic ratios of related molecular markers revealing that BB and coal combustion were dominant under the influence of long-range transport air mass, while vehicle emissions were dominant under local/median-range air mass influence (Q. Wang et al., 2020). By employing multiple linear regression with molecular markers of OA (e.g., hydroxy benzonic acids and 2,3-dihydroxy-4-oxopentanoic acid), anthropogenic organic compounds, especially aromatics, presumably were the most likely precursors of LO-OOA1 (Yao et al., 2022). Meanwhile, these studies underscore the importance of high time resolution organic markers for discerning OA pollution sources, providing observational evidence to support their use in source apportionment.

CONTROL STRATEGIES ON URBAN AIR POLLUTION

Potential Control Measures for O_3 Pollution

O_3-Precursor Relationship

Figure 10.17 shows the RIR of anthropogenic VOCs (AVOCs or AHC), biogenic VOCs (BVOCs or BHC), CO and NO_x, as a measure of the sensitivity of O_3 formation to the changes of the precursors (Cardelino & Chameides,

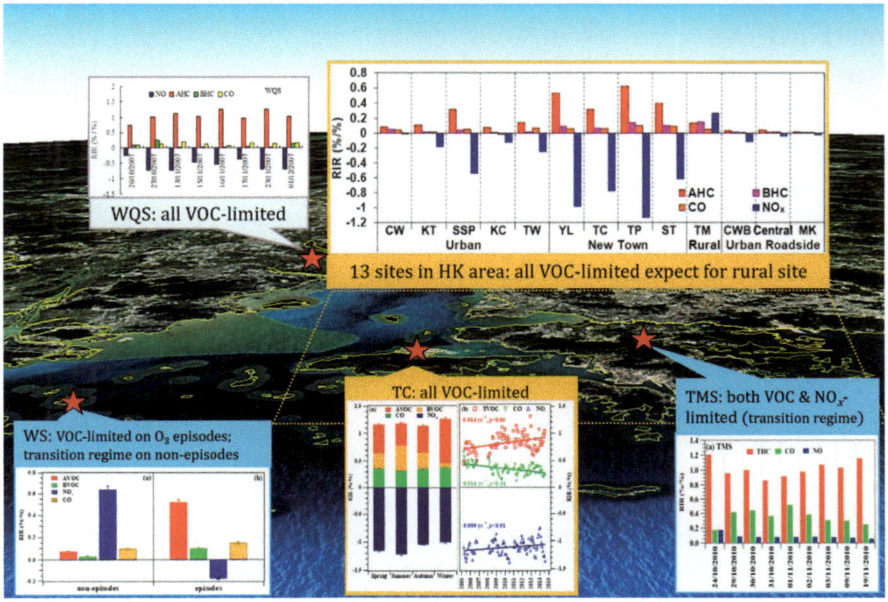

FIGURE 10.17 RIR values of VOC groups, CO and NO_x at different sites in Hong Kong and its surrounding areas.

1995). The VOC groups and CO had positive RIR values at all sites, and the RIR values of VOC groups were higher than that of CO, indicating that O_3 production was VOC-limited (except WS, shown below). The RIR values of AVOCs were mostly the highest, followed by BVOCs and CO. In contrast, the average RIR for NO_x was negative at urban and suburban sites, suggesting that cutting NO_x led to O_3 increase. Different from other sites where O_3 formation was limited by VOCs, the RIR of NO_x at WS and TMS were positive, same as VOCs and CO. Since the RIR of NO at TMS was much lower than that of VOCs, O_3 production was still VOC-limited at TMS. In contrast, the average RIR for NO_x was much larger than those of VOCs and CO on non O_3 episodes, indicating that O_3 formation was limited by NO_x. To understand the dominant VOC groups/species responsible for O_3 formation, more RIR values of VOC groups/species at different types of sites can be found in our previous studies (Cheng, Guo, Wang, et al., 2010; Lyu, Liu, et al., 2016; Y. Wang et al., 2018) and detailed descriptions of 13 sites in Hong Kong are provided in Table 10.1 and in Lyu, Liu, et al. (2016).

O_3 Control Measures on Cutting Ratios of VOCs and NO_x

The isopleths of net O_3 production rate in the five cities, based upon the percentages of OH reactivity values with VOCs (OH reactivity VOCs) and NO_x (OH reactivity NO_x), are plotted in Figure 10.15. In the five cities, the average OH reactivity VOCs on each sampling day were within the range of 48–193% of the average OH reactivity VOCs of all VOC sampling days. For OH reactivity NO_x, the range was

TABLE 10.1
Description of the 13 AQMS Sites in Hong Kong

Type	Site	No.	Description
Urban	Central Western (CW)	10	General urban air quality monitoring stations in residential areas and some mixed with commercial and/or industrial areas.
	Kwun Tong (KT)	21	
	Sham Shui PO (SSP)	22	
	Kwai Chung (KC)	13	
	Tsuen Wan (TW)	7	
Suburban/New Town	Yuen Long (YL)	6	Mainly residential areas in new territories.
	Tung Chung (TC)	9	
	Tai Po (TP)	29	
	Sha Tin (ST)	18	
Rural	Tap Mun (TM)	8	Remote island off the northeast coast of Hong Kong
Roadside	Censeway (CWB)	2	Commercial with high-rise buildings along the streets, high traffic and pedestrian flow, and frequent trafbc jam in rush hours
	Central	3	
	Mong Kou (MK)	1	

37–173%. Based on a series of hypothetical scenarios with different OH reactivity values of VOCs and NO_x, we simulated the net O_3 production rates in each city using the PBM-MCM model. To include the OH reactivity VOCs and OH reactivity NO_x values on all sampling days, factors between 0 and 200% with the interval of 10% were implemented to the average diurnal profiles of both VOCs and NO_x in each city. In each scenario, the concentrations of VOCs and NO_x were constrained to these scaled factors. Other air pollutants and meteorological conditions were consistent with the base case, which had 100% input of both VOCs and NO_x.

In Chengdu and Shanghai, the results suggested controlling VOCs be effective to mitigate O_3 formation, whereas reducing NO_x too fast would result in more local O_3 production. However, cutting either VOCs or NO_x would effectively lower O_3 production in Beijing and Wuhan, while reducing NO_x would be more effective than cutting VOCs in Lanzhou. Further, it was found that the local O_3 formation on O_3 episode days (*i.e.*, blue blocks in Figure 10.15) were in either transitional or NO_x-limited regime in Beijing, Wuhan, and Lanzhou, and in VOC-limited regime in Chengdu, basically in line with the average profile in each city. However, the net O_3 production rates on non O_3 episode days were not consistent in each city, which deserves more attention when making local control policies (X. Liu et al., 2021).

Source Contribution to VOCs and O_3 Production

Figure 10.18 presents the contribution of each source to ambient VOC mixing ratios and to O_3 production in the five cities. Note that the uncertainty of each value was at a considerable level (*i.e.*, at an average of 31.0% of their mean values), which might be due to the limited number of VOC samples collected in each city. Among the six

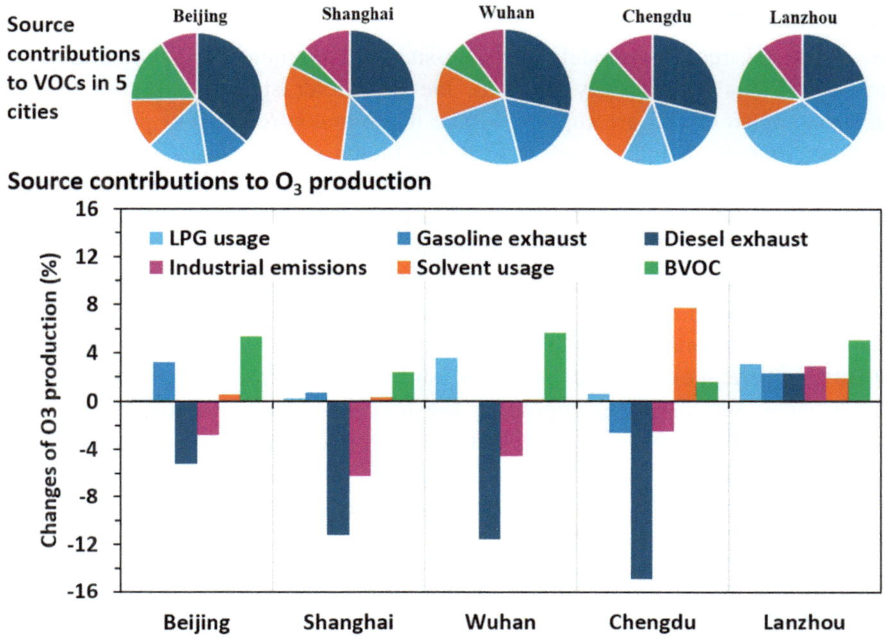

FIGURE 10.18 Contribution of each source to ambient VOC mixing ratios and to O_3 production in the five cities.

sources, vehicular exhausts, including gasoline exhaust and diesel exhaust, contributed the most to VOC mixing ratios in all the five cities (39.1–53.9%). Diesel exhaust ranked first (26.5–34.0%) in the other four cities except Lanzhou, where both gasoline exhaust (23.9 ± 7.0%) and industrial emissions (23.1 ± 7.8%) contributed the most to VOCs. Apart from traffic related sources, solvent usage was the second VOC source in Shanghai, Chengdu and Beijing (15.4–22.5%), and LPG usage had a considerable contribution in Wuhan (19.6 ± 8.0%). In comparison, Lanzhou had the largest industrial emissions (23.1 ± 7.8%) among the five cities, consistent with the contribution of 23.2% reported by Jia et al. (2016). Compared to anthropogenic sources, the contribution of biogenic emissions (4.6–7.2%) was the smallest. The contributions were consistent with the findings of previous studies and more details can be found in X. Liu et al. (2021).

Regarding the contribution of the six sources to O_3 production in each city, the biogenic emission in the cities except Chengdu was more obvious than other sources, likely due to the high loadings of NO_x, which led to negative contributions of vehicular exhausts and industrial sources to O_3 production in the cities except Lanzhou, where the O_3 production seemed to be controlled by NO_x. In addition, lower photochemical reactivity of VOC species in LPG usage and solvent usage than that in biogenic emission was the reason of less contribution to O_3 production. Furthermore, among the anthropogenic sources, gasoline exhaust in Beijing, LPG usage in Wuhan, solvent usage in Chengdu, and both LPG usage and industrial emissions in Lanzhou were the top contributors to local O_3 production.

Aerosol Control Measures

Based on the receptor modeling and in-situ measurements, it is concluded that while PM has dramatically reduced due to stricter standards during the last decades, much efforts are supposed to be made to control aerosol pollution from various sources including vehicles, ships, biomass burning and even cooking. To tackle this problem, it is important to reduce the sources of atmospheric pollutants, such as SO_2, NO_x, and VOCs. For example, the reduction in SO_2 in Hong Kong during the last decade was mainly attributed to stricter standards for the use of sulfur fuel oil (LSFO) (Euro IV standard) and the phase-out of old diesel vehicles. Meanwhile, more stringent control on the types of transportation should also be considered. Furthermore, it is also important to increase the sink of atmospheric pollutants in order to inhibit the formation of particles in the atmosphere. For instance, improving the filer for vehicles, ships, chimneys as well as kitchen hoods is necessary. As for essential emission activities, meteorological conditions should be considered since meteorological parameters (i.e., solar irradiation, wind speed, humidity, temperature) are key factors of particle formation and growth.

REFERENCES

Brean, J., Dall'Osto, M., Simo, R., Shi, Z. B., Beddows, D. C. S., & Harrison, R. M. (2021). Open ocean and coastal new particle formation from sulfuric acid and amines around the Antarctic Peninsula. *Nature Geoscience*, 14(6), 383–388. 10.1038/s41561-021-00751-y

Byun, D., & Schere, K. L. (2006). Review of the governing equations, computational algorithms, and other components of the Models-3 Community Multiscale Air Quality (CMAQ) modeling system. *Applied mechanics reviews*, 59(2), 51–77.

Canagaratna, M. R., Jayne, J. T., Jimenez, J. L., Allan, J. D., Alfarra, M. R., Zhang, Q., Onasch, T. B., Drewnick, F., Coe, H., Middlebrook, A., Delia, A., Williams, L. R., Trimborn, A. M., Northway, M. J., Decarlo, P. F., Kolb, C. E., Davidovits, P., & Worsnop, D. R. (2007). Chemical and microphysical characterization of ambient aerosols with the aerodyne aerosol mass spectrometer. *Mass spectrometry reviews*, 26(2), 185–222. 10.1002/mas.20115

Cardelino, C. A., & Chameides, W. L. (1995). An observation-based model for analyzing ozone precursor relationships in the urban atmosphere. *Journal of the Air & Waste Management Association*, 45(3), 161–180. 10.1080/10473289.1995.10467356

Carracedo, L. G., Lehtipalo, K., Ahonen, L. R., Sarnela, N., Holm, S., Kangasluoma, J., Kulmala, M., Winkler, P. M., & Stolzenburg, D. (2022). On the relation between apparent ion and total particle growth rates in the boreal forest and related chamber experiments. *Atmospheric Chemistry and Physics*, 22(19), 13153–13166. 10.5194/acp-22-13153-2022

Chelani, A. B. (2013). Statistical characteristics of ambient PM2.5 concentration at a traffic site in delhi: Source identification using persistence analysis and nonparametric wind regression. *Aerosol and Air Quality Research*, 13(6), 1768–1778. 10.4209/aaqr.2012.09.0243

Cheng, H. R., Guo, H., Saunders, S. M., Lam, S. H. M., Jiang, F., Wang, X. M., Simpson, I. J., Blake, D. R., Louie, P. K. K., & Wang, T. J. (2010). Assessing photochemical ozone formation in the Pearl River Delta with a photochemical trajectory model. *Atmospheric Environment*, 44(34), 4199–4208. 10.1016/j.atmosenv.2010.07.019

Cheng, H. R., Guo, H., Wang, X. M., Saunders, S. M., Lam, S. H., Jiang, F., Wang, T. J., Ding, A. J., Lee, S. C., & Ho, K. F. (2010). On the relationship between ozone and its precursors in the Pearl River Delta: application of an observation-based model (OBM). *Environ Sci Pollut Res Int*, 17(3), 547–560. 10.1007/s11356-009-0247-9

Cheung, H. C., Morawska, L., & Ristovski, Z. D. (2011). Observation of new particle formation in subtropical urban environment. *Atmospheric Chemistry and Physics*, 11(8), 3823–3833. 10.5194/acp-11-3823-2011

Cheung, K., Ling, Z. H., Wang, D. W., Wang, Y., Guo, H., Lee, B., Li, Y. J., & Chan, C. K. (2015). Characterization and source identification of sub-micron particles at the HKUST Supersite in Hong Kong. *Science of the Total Environment*, 527-528, 287–296. 10.1016/j.scitotenv.2015.04.087

CMA. (2019). China climate bulletin of 2018. Retrieved from http://zwgk.cma.gov.cn/zfxxgk/gknr/qxbg/201903/t20190319_1709281.html

DeCarlo, P. F., Kimmel, J. R., Trimborn, A., Northway, M. J., Jayne, J. T., Aiken, A. C., Gonin, M., Fuhrer, K., Horvath, T., Docherty, K. S., Worsnop, D. R., & Jimenez, J. L. (2006). Field-Deployable, high-resolution, time-of-flight aerosol mass spectrometer. *Analytical Chemistry*, 78(24), 8281–8289. 10.1021/ac061249n

Ding, A. J., Fu, C. B., Yang, X. Q., Sun, J. N., Zheng, L. F., Xie, Y. N., Herrmann, E., Nie, W., Petäjä, T., Kerminen, V. M., & Kulmala, M. (2013). Ozone and fine particle in the western Yangtze River Delta: an overview of 1 yr data at the SORPES station. *Atmospheric Chemistry and Physics*, 13(11), 5813–5830. 10.5194/acp-13-5813-2013

Ding, A. J., Wang, T., & Fu, C. B. (2013). Transport characteristics and origins of carbon monoxide and ozone in Hong Kong, South China. *Journal of Geophysical Research-Atmospheres*, 118(16), 9475–9488. 10.1002/jgrd.50714

Ding, A. J., Wang, T., Zhao, M., Wang, T. J., & Li, Z. K. (2004). Simulation of sea-land breezes and a discussion of their implications on the transport of air pollution during a multi-day ozone episode in the Pearl River Delta of China. *Atmospheric Environment*, 38(39), 6737–6750. 10.1016/j.atmosenv.2004.09.017

Ding, X., Wang, X.-M., Gao, B., Fu, X.-X., He, Q.-F., Zhao, X.-Y., Yu, J.-Z., & Zheng, M. (2012). Tracer-based estimation of secondary organic carbon in the Pearl River Delta, south China. *Journal of Geophysical Research: Atmospheres*, 117, D05313. 10.1029/2011jd016596

Ding, X., Wang, X., Xie, Z., Zhang, Z., & Sun, L. (2013). Impacts of Siberian biomass burning on organic aerosols over the North Pacific Ocean and the Arctic: Primary and secondary organic tracers. *Environ Sci Technol*, 47(7), 3149–3157. 10.1021/es3037093

Draxler, R. R., & Rolph, G. D. (2003). HYSPLIT (HYbrid Single-Particle Lagrangian Integrated Trajectory) Model access via NOAA ARL READY Website (http://www.arl.noaa.gov/ready/hysplit4.html).

Engelbrecht, J. P., Menéndez, I., & Derbyshire, E. (2014). Sources of PM2.5 impacting on Gran Canaria, Spain. *CATENA*, 117, 119–132. 10.1016/j.catena.2013.06.017

Fu, X., Guo, H., Wang, X., Ding, X., He, Q., Liu, T., & Zhang, Z. (2015). PM2.5 acidity at a background site in the Pearl River Delta region in fall-winter of 2007–2012. *Journal of Hazardous Materials*, 286, 484–492. 10.1016/j.jhazmat.2015.01.022

Fung, J. C. H., Lau, A. K. H., Lam, J. S. L., & Yuan, Z. (2005). Observational and modeling analysis of a severe air pollution episode in western Hong Kong. *Journal of Geophysical Research: Atmospheres*, 110(D9). 10.1029/2004JD005105

Goldstein, A. H., Koven, C. D., Heald, C. L., & Fung, I. Y. (2009). Biogenic carbon and anthropogenic pollutants combine to form a cooling haze over the southeastern United States. *Proceedings of the National Academy of Sciences*, 106(22), 8835–8840. 10.1073/pnas.0904128106

Gordon, H., Sengupta, K., Rap, A., Duplissy, J., Frege, C., Williamson, C., Heinritzi, M., Simon, M., Yan, C., Almeida, J., Trostl, J., Nieminen, T., Ortega, I. K., Wagner, R., Dunne, E. M., Adamov, A., Amorim, A., Bernhammer, A. K., Bianchi, F., … Carslaw, K. S. (2016). Reduced anthropogenic aerosol radiative forcing caused by biogenic new particle formation. *Proceedings of the National Academy of Sciences of the United States of America*, 113(43), 12053–12058. 10.1073/pnas.1602360113

Grell, G. A., Peckham, S. E., Schmitz, R., McKeen, S. A., Frost, G., Skamarock, W. C., & Eder, B. (2005). Fully coupled "online" chemistry within the WRF model. *Atmospheric Environment*, 39(37), 6957–6975. 10.1016/j.atmosenv.2005.04.027

Guenther, A., Karl, T., Harley, P., Wiedinmyer, C., Palmer, P. I., & Geron, C. (2006). Estimates of global terrestrial isoprene emissions using MEGAN (Model of Emissions of Gases and Aerosols from Nature). *Atmos. Chem. Phys.*, 6(11), 3181–3210. 10.5194/acp-6-3181-2006

Guo, H., Cheng, H. R., Ling, Z. H., Louie, P. K., & Ayoko, G. A. (2011). Which emission sources are responsible for the volatile organic compounds in the atmosphere of Pearl River Delta? *J Hazard Mater*, 188(1-3), 116–124. 10.1016/j.jhazmat.2011.01.081

Guo, H., Ding, A., Morawska, L., He, C., Ayoko, G., Li, Y.-s., & Hung, W.-t. (2008). Size distribution and new particle formation in subtropical eastern Australia. *Environmental Chemistry*, 5(6), 382–390. 10.1071/EN08058

Guo, H., Ding, A. J., So, K. L., Ayoko, G., Li, Y. S., & Hung, W. T. (2009). Receptor modeling of source apportionment of Hong Kong aerosols and the implication of urban and regional contribution. *Atmospheric Environment*, 43(6), 1159–1169. 10.1016/j.atmosenv.2008.04.046

Guo, H., Jiang, F., Cheng, H. R., Simpson, I. J., Wang, X. M., Ding, A. J., Wang, T. J., Saunders, S. M., Wang, T., Lam, S. H. M., Blake, D. R., Zhang, Y. L., & Xie, M. (2009). Concurrent observations of air pollutants at two sites in the Pearl River Delta and the implication of regional transport. *Atmos. Chem. Phys.*, 9(19), 7343–7360. 10.5194/acp-9-7343-2009

Guo, H., Lee, S. C., Ho, K. F., Wang, X. M., & Zou, S. C. (2003). Particle-associated polycyclic aromatic hydrocarbons in urban air of Hong Kong. *Atmospheric Environment*, 37(38), 5307–5317. 10.1016/j.atmosenv.2003.09.011

Guo, H., Ling, Z. H., Cheng, H. R., Simpson, I. J., Lyu, X. P., Wang, X. M., Shao, M., Lu, H. X., Ayoko, G., Zhang, Y. L., Saunders, S. M., Lam, S. H. M., Wang, J. L., & Blake, D. R. (2017). Tropospheric volatile organic compounds in China. *Science of The Total Environment*, 574, 1021–1043. 10.1016/j.scitotenv.2016.09.116

Guo, H., Ling, Z. H., Cheung, K., Jiang, F., Wang, D. W., Simpson, I. J., Barletta, B., Meinardi, S., Wang, T. J., Wang, X. M., Saunders, S. M., & Blake, D. R. (2013). Characterization of photochemical pollution at different elevations in mountainous areas in Hong Kong. *Atmospheric Chemistry and Physics*, 13(8), 3881–3898. 10.5194/acp-13-3881-2013

Guo, H., Ling, Z. H., Cheung, K., Wang, D. W., Simpson, I. J., & Blake, D. R. (2013). Acetone in the atmosphere of Hong Kong: Abundance, sources and photochemical precursors. *Atmospheric Environment*, 65, 80–88. 10.1016/j.atmosenv.2012.10.027

Guo, H., So, K. L., Simpson, I. J., Barletta, B., Meinardi, S., & Blake, D. R. (2007). C1–C8 volatile organic compounds in the atmosphere of Hong Kong: Overview of atmospheric processing and source apportionment. *Atmospheric Environment*, 41(7), 1456–1472. 10.1016/j.atmosenv.2006.10.011

Guo, H., Wang, D. W., Cheung, K., Ling, Z. H., Chan, C. K., & Yao, X. H. (2012). Observation of aerosol size distribution and new particle formation at a mountain site in subtropical Hong Kong. *Atmospheric Chemistry and Physics*, 12(20), 9923–9939. 10.5194/acp-12-9923-2012

Guo, H., Wang, T., Blake, D. R., Simpson, I. J., Kwok, Y. H., & Li, Y. S. (2006). Regional and local contributions to ambient non-methane volatile organic compounds at a polluted rural/coastal site in Pearl River Delta, China. *Atmospheric Environment*, 40(13), 2345–2359. 10.1016/j.atmosenv.2005.12.011

Guo, S., Hu, M., Zamora, M. L., Peng, J., Shang, D., Zheng, J., Du, Z., Wu, Z., Shao, M., Zeng, L., Molina, M. J., & Zhang, R. (2014). Elucidating severe urban haze formation in China. *Proceedings of the National Academy of Sciences*, 111(49), 17373–17378. 10.1073/pnas.1419604111

He, J., Gong, S., Yu, Y., Yu, L., Wu, L., Mao, H., Song, C., Zhao, S., Liu, H., Li, X., & Li, R. (2017). Air pollution characteristics and their relation to meteorological conditions during 2014–2015 in major Chinese cities. *Environ Pollut*, 223, 484–496. 10.1016/j.envpol.2017.01.050

HKEPD. (2015). Air Quality in Hong Kong 2014. http://www.aqhi.gov.hk/api_history/english/report/files/AQR2014e_Update0616.pdf.

Ho, K. F., Lee, S. C., Ho, W. K., Blake, D. R., Cheng, Y., Li, Y. S., Ho, S. S. H., Fung, K., Louie, P. K. K., & Park, D. (2009). Vehicular emission of volatile organic compounds (VOCs) from a tunnel study in Hong Kong. *Atmos. Chem. Phys.*, 9(19), 7491–7504. 10.5194/acp-9-7491-2009

Holloway, A. M., & Wayne, R. P. (2010). Atmospheric chemistry. *Royal Society of Chemistry.*

Hoyle, C. R., Boy, M., Donahue, N. M., Fry, J. L., Glasius, M., Guenther, A., Hallar, A. G., Huff Hartz, K., Petters, M. D., Petäjä, T., Rosenoern, T., & Sullivan, A. P. (2011). A review of the anthropogenic influence on biogenic secondary organic aerosol. *Atmospheric Chemistry and Physics*, 11(1), 321–343. 10.5194/acp-11-321-2011

Huang, R.-J., Zhang, Y., Bozzetti, C., Ho, K.-F., Cao, J.-J., Han, Y., Daellenbach, K. R., Slowik, J. G., Platt, S. M., Canonaco, F., Zotter, P., Wolf, R., Pieber, S. M., Bruns, E. A., Crippa, M., Ciarelli, G., Piazzalunga, A., Schwikowski, M., Abbaszade, G., … Prévôt, A. S. H. (2014). High secondary aerosol contribution to particulate pollution during haze events in China. *Nature*, 514(7521), 218–222. 10.1038/nature13774

Hunter, J. F., Day, D. A., Palm, B. B., Yatavelli, R. L. N., Chan, A. W. H., Kaser, L., Cappellin, L., Hayes, P. L., Cross, E. S., Carrasquillo, A. J., Campuzano-Jost, P., Stark, H., Zhao, Y., Hohaus, T., Smith, James N., Hansel, A., Karl, T., Goldstein, A. H., Guenther, A., … Kroll, J. H. (2017). Comprehensive characterization of atmospheric organic carbon at a forested site. *Nature Geoscience*, 10(10), 748–753. 10.1038/ngeo3018

Jacob, D. J. (1999). *Introduction to atmospheric chemistry*. Princeton University Press.

Jaeckels, J. M., Bae, M.-S., & Schauer, J. J. (2007). Positive Matrix Factorization (PMF) Analysis of Molecular Marker Measurements to Quantify the Sources of Organic Aerosols. *Environ Sci Technol*, 41(16), 5763–5769. 10.1021/es062536b

Jenkin, M. E., Saunders, S. M., Wagner, V., & Pilling, M. J. (2003). Protocol for the development of the Master Chemical Mechanism, MCM v3 (Part B): tropospheric degradation of aromatic volatile organic compounds. *Atmospheric Chemistry and Physics*, 3, 181–193. 10.5194/acp-3-161-2003

Jenkin, M. E., Young, J. C., & Rickard, A. R. (2015). The MCM v3.3.1 degradation scheme for isoprene. *Atmos. Chem. Phys.*, 15(20), 11433–11459. 10.5194/acp-15-11433-2015

Jia, C., Mao, X., Huang, T., Liang, X., Wang, Y., Shen, Y., Jiang, W., Wang, H., Bai, Z., Ma, M., Yu, Z., Ma, J., & Gao, H. (2016). Non-methane hydrocarbons (NMHCs) and their contribution to ozone formation potential in a petrochemical industrialized city, Northwest China. *Atmospheric Research*, 169, 225–236. 10.1016/j.atmosres.2015.10.006

Jiang, F., Guo, H., Wang, T. J., Cheng, H. R., Wang, X. M., Simpson, I. J., Ding, A. J., Saunders, S. M., Lam, S. H. M., & Blake, D. R. (2010). An ozone episode in the Pearl River Delta: Field observation and model simulation. *Journal of Geophysical Research*, 115(D22). 10.1029/2009jd013583

Jimenez, J. L., Canagaratna, M. R., Donahue, N. M., Prevot, A. S. H., Zhang, Q., Kroll, J. H., DeCarlo, P. F., Allan, J. D., Coe, H., Ng, N. L., Aiken, A. C., Docherty, K. S., Ulbrich, I. M., Grieshop, A. P., Robinson, A. L., Duplissy, J., Smith, J. D., Wilson, K. R., Lanz, V. A., … Worsnop, D. R. (2009). evolution of organic aerosols in the atmosphere. *Science*, 326(5959), 1525–1529. 10.1126/science.1180353

Jokinen, T., Lehtipalo, K., Thakur, R. C., Ylivinkka, I., Neitola, K., Sarnela, N., Laitinen, T., Kulmala, M., Petaja, T., & Sipila, M. (2022). Measurement report: Long-term measurements of aerosol precursor concentrations in the Finnish subarctic boreal forest. *Atmospheric Chemistry and Physics*, 22(4), 2237–2254. 10.5194/acp-22-2237-2022

Jones, M. (2008). Kernelization, Max-Cut Above Tight Bounds. In M.-Y. Kao (Ed.), *Encyclopedia of Algorithms* (pp. 1–5). Springer US. 10.1007/978-3-642-27848-8_531-1

Jorga, S. D., Florou, K., Patoulias, D., & Pandis, S. N. (2023). New particle formation and growth during summer in an urban environment: a dual chamber study. *Atmospheric Chemistry and Physics*, 23(1), 85–97. 10.5194/acp-23-85-2023

Jorquera, H., & Rappenglück, B. (2004). Receptor modeling of ambient VOC at Santiago, Chile. *Atmospheric Environment*, 38(25), 4243–4263. 10.1016/j.atmosenv.2004.04.030

Kerminen, V. M., Chen, X. M., Vakkari, V., Petaja, T., Kulmala, M., & Bianchi, F. (2018). Atmospheric new particle formation and growth: Review of field observations. *Environmental Research Letters*, 13(10), Article 103003. 10.1088/1748-9326/aadf3c

Kerminen, V. M., Petäjä, T., Manninen, H. E., Paasonen, P., Nieminen, T., Sipilä, M., Junninen, H., Ehn, M., Gagné, S., Laakso, L., Riipinen, I., Vehkamäki, H., Kurten, T., Ortega, I. K., Dal Maso, M., Brus, D., Hyvärinen, A., Lihavainen, H., Leppä, J., … Kulmala, M. (2010). Atmospheric nucleation: highlights of the EUCAARI project and future directions. *Atmospheric Chemistry and Physics*, 10(22), 10829–10848. 10.5194/acp-10-10829-2010

Kleindienst, T. E., Jaoui, M., Lewandowski, M., Offenberg, J. H., Lewis, C. W., Bhave, P. V., & Edney, E. O. (2007). Estimates of the contributions of biogenic and anthropogenic hydrocarbons to secondary organic aerosol at a southeastern US location. *Atmospheric Environment*, 41(37), 8288–8300. 10.1016/j.atmosenv.2007.06.045

Knighton, W. B., Herndon, S. C., Wood, E. C., Fortner, E. C., Onasch, T. B., Wormhoudt, J., Kolb, C. E., Lee, B. H., Zavala, M., Molina, L., & Jones, M. (2012). Detecting fugitive emissions of 1,3-butadiene and styrene from a petrochemical facility: An application of a mobile laboratory and a modified proton transfer reaction mass spectrometer. *Industrial & Engineering Chemistry Research*, 51(39), 12706–12711. 10.1021/ie202794j

Kreisberg, N. M., Hering, S. V., Williams, B. J., Worton, D. R., & Goldstein, A. H. (2009). Quantification of hourly speciated organic compounds in atmospheric aerosols, measured by an in-situ thermal desorption aerosol gas chromatograph (TAG). *Aerosol Science and Technology*, 43(1), 38–52. 10.1080/02786820802459583

Kulmala, M., Kontkanen, J., Junninen, H., Lehtipalo, K., Manninen, H. E., Nieminen, T., Petäjä, T., Sipilä, M., Schobesberger, S., Rantala, P., Franchin, A., Jokinen, T., Järvinen, E., Äijälä, M., Kangasluoma, J., Hakala, J., Aalto, P. P., Paasonen, P., Mikkilä, J., … Worsnop, D. R. (2013). Direct observations of atmospheric aerosol nucleation. *Science*, 339(6122), 943–946. 10.1126/science.1227385

Kulmala, M., Lehtinen, K. E. J., & Laaksonen, A. (2006). Cluster activation theory as an explanation of the linear dependence between formation rate of 3nm particles and sulphuric acid concentration. *Atmospheric Chemistry and Physics*, 6(3), 787–793. 10.5194/acp-6-787-2006

Kulmala, M., Maso, M. D., Mäkela, J. M., Pirjola, L., Väkevä, M., Aalto, P., Miikkulainen, P., Hämeri, K., & O'Dowd, C. D. (2001). On the formation, growth and composition of nucleation mode particles. *Tellus B: Chemical and Physical Meteorology*, 53(4), 479. 10.3402/tellusb.v53i4.16622

Kulmala, M., Petäjä, T., Mönkkönen, P., Koponen, I. K., Dal Maso, M., Aalto, P. P., Lehtinen, K. E. J., & Kerminen, V. M. (2005). On the growth of nucleation mode particles: source rates of condensable vapor in polluted and clean environments. *Atmospheric Chemistry and Physics*, 5(2), 409–416. 10.5194/acp-5-409-2005

Kulmala, M., Petäjä, T., Nieminen, T., Sipilä, M., Manninen, H. E., Lehtipalo, K., Dal Maso, M., Aalto, P. P., Junninen, H., Paasonen, P., Riipinen, I., Lehtinen, K. E. J., Laaksonen, A., & Kerminen, V.-M. (2012). Measurement of the nucleation of atmospheric aerosol particles. *Nature Protocols*, 7(9), 1651–1667. 10.1038/nprot.2012.091

Kulmala, M., Pirjola, L., & Mäkelä, J. M. (2000). Stable sulphate clusters as a source of new atmospheric particles. *Nature*, 404(6773), 66–69. 10.1038/35003550

Kulmala, M., Vehkamäki, H., Petäjä, T., Dal Maso, M., Lauri, A., Kerminen, V. M., Birmili, W., & McMurry, P. H. (2004). Formation and growth rates of ultrafine atmospheric particles: A review of observations. *Journal of Aerosol Science*, 35(2), 143–176. 10.1016/j.jaerosci.2003.10.003

Kumar, R., Naja, M., Pfister, G. G., Barth, M. C., Wiedinmyer, C., & Brasseur, G. P. (2012). Simulations over South Asia using the Weather Research and Forecasting model with Chemistry (WRF-Chem): Chemistry evaluation and initial results. *Geoscientific Model Development*, 5(3), 619–648. 10.5194/gmd-5-619-2012

Kwok, R. H. F., Baker, K. R., Napelenok, S. L., & Tonnesen, G. S. (2015). Photochemical grid model implementation and application of VOC, NOx, and O3 source apportionment. *Geoscientific Model Development*, 8(1), 99–114. 10.5194/gmd-8-99-2015

Laakso, L., Petäjä, T., Lehtinen, K. E. J., Kulmala, M., Paatero, J., Hõrrak, U., Tammet, H., & Joutsensaari, J. (2004). Ion production rate in a boreal forest based on ion, particle and radiation measurements. *Atmospheric Chemistry and Physics*, 4(7), 1933–1943. 10.5194/acp-4-1933-2004

Lam, S. H. M., Saunders, S. M., Guo, H., Ling, Z. H., Jiang, F., Wang, X. M., & Wang, T. J. (2013). Modelling VOC source impacts on high ozone episode days observed at a mountain summit in Hong Kong under the influence of mountain-valley breezes. *Atmospheric Environment*, 81, 166–176. 10.1016/j.atmosenv.2013.08.060

Lee, B. P., Li, Y. J., Yu, J. Z., Louie, P. K. K., & Chan, C. K. (2013). Physical and chemical characterization of ambient aerosol by HR-ToF-AMS at a suburban site in Hong Kong during springtime 2011. *Journal of Geophysical Research: Atmospheres*, 118(15), 8625–8639. 10.1002/jgrd.50658

Lee, Y. C., Calori, G., Hills, P., & Carmichael, G. R. (2002). Ozone episodes in urban Hong Kong 1994–1999. *Atmospheric Environment*, 36(12), 1957–1968. 10.1016/S1352-2310(02)00150-4

Lee, Y. C., Shindell, D. T., Faluvegi, G., Wenig, M., Lam, Y. F., Ning, Z., Hao, S., & Lai, C. S. (2014). Increase of ozone concentrations, its temperature sensitivity and the precursor factor in South China. *Tellus B*, 66(23455). 10.3402/tellusb.v66.23455

Lehtipalo, K., Yan, C., Dada, L., Bianchi, F., Xiao, M., Wagner, R., Stolzenburg, D., Ahonen, L. R., Amorim, A., Baccarini, A., Bauer, P. S., Baumgartner, B., Bergen, A., Bernhammer, A.-K., Breitenlechner, M., Brilke, S., Buchholz, A., Mazon, S. B., Chen, D., … Worsnop, D. R. (2018). Multicomponent new particle formation from sulfuric acid, ammonia, and biogenic vapors. *Science Advances*, 4(12), eaau5363. doi:10.1126/sciadv.aau5363

Leino, K., Lampilahti, J., Poutanen, P., Vaananen, R., Manninen, A., Mazon, S. B., Dada, L., Franck, A., Wimmer, D., Aalto, P. P., Ahonen, L. R., Enroth, J., Kangasluoma, J., Keronen, P., Korhonen, F., Laakso, H., Matilainen, T., Siivola, E., Manninen, H. E., … Kulmala, M. (2019). Vertical profiles of sub-3nm particles over the boreal forest. *Atmospheric Chemistry and Physics*, 19(6), 4127–4138. 10.5194/acp-19-4127-2019

Li, K., Jacob, D. J., Shen, L., Lu, X., De Smedt, I., & Liao, H. (2020). Increases in surface ozone pollution in China from 2013 to 2019: anthropogenic and meteorological influences. *Atmospheric Chemistry and Physics*, 20(19), 11423–11433. 10.5194/acp-20-11423-2020

Li, M., Zhang, Q., Kurokawa, J.-i., Woo, J.-H., He, K., Lu, Z., Ohara, T., Song, Y., Streets, D. G., Carmichael, G. R., Cheng, Y., Hong, C., Huo, H., Jiang, X., Kang, S., Liu, F., Su, H., & Zheng, B. (2017). MIX: a mosaic Asian anthropogenic emission inventory under the international collaboration framework of the MICS-Asia and HTAP. *Atmospheric Chemistry and Physics*, 17(2), 935–963. 10.5194/acp-17-935-2017

Li, Y., Lau, A. K. H., Fung, J. C. H., Zheng, J. Y., Zhong, L. J., & Louie, P. K. K. (2012). Ozone source apportionment (OSAT) to differentiate local regional and super-regional source contributions in the Pearl River Delta region, China. *Journal of Geophysical Research-Atmospheres*, 117. 10.1029/2011JD017340

Li, Y., Pöschl, U., & Shiraiwa, M. (2016). Molecular corridors and parameterizations of volatility in the chemical evolution of organic aerosols. *Atmospheric Chemistry and Physics*, 16(5), 3327–3344. 10.5194/acp-16-3327-2016

Li, Y. J., Lee, B. P., Su, L., Fung, J. C. H., & Chan, C. K. (2015). Seasonal characteristics of fine particulate matter (PM) based on high-resolution time-of-flight aerosol mass spectrometric (HR-ToF-AMS) measurements at the HKUST Supersite in Hong Kong. *Atmospheric Chemistry and Physics*, 15(1), 37–53. 10.5194/acp-15-37-2015

Li, Z., Lau, W. K. M., Ramanathan, V., Wu, G., Ding, Y., Manoj, M. G., Liu, J., Qian, Y., Li, J., Zhou, T., Fan, J., Rosenfeld, D., Ming, Y., Wang, Y., Huang, J., Wang, B., Xu, X., Lee, S. S., Cribb, M., … Brasseur, G. P. (2016). Aerosol and monsoon climate interactions over Asia. *Reviews of Geophysics*, 54(4), 866–929. 10.1002/2015rg000500

Liao, H., Chang, W., & Yang, Y. (2015). Climatic effects of air pollutants over China: A review. *Advances in Atmospheric Sciences*, 32(1), 115–139. 10.1007/s00376-014-0013-x

Ling, Z. H., & Guo, H. (2014). Contribution of VOC sources to photochemical ozone formation and its control policy implication in Hong Kong. *Environmental Science & Policy*, 38, 180–191. 10.1016/j.envsci.2013.12.004

Ling, Z. H., Guo, H., Cheng, H. R., & Yu, Y. F. (2011). Sources of ambient volatile organic compounds and their contributions to photochemical ozone formation at a site in the Pearl River Delta, southern China. *Environ Pollut*, 159(10), 2310–2319. 10.1016/j.envpol.2011.05.001

Ling, Z. H., Guo, H., Lam, S. H. M., Saunders, S. M., & Wang, T. (2014). Atmospheric photochemical reactivity and ozone production at two sites in Hong Kong: Application of a Master Chemical Mechanism-photochemical box model. *Journal of Geophysical Research-Atmospheres*, 119(17), 10567–10582. 10.1002/2014jd021794

Liu, L., Yu, F., Du, L., Yang, Z., Francisco, J. S., & Zhang, X. (2021). Rapid sulfuric acid–dimethylamine nucleation enhanced by nitric acid in polluted regions. *Proceedings of the National Academy of Sciences*, 118(35), e2108384118. 10.1073/pnas.2108384118

Liu, X., Guo, H., Zeng, L., Lyu, X., Wang, Y., Zeren, Y., Yang, J., Zhang, L., Zhao, S., Li, J., & Zhang, G. (2021). Photochemical ozone pollution in five Chinese megacities in summer 2018. *Science of The Total Environment*, 801, 149603. 10.1016/j.scitotenv.2021.149603

Liu, X., Lyu, X., Wang, Y., Jiang, F., & Guo, H. (2019). Intercomparison of O3 formation and radical chemistry in the past decade at a suburban site in Hong Kong. *Atmos. Chem. Phys.*, 19(7), 5127–5145. 10.5194/acp-19-5127-2019

Liu, Y., Shao, M., Fu, L., Lu, S., Zeng, L., & Tang, D. (2008). Source profiles of volatile organic compounds (VOCs) measured in China: Part I. *Atmospheric Environment*, 42(25), 6247–6260. 10.1016/j.atmosenv.2008.01.070

Liu, Y., Shao, M., Lu, S., Chang, C.-C., Wang, J.-L., & Fu, L. (2008). Source apportionment of ambient volatile organic compounds in the Pearl River Delta, China: Part II. *Atmospheric Environment*, 42(25), 6261–6274. 10.1016/j.atmosenv.2008.02.027

Liu, Y., & Wang, T. (2020). Worsening urban ozone pollution in China from 2013 to 2017 – Part 2: The effects of emission changes and implications for multi-pollutant control. *Atmos. Chem. Phys.*, 20(11), 6323–6337. 10.5194/acp-20-6323-2020

Lo, J. C. F., Lau, A. K. H., Fung, J. C. H., & Chen, F. (2006). Investigation of enhanced cross-city transport and trapping of air pollutants by coastal and urban land-sea breeze circulations. *Journal of Geophysical Research: Atmospheres*, 111(D14). 10.1029/2005jd006837

Lu, H., Huang, F., & Guo, H. (2021). Differential removal of nanoparticles on the surface of a thin film substrate. *ACS Omega*, 6(25), 16280–16287. 10.1021/acsomega.1c00334

Lu, H., Wang, G., & Guo, H. (2022). Ambient acidic ultrafine particles in different land-use areas in two representative Chinese cities. *Science of The Total Environment*, 830, 154774. 10.1016/j.scitotenv.2022.154774

Lu, X., Chow, K.-C., Yao, T., Lau, A. K. H., & Fung, J. C. H. (2009). Effects of urbanization on the land sea breeze circulation over the Pearl River Delta region in winter. *International Journal of Climatology*, n/a-n/a. 10.1002/joc.1947

Lu, X., Zhang, L., Wang, X., Gao, M., Li, K., Zhang, Y., Yue, X., & Zhang, Y. (2020). Rapid Increases in Warm-Season Surface Ozone and Resulting Health Impact in China Since 2013. *Environmental Science & Technology Letters*. 10.1021/acs.estlett.0c00171

Lyu, X., Chen, N., Guo, H., Zeng, L., Zhang, W., Shen, F., Quan, J., & Wang, N. (2016). Chemical characteristics and causes of airborne particulate pollution in warm seasons in Wuhan, central China. *Atmos. Chem. Phys.*, 16(16), 10671–10687. 10.5194/acp-16-10671-2016

Lyu, X., Guo, H., Yao, D., Lu, H., Huo, Y., Xu, W., Kreisberg, N., Goldstein, A. H., Jayne, J., Worsnop, D., Tan, Y., Lee, S.-C., & Wang, T. (2020). In situ measurements of molecular markers facilitate understanding of dynamic sources of atmospheric organic aerosols. *Environ Sci Technol*, 54(18), 11058–11069. 10.1021/acs.est.0c02277

Lyu, X., Wang, N., Guo, H., Xue, L., Jiang, F., Zeren, Y., Cheng, H., Cai, Z., Han, L., & Zhou, Y. (2019). Causes of a continuous summertime O3 pollution event in Jinan, a central city in the North China Plain. *Atmos. Chem. Phys.*, 19(5), 3025–3042. 10.5194/acp-19-3025-2019

Lyu, X. P., Chen, N., Guo, H., Zhang, W. H., Wang, N., Wang, Y., & Liu, M. (2015). Ambient volatile organic compounds and their effect on ozone production in Wuhan, central China. *Sci Total Environ*, 541, 200–209. 10.1016/j.scitotenv.2015.09.093

Lyu, X. P., Guo, H., Cheng, H. R., & Wang, D. W. (2018). New particle formation and growth at a suburban site and a background site in Hong Kong. *Chemosphere*, 193, 664–674. 10.1016/j.chemosphere.2017.11.060

Lyu, X. P., Guo, H., Cheng, H. R., Wang, X. M., Ding, X., Lu, H. X., Yao, D. W., & Xu, C. (2017). Observation of SOA tracers at a mountainous site in Hong Kong: Chemical characteristics, origins and implication on particle growth. *Science of The Total Environment*, 605, 180–189. 10.1016/j.scitotenv.2017.06.161

Lyu, X. P., Guo, H., Simpson, I. J., Meinardi, S., Louie, P. K. K., Ling, Z. H., Wang, Y., Liu, M., Luk, C. W. Y., Wang, N., & Blake, D. R. (2016). Effectiveness of replacing catalytic converters in LPG-fueled vehicles in Hong Kong. *Atmos. Chem. Phys.*, 16(10), 6609–6626. 10.5194/acp-16-6609-2016

Lyu, X. P., Liu, M., Guo, H., Ling, Z. H., Wang, Y., Louie, P. K. K., & Luk, C. W. Y. (2016). Spatiotemporal variation of ozone precursors and ozone formation in Hong Kong: Grid field measurement and modelling study. *Science of The Total Environment*, 569, 1341–1349. 10.1016/j.scitotenv.2016.06.214

Lyu, X. P., Zeng, L. W., Guo, H., Simpson, I. J., Ling, Z. H., Wang, Y., Murray, F., Louie, P. K. K., Saunders, S. M., Lam, S. H. M., & Blake, D. R. (2017). Evaluation of the effectiveness of air pollution control measures in Hong Kong. *Environ Pollut*, 220, Part A, 87–94. http://dx.doi.org/10.1016/j.envpol.2016.09.025

Ma, Z., Xu, J., Quan, W., Zhang, Z., Lin, W., & Xu, X. (2016). Significant increase of surface ozone at a rural site, north of eastern China. *Atmospheric Chemistry and Physics*, 16(6), 3969–3977. 10.5194/acp-16-3969-2016

Mikkonen, S., Korhonen, H., Romakkaniemi, S., Smith, J. N., Joutsensaari, J., Lehtinen, K. E. J., Hamed, A., Breider, T. J., Birmili, W., Spindler, G., Plass-Duelmer, C., Facchini, M. C., & Laaksonen, A. (2011). Meteorological and trace gas factors affecting the number concentration of atmospheric Aitken (Dp=50 nm) particles in the continental boundary layer: parameterization using a multivariate mixed effects model. *Geoscientific Model Development*, 4(1), 1–13. 10.5194/gmd-4-1-2011

Modini, R. L., Ristovski, Z. D., Johnson, G. R., He, C., Surawski, N., Morawska, L., Suni, T., & Kulmala, M. (2009). New particle formation and growth at a remote, sub-tropical coastal location. *Atmospheric Chemistry and Physics*, 9(19), 7607–7621. 10.5194/acp-9-7607-2009

Monks, P. S., Archibald, A. T., Colette, A., Cooper, O., Coyle, M., Derwent, R., Fowler, D., Granier, C., Law, K. S., Mills, G. E., Stevenson, D. S., Tarasova, O., Thouret, V., von Schneidemesser, E., Sommariva, R., Wild, O., & Williams, M. L. (2015). Tropospheric ozone and its precursors from the urban to the global scale from air quality to short-lived climate forcer. *Atmos. Chem. Phys.*, 15(15), 8889–8973. 10.5194/acp-15-8889-2015

NARSTO. (2000). An assessment of tropospheric ozone pollution: A North American perspective.

O'Dowd, C. D., Hämeri, K., Mäkelä, J., Väkeva, M., Aalto, P., de Leeuw, G., Kunz, G. J., Becker, E., Hansson, H.-C., Allen, A. G., Harrison, R. M., Berresheim, H., Kleefeld, C., Geever, M., Jennings, S. G., & Kulmala, M. (2002). Coastal new particle formation: Environmental conditions and aerosol physicochemical characteristics during nucleation bursts. *Journal of Geophysical Research: Atmospheres*, 107(D19), PAR 12-11-PAR 12-17. 10.1029/2000JD000206

Ou, J. M., Guo, H., Zheng, J. Y., Cheung, K., Louie, P. K. K., Ling, Z. H., & Wang, D. W. (2015). Concentrations and sources of non-methane hydrocarbons (NMHCs) from 2005 to 2013 in Hong Kong: A multi-year real-time data analysis. *Atmospheric Environment*, 103, 196–206. 10.1016/j.atmosenv.2014.12.048

Pope, C. A., & Dockery, D. W. (2006). Health effects of fine particulate air pollution: Lines that connect. *Journal of the Air & Waste Management Association*, 56(6), 709–742. 10.1080/10473289.2006.10464485

Qian, S., Sakurai, H., & McMurry, P. H. (2007). Characteristics of regional nucleation events in urban East St. Louis. *Atmospheric Environment*, 41(19), 4119–4127. 10.1016/j.atmosenv.2007.01.011

Quelever, L. L. J., Dada, L., Asmi, E., Lampilahti, J., Chan, T., Ferrara, J. E., Copes, G. E., Perez-Fogwill, G., Barreira, L., Aurela, M., Worsnop, D. R., Jokinen, T., & Sipila, M. (2022). Investigation of new particle formation mechanisms and aerosol processes at Marambio Station, Antarctic Peninsula. *Atmospheric Chemistry and Physics*, 22(12), 8417–8437. 10.5194/acp-22-8417-2022

Rao, S. T., & Zurbenko, I. G. (1994). Detecting and tracking changes in ozone air quality. *Air & Waste*, 44(9), 1089–1092. 10.1080/10473289.1994.10467303

Rao, S. T., Zurbenko, I. G., Neagu, R., Porter, P. S., Ku, J. Y., & Henry, R. F. (1997). Space and time scales in ambient ozone data. *Bulletin of the American Meteorological Society*, 78(10), 2153–2166. 10.1175/1520-0477(1997)078<2153:SATSIA>2.0.CO;2

Rosenfeld, D., Lohmann, U., Raga, G. B., O'Dowd, C. D., Kulmala, M., Fuzzi, S., Reissell, A., & Andreae, M. O. (2008). Flood or drought: How do aerosols affect precipitation? *Science*, 321(5894), 1309–1313. 10.1126/science.1160606

Sahoo, D., Petersen, B., & Miles, P. C. (2011). Measurement of equivalence ratio in a light-duty low temperature combustion diesel engine by planar laser induced fluorescence of a fuel tracer. *SAE International Journal of Engines*, 4(2), 2312–2325. 10.4271/2011-24-0064

Sánchez-Ccoyllo, O. R., Ynoue, R. Y., Martins, L. D., & de Fátima Andrade, M. (2006). Impacts of ozone precursor limitation and meteorological variables on ozone concentration in São Paulo, Brazil. *Atmospheric Environment*, 40, 552–562. 10.1016/j.atmosenv.2006.04.069

Saunders, S. M., Jenkin, M. E., Derwent, R. G., & Pilling, M. J. (2003). Protocol for the development of the Master Chemical Mechanism, MCM v3 (Part A): Tropospheric degradation of non-aromatic volatile organic compounds. *Atmospheric Chemistry and Physics*, 3, 161–180. 10.5194/acp-3-181-2003

Shao, M., Zhang, Y., Zeng, L., Tang, X., Zhang, J., Zhong, L., & Wang, B. (2009). Ground-level ozone in the Pearl River Delta and the roles of VOC and NOx in its production. *J Environ Manage*, 90(1), 512–518. 10.1016/j.jenvman.2007.12.008

Shao, P., An, J., Xin, J., Wu, F., Wang, J., Ji, D., & Wang, Y. (2016). Source apportionment of VOCs and the contribution to photochemical ozone formation during summer in the typical industrial area in the Yangtze River Delta, China. *Atmospheric Research*, 176-177, 64–74. 10.1016/j.atmosres.2016.02.015

Sihto, S. L., Kulmala, M., Kerminen, V. M., Dal Maso, M., Petäjä, T., Riipinen, I., Korhonen, H., Arnold, F., Janson, R., Boy, M., Laaksonen, A., & Lehtinen, K. E. J. (2006). Atmospheric sulphuric acid and aerosol formation: Implications from atmospheric measurements for nucleation and early growth mechanisms. *Atmospheric Chemistry and Physics*, 6(12), 4079–4091. 10.5194/acp-6-4079-2006

Sillman, S. (1999). The relation between ozone, NOx and hydrocarbons in urban and polluted rural environments. *Atmospheric Environment*, 33(12), 1821–1845. 10.1016/S1352-2310(98)00345-8

Simpson, I. J., Blake, N. J., Barletta, B., Diskin, G. S., Fuelberg, H. E., Gorham, K., Huey, L. G., Meinardi, S., Rowland, F. S., Vay, S. A., Weinheimer, A. J., Yang, M., & Blake, D. R. (2010). Characterization of trace gases measured over Alberta oil sands mining operations: 76 speciated C2-C10 volatile organic compounds (VOCs), CO_2, CH_4, CO, NO, NO_2, NO_y, O3 and SO_2. *Atmospheric Chemistry and Physics*, 10(23), 11931–11954. 10.5194/acp-10-11931-2010

Sipilä, M., Berndt, T., Petäjä, T., Brus, D., Vanhanen, J., Stratmann, F., Patokoski, J., Mauldin, R. L., Hyvärinen, A.-P., Lihavainen, H., & Kulmala, M. (2010). The role of sulfuric acid in atmospheric nucleation. *Science*, 327(5970), 1243–1246. 10.1126/science.1180315

Song, M., Tan, Q., Feng, M., Qu, Y., Liu, X., An, J., & Zhang, Y. (2018). Source apportionment and secondary transformation of atmospheric nonmethane hydrocarbons in Chengdu, Southwest China. *Journal of Geophysical Research: Atmospheres*, 123(17), 9741–9763. 10.1029/2018JD028479

Stein, A. F., Ngan, F., Draxler, R. R., & Chai, T. (2015). Potential use of transport and dispersion model ensembles for forecasting applications. *Weather and Forecasting*, 30(3), 639–655. 10.1175/waf-d-14-00153.1

Sun, Y. L., Zhang, Q., Schwab, J. J., Demerjian, K. L., Chen, W. N., Bae, M. S., Hung, H. M., Hogrefe, O., Frank, B., Rattigan, O. V., & Lin, Y. C. (2011). Characterization of the sources and processes of organic and inorganic aerosols in New York city with a high-resolution time-of-flight aerosol mass apectrometer. *Atmospheric Chemistry and Physics*, 11(4), 1581–1602. 10.5194/acp-11-1581-2011

Tie, X., Brasseur, G., & Ying, Z. (2010). Impact of model resolution on chemical ozone formation in Mexico City: application of the WRF-Chem model. *Atmospheric Chemistry and Physics*, 10(18), 8983–8995. 10.5194/acp-10-8983-2010

Tröstl, J., Chuang, W. K., Gordon, H., Heinritzi, M., Yan, C., Molteni, U., Ahlm, L., Frege, C., Bianchi, F., Wagner, R., Simon, M., Lehtipalo, K., Williamson, C., Craven, J. S., Duplissy, J., Adamov, A., Almeida, J., Bernhammer, A.-K., Breitenlechner, M., … Baltensperger, U. (2016). The role of low-volatility organic compounds in initial particle growth in the atmosphere. *Nature*, 533(7604), 527–531. 10.1038/nature18271

Vedantham, R., Landis, M. S., Olson, D., & Pancras, J. P. (2014). Source identification of PM2.5 in Steubenville, Ohio using a hybrid method for highly time-resolved data. *Environ Sci Technol*, 48(3), 1718–1726. 10.1021/es402704n

Wang, D., Guo, H., & Chan, C. K. (2014). Diffusion Sampler for Measurement of Acidic Ultrafine Particles in the Atmosphere. *Aerosol Science and Technology*, 48(12), 1236–1246. 10.1080/02786826.2014.978937

Wang, D., Guo, H., Cheung, K., & Gan, F. (2014). Observation of nucleation mode particle burst and new particle formation events at an urban site in Hong Kong. *Atmospheric Environment*, 99, 196–205. 10.1016/j.atmosenv.2014.09.074

Wang, H., Lyu, X., Guo, H., Wang, Y., Zou, S., Ling, Z., Wang, X., Jiang, F., Zeren, Y., Pan, W., Huang, X., & Shen, J. (2018). Ozone pollution around a coastal region of South China Sea: interaction between marine and continental air. *Atmos. Chem. Phys.*, 18(6), 4277–4295. 10.5194/acp-18-4277-2018

Wang, J.-L., Wang, C.-H., Lai, C.-H., Chang, C.-C., Liu, Y., Zhang, Y., Liu, S., & Shao, M. (2008). Characterization of ozone precursors in the Pearl River Delta by time series observation of non-methane hydrocarbons. *Atmospheric Environment*, 42(25), 6233–6246. 10.1016/j.atmosenv.2008.01.050

Wang, M., Kong, W., Marten, R., He, X.-C., Chen, D., Pfeifer, J., Heitto, A., Kontkanen, J., Dada, L., Kürten, A., Yli-Juuti, T., Manninen, H. E., Amanatidis, S., Amorim, A., Baalbaki, R., Baccarini, A., Bell, D. M., Bertozzi, B., Bräkling, S., … Donahue, N. M. (2020). Rapid growth of new atmospheric particles by nitric acid and ammonia condensation. *Nature*, 581(7807), 184–189. 10.1038/s41586-020-2270-4

Wang, N., Guo, H., Jiang, F., Ling, Z. H., & Wang, T. (2015). Simulation of ozone formation at different elevations in mountainous area of Hong Kong using WRF-CMAQ model. *Sci Total Environ*, 505, 939–951. 10.1016/j.scitotenv.2014.10.070

Wang, Q., He, X., Huang, X. H. H., Griffith, S. M., Feng, Y., Zhang, T., Zhang, Q., Wu, D., & Yu, J. Z. (2017). Impact of secondary organic aerosol tracers on tracer-based source apportionment of organic carbon and PM2.5: A case study in the Pearl River Delta, China. *ACS Earth and Space Chemistry*, 1(9), 562–571. 10.1021/acsearthspacechem.7b00088

Wang, Q., He, X., Zhou, M., Huang, D. D., Qiao, L., Zhu, S., Ma, Y.-g., Wang, H.-l., Li, L., Huang, C., Huang, X. H. H., Xu, W., Worsnop, D., Goldstein, A. H., Guo, H., & Yu, J. Z. (2020). Hourly measurements of organic molecular markers in Urban Shanghai, China: Primary organic aerosol source identification and observation of cooking aerosol aging. *ACS Earth and Space Chemistry*, 4(9), 1670–1685. 10.1021/acsearthspacechem.0c00205

Wang, T., Dai, J., Lam, K. S., Nan Poon, C., & Brasseur, G. P. (2019). Twenty-five years of lower tropospheric ozone observations in tropical East Asia: The influence of emissions and weather patterns. *Geophysical Research Letters*, 46(20), 11463–11470. 10.1029/2019gl084459

Wang, T., Poon, C. N., Kwok, Y. H., & Li, Y. S. (2003). Characterizing the temporal variability and emission patterns of pollution plumes in the Pearl River Delta of China. *Atmospheric Environment*, 37(25), 3539–3550. 10.1016/S1352-2310(03)00363-7

Wang, T., Wei, X. L., Ding, A. J., Poon, C. N., Lam, K. S., Li, Y. S., Chan, L. Y., & Anson, M. (2009). Increasing surface ozone concentrations in the background atmosphere of Southern China, 1994-2007. *Atmospheric Chemistry and Physics*, 9(16), 6217–6227. 10.5194/acp-9-6217-2009

Wang, T., Xue, L., Brimblecombe, P., Lam, Y. F., Li, L., & Zhang, L. (2017). Ozone pollution in China: A review of concentrations, meteorological influences, chemical precursors, and effects. *Science of The Total Environment*, 575, 1582–1596. 10.1016/j.scitotenv.2016.10.081

Wang, Y., Guo, H., Lyu, X., Zhang, L., Zeren, Y., Zou, S., & Ling, Z. (2019). Photochemical evolution of continental air masses and their influence on ozone formation over the South China Sea. *Science of The Total Environment*, 673, 424–434. 10.1016/j.scitotenv.2019.04.075

Wang, Y., Guo, H., Zou, S., Lyu, X., Ling, Z., Cheng, H., & Zeren, Y. (2018). Surface O3 photochemistry over the South China Sea: Application of a near-explicit chemical mechanism box model. *Environ Pollut*, 234, 155–166. 10.1016/j.envpol.2017.11.001

Wang, Y., Wan, Q., Meng, W., Liao, F., Tan, H., & Zhang, R. (2011). Long-term impacts of aerosols on precipitation and lightning over the Pearl River Delta megacity area in China. *Atmospheric Chemistry and Physics*, 11(23), 12421–12436. 10.5194/acp-11-12421-2011

Wang, Y., Wang, H., Guo, H., Lyu, X., Cheng, H., Ling, Z., Louie, P. K. K., Simpson, I. J., Meinardi, S., & Blake, D. R. (2017). Long-term O3–precursor relationships in Hong Kong: field observation and model simulation. *Atmos. Chem. Phys.*, 17(18), 10919–10935. 10.5194/acp-17-10919-2017

Williams, B. J., Goldstein, A. H., Kreisberg, N. M., & Hering, S. V. (2010). In situ measurements of gas/particle-phase transitions for atmospheric semivolatile organic compounds. *Proceedings of the National Academy of Sciences*, 107(15), 6676–6681. 10.1073/pnas.0911858107

Williams, B. J., Jayne, J. T., Lambe, A. T., Hohaus, T., Kimmel, J. R., Sueper, D., Brooks, W., Williams, L. R., Trimborn, A. M., Martinez, R. E., Hayes, P. L., Jimenez, J. L., Kreisberg, N. M., Hering, S. V., Worton, D. R., Goldstein, A. H., & Worsnop, D. R. (2014). The first combined Thermal Desorption Aerosol Gas Chromatograph—Aerosol Mass Spectrometer (TAG-AMS). *Aerosol Science and Technology*, 48(4), 358–370. 10.1080/02786826.2013.875114

Wu, F., Yu, Y., Sun, J., Zhang, J., Wang, J., Tang, G., & Wang, Y. (2016). Characteristics, source apportionment and reactivity of ambient volatile organic compounds at Dinghu Mountain in Guangdong Province, China. *Science of the Total Environment*, 548-549, 347–359. 10.1016/j.scitotenv.2015.11.069

Wu, R., Li, J., Hao, Y., Li, Y., Zeng, L., & Xie, S. (2016). Evolution process and sources of ambient volatile organic compounds during a severe haze event in Beijing, China. *Science of the Total Environment*, 560-561, 62–72. 10.1016/j.scitotenv.2016.04.030

Xu, X., Lin, W., Wang, T., Yan, P., Tang, J., Meng, Z., & Wang, Y. (2008). Long-term trend of surface ozone at a regional background station in eastern China 1991-2006: enhanced variability. *Atmospheric Chemistry and Physics*, 8(10), 2595–2607. 10.5194/acp-8-2595-2008

Xue, L. K., Wang, T., Louie, P. K., Luk, C. W., Blake, D. R., & Xu, Z. (2014). Increasing external effects negate local efforts to control ozone air pollution: a case study of Hong Kong and implications for other Chinese cities. *Environ Sci Technol*, 48(18), 10769–10775. 10.1021/es503278g

Yang, L., Luo, H., Yuan, Z., Zheng, J., Huang, Z., Li, C., Lin, X., Louie, P. K. K., Chen, D., & Bian, Y. (2019). Quantitative impacts of meteorology and precursor emission changes on the long-term trend of ambient ozone over the Pearl River Delta, China, and implications for ozone control strategy. *Atmospheric Chemistry and Physics*, 19(20), 12901–12916. 10.5194/acp-19-12901-2019

Yao, D., Guo, H., Lyu, X., Lu, H., & Huo, Y. (2022). Secondary organic aerosol formation at an urban background site on the coastline of South China: Precursors and aging processes. *Environ Pollut*, 309, 119778. 10.1016/j.envpol.2022.119778

Yao, D., Lyu, X., Lu, H., Zeng, L., Liu, T., Chan, C. K., & Guo, H. (2021). Characteristics, sources and evolution processes of atmospheric organic aerosols at a roadside site in Hong Kong. *Atmospheric Environment*, 252, 118298. 10.1016/j.atmosenv.2021.118298

Yao, D., Lyu, X., Murray, F., Morawska, L., Yu, W., Wang, J., & Guo, H. (2019). Continuous effectiveness of replacing catalytic converters on liquified petroleum gas-fueled vehicles in Hong Kong. *Science of The Total Environment*, 648, 830–838. 10.1016/j.scitotenv.2018.08.191

Yarwood, G., Rao, S., Yocke, M., & Whitten, G. (2005). Updates to the carbon bond chemical mechanism: CB05. Final report to the US EPA, RT-0400675, 8.

Yin, R. J., Yan, C., Cai, R. L., Li, X. X., Shen, J. W., Lu, Y. Q., Schobesberger, S., Fu, Y. Y., Deng, C. J., Wang, L., Liu, Y. C., Zheng, J., Xie, H. B., Bianchi, F., Worsnop, D. R., Kulmala, M., & Jiang, J. K. (2021). Acid-base clusters during atmospheric new particle formation in Urban Beijing. *Environ Sci Technol*, 55(16), 10994–11005. 10.1021/acs.est.1c02701

Yu, H., Ren, L., Huang, X., Xie, M., He, J., & Xiao, H. (2019). Iodine speciation and size distribution in ambient aerosols at a coastal new particle formation hotspot in China. *Atmospheric Chemistry and Physics*, 19(6), 4025–4039. 10.5194/acp-19-4025-2019

Yuan, B., Shao, M., Lu, S., & Wang, B. (2010). Source profiles of volatile organic compounds associated with solvent use in Beijing, China. *Atmospheric Environment*, 44(15), 1919–1926. 10.1016/j.atmosenv.2010.02.014

Zeng, L., Yang, J., Guo, H., & Lyu, X. (2022). Impact of NOx reduction on long-term surface ozone pollution in roadside and suburban Hong Kong: Field measurements and model simulations. *Chemosphere*, 302, 134816. 10.1016/j.chemosphere.2022.134816

Zeren, Y., Guo, H., Lyu, X., Jiang, F., Wang, Y., Liu, X., Zeng, L., Li, M., & Li, L. (2019). An ozone "pool" in South China: Investigations on atmospheric dynamics and photochemical processes over the Pearl River Estuary. *Journal of Geophysical Research: Atmospheres*, 124(22), 12340–12355. 10.1029/2019jd030833

Zeren, Y., Guo, H., Lyu, X., Zhou, B., Liu, X., Yang, L., Yuan, Z., & Wang, Y. (2022). Remarkable spring increase overwhelmed hard-earned autumn decrease in ozone pollution from 2005 to 2017 at a suburban site in Hong Kong, South China. *Science of The Total Environment*, 831, 154788. 10.1016/j.scitotenv.2022.154788

Zeren, Y., Zhou, B., Zheng, Y., Jiang, F., Lyu, X., Xue, L., Wang, H., Liu, X., & Guo, H. (2022). Does ozone pollution share the same formation mechanisms in the bay areas of China? *Environ Sci Technol*, 56(20), 14326–14337. 10.1021/acs.est.2c05126

Zhang, F., Guo, H., Chen, Y., Matthias, V., Zhang, Y., Yang, X., & Chen, J. (2020). Size-segregated characteristics of organic carbon (OC), elemental carbon (EC) and organic matter in particulate matter (PM) emitted from different types of ships in China. *Atmos. Chem. Phys.*, 20(3), 1549–1564. 10.5194/acp-20-1549-2020

Zhang, J., Wang, T., Chameides, W. L., Cardelino, C., Kwok, J., Blake, D. R., Ding, A., & So, K. L. (2007). Ozone production and hydrocarbon reactivity in Hong Kong, Southern China. *Atmospheric Chemistry and Physics*, 7, 557–573. 10.5194/acp-7-557-2007

Zhang, Q., Jimenez, J. L., Canagaratna, M. R., Ulbrich, I. M., Ng, N. L., Worsnop, D. R., & Sun, Y. (2011). Understanding atmospheric organic aerosols via factor analysis of aerosol mass spectrometry: a review. *Analytical and Bioanalytical Chemistry*, 401(10), 3045–3067. 10.1007/s00216-011-5355-y

Zhang, Q., Streets, D. G., Carmichael, G. R., He, K., Huo, H., Kannari, A., Klimont, Z., Park, I., Reddy, S., & Fu, J. (2009). Asian emissions in 2006 for the NASA INTEX-B mission. *Atmospheric Chemistry and Physics*, 9(14), 5131–5153.

Zhang, Q., Yuan, B., Shao, M., Wang, X., Lu, S., Lu, K., Wang, M., Chen, L., Chang, C. C., & Liu, S. C. (2014). Variations of ground-level O3 and its precursors in Beijing in summertime between 2005 and 2011. *Atmospheric Chemistry and Physics*, 14(12), 6089–6101. 10.5194/acp-14-6089-2014

Zhang, R. (2010). Getting to the critical nucleus of aerosol formation. *Science*, 328(5984), 1366–1367. doi:10.1126/science.1189732

Zhang, R., Khalizov, A., Wang, L., Hu, M., & Xu, W. (2012). Nucleation and growth of nanoparticles in the atmosphere. *Chemical Reviews*, 112(3), 1957–2011. 10.1021/cr2001756

Zhao, Y., Kreisberg, N. M., Worton, D. R., Isaacman, G., Gentner, D. R., Chan, A. W. H., Weber, R. J., Liu, S., Day, D. A., Russell, L. M., Hering, S. V., & Goldstein, A. H. (2013). Sources of organic aerosol investigated using organic compounds as tracers measured during CalNex in Bakersfield. *Journal of Geophysical Research: Atmospheres*, 118(19), 11,388–311,398. 10.1002/jgrd.50825

Zhao, Y., Kreisberg, N. M., Worton, D. R., Isaacman, G., Weber, R. J., Liu, S., Day, D. A., Russell, L. M., Markovic, M. Z., VandenBoer, T. C., Murphy, J. G., Hering, S. V., & Goldstein, A. H. (2013). Insights into secondary organic aerosol formation mechanisms from measured gas/particle partitioning of specific organic tracer compounds. *Environ Sci Technol*, 47(8), 3781–3787. 10.1021/es304587x

Zheng, B., Tong, D., Li, M., Liu, F., Hong, C., Geng, G., Li, H., Li, X., Peng, L., Qi, J., Yan, L., Zhang, Y., Zhao, H., Zheng, Y., He, K., & Zhang, Q. (2018). Trends in China's anthropogenic emissions since 2010 as the consequence of clean air actions. *Atmospheric Chemistry and Physics*, 18(19), 14095–14111. 10.5194/acp-18-14095-2018

Zheng, J., Zhong, L., Wang, T., Louie, P. K. K., & Li, Z. (2010). Ground-level ozone in the Pearl River Delta region: Analysis of data from a recently established regional air quality monitoring network. *Atmospheric Environment*, 44(6), 814–823. 10.1016/j.atmosenv.2009.11.032

Zhou, S., Guo, F., Chao, C.-Y., Yoon, S., Alvarez, S. L., Shrestha, S., Flynn, J. H., III, Usenko, S., Sheesley, R. J., & Griffin, R. J. (2023). Marine submicron aerosols from the Gulf of Mexico: Polluted and acidic with rapid production of sulfate and organosulfates. *Environ Sci Technol*, 57(13), 5149–5159. 10.1021/acs.est.2c05469

11 Geospatial Analysis of Emerging Drought Risk in a Warming Climate

Shuo Wang and Yamin Qing

INTRODUCTION

Extreme weather events are recognized as a significant threat to human society and the environment, which have devastating impacts on infrastructure, ecosystems, and human lives (Abbass et al., 2022). As the Earth's climate continues to warm, the frequency and severity of extreme weather events, such as droughts, floods, heatwaves, and wildfires, are expected to increase (AghaKouchak et al., 2020; Ebi et al., 2021; You and Wang, 2021). In recent years, an emerging type of drought, known as "flash drought", has attracted increasing attention from the scientific communities. Flash drought is characterized by its rapid onset and devastating impacts, and has become a pressing concern for policymakers, water managers, and agricultural producers (Svoboda et al., 2002; Otkin et al., 2015; Crausbay et al., 2017; Otkin et al., 2018). For example, the historical flash drought that occurred over the Central United States without any early warning caused a tremendous impact on agricultural production and the economy, with $12 billion losses attributed to this event (Hoerling et al., 2014). The flash drought that occurred over Southern Queensland, Australia in 2018 de-vegetated the landscape and drove livestock numbers to the lowest level in the country (Nguyen et al., 2019). Under the influence of global warming, flash droughts have occurred more frequently, including those in southern China (Wang et al., 2016; Yuan et al., 2019), USA (Hoerling et al., 2014; Gerken et al., 2018), and Africa (Yuan et al., 2018). Therefore, there is an urgent need to investigate flash drought risk and its underlying drivers for enhancing climate adaptation and resilience to the emerging drought event.

The complex nature of flash droughts makes it challenging to predict and identify their onset and duration, and there is a growing need for innovative approaches to assess and monitor the emerging drought risk. The improvement of geospatial analysis techniques has brought significant advantages to the emerging drought risk analysis (Jeyaseelan, 2003; Belal et al., 2014; Arabameri et al., 2022). For example, geospatial analysis enables the identification of critical areas affected by flash drought and helps in the assessment of the degree of vulnerability and exposure to flash drought risk (Al-Hemoud et al., 2023; Dunne & Kuleshov, 2023).

 DOI: 10.1201/9781003244561-14

Thus, geospatial analysis can provide an in-depth understanding of the impact of flash drought on various sectors, such as agriculture, water resources, and ecosystems, which assists in the development of effective mitigation and adaptation measures. On the other hand, the use of remote sensing data in geospatial analysis allows for the monitoring of vegetation health, soil moisture levels, and water availability, providing critical information for flash drought forecasting and early warning systems (Liou & Mulualem, 2019; Mullapudi et al., 2022). Furthermore, geospatial analysis techniques can integrate various datasets, including meteorological, hydrological, and socio-economic data, to provide a comprehensive assessment of flash drought risk (Hoque et al., 2020; Arabameri et al., 2022). Overall, the improvement of geospatial analysis techniques has facilitated more informed and effective means to drought risk assessment and has the potential to reduce the economic, social, and environmental impacts of emerging droughts.

In this chapter, we aim to identify and assess the emerging drought risks using geospatial analysis, particularly for flash droughts and drought-related compound events. Using the remote sensing data and climate models, we will be able to gain a better understanding of the evolution and underlying mechanism of emerging droughts under a warming climate and develop effective strategies to mitigate the impacts of drought hazards on society and the environment.

SPATIOTEMPORAL EVOLUTION OF FLASH DROUGHT

The spatial distribution of flash drought that occurred globally during the period of 2000–2020 has been analyzed using satellite-based datasets (Table 11.1). As shown in Figure 11.1(a), flash droughts are likely to occur in humid and semi-humid regions: the North Asia, East Asia, Southeast Asia, Eastern North America, and Central North America (Qing et al., 2022; Yuan et al., 2023). On the contrary, the water-limited areas, including South Asia, Western Africa, Eastern Africa, Southern Africa, and Australia, have experienced fewer flash drought events during the period of 2000–2020. Additionally, the identified flash drought events present a sharp decrease in soil moisture (SM) during the onset phase, on average, from the 52.76–64.12th percentile to the 25.40–39.37th percentile (from the highest point to the lowest point), with a decline rate of larger than the 5th percentile per pentad (Figure 11.1(b)) (Qing et al., 2022). Compared with the slowly evolving droughts,

TABLE 11.1
Summary of the Datasets Used in This Chapter

Dataset	Source	Time Resolution	Spatial Resolution	Time Period
GLDAS	https://disc.gsfc.nasa.gov/datasets?keywords=GLDAS&page=1/	3hours/daily	0.25° × 0.25°	2000–2020
ERA5	https://cds.climate.copernicus.eu/	daily	0.25° × 0.25°	1980–2020
MERRA-2	https://gmao.gsfc.nasa.gov/reanalysis/MERRA-2/	3hours	0.5° × 0.625°	1980–2020
GLEAM	https://www.gleam.eu/	daily	0.25°× 0.25°	1980–2020

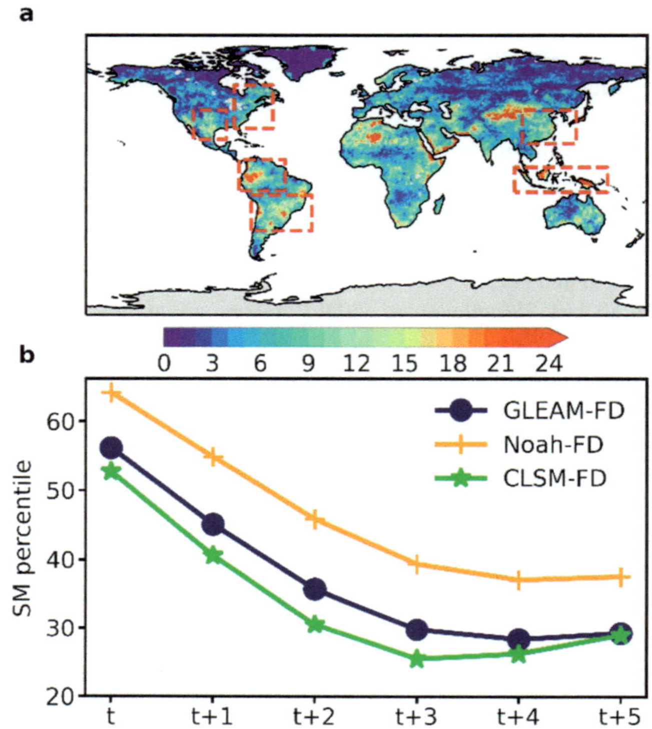

FIGURE 11.1 (a) Frequencies of occurrence of flash droughts during the period of 2000–2020. (b) Temporal variations in SM percentiles for flash drought events of all grid points.

flash droughts have a rapid onset accompanied by a rapid intensification rate. Such a rapid onset phase may not provide sufficient time for acclimatory responses to occur, resulting in more severe physiological responses.

On the other hand, flash droughts do not appear to be occurring more frequently in most global regions, just coming on faster (Qing et al., 2022). Among all five types of flash drought, including 1 pentad, 2 pentads, 3 pentads, 4 pentads, and 5 pentads, 33.64% – 46.18% of flash droughts developed within 1 pentad, particularly in South Asia, Central Asia, Western Africa, and the Amazon Basin. Approximately 20.12% to 28.17% and 19.32% to 32.53% of flash droughts occurred in 2 and 3 pentads, respectively. In contrast, only 5.34% to 11.22% and 0.99% to 2.26% of flash droughts developed in 4 and 5 pentads, respectively. Overall, over 70% of flash droughts developed within half a month (≤3 pentad), and more than 30% of flash droughts developed within 5 days (≤1 pentad), accompanied by a high intensification rate (≥5th percentile) (Figure 11.2). These indicate the rapid onset and development of flash droughts, which contrasts with traditional droughts that may take 5–6 months to develop due to the cumulative effects of climate factors. The proportion of flash droughts with a 1-pentad onset of drying also shows a significant ($P < 0.01$) increasing trend, the estimated slope suggests a rise of 3.23% to 19.03% in the proportion of flash droughts with 1-pentad onset during the period of 2000–2020,

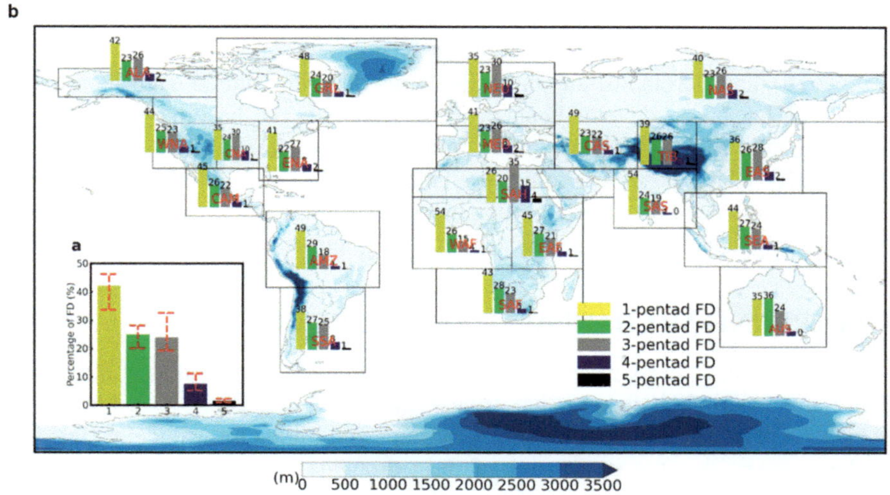

FIGURE 11.2 (a) Global proportion of flash droughts at different onset times relative to all flash droughts. The red dashed lines at the top of each bar represent the range of uncertainty in three different datasets. (b) Proportion of flash droughts at different onset times relative to all flash droughts over 21 regions.

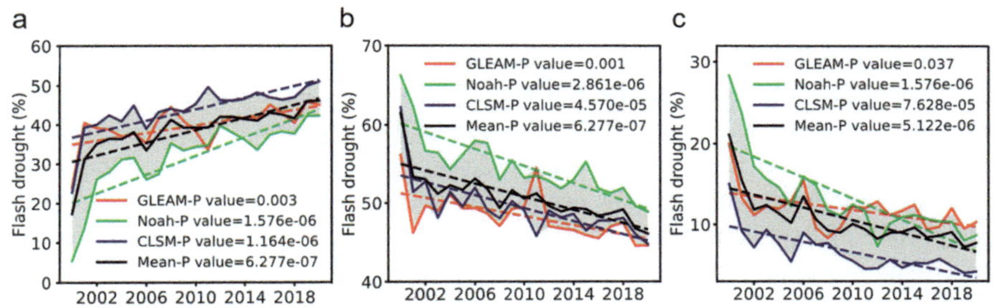

FIGURE 11.3 Temporal evolutions of percentages of flash droughts with the onset times of 1 (a), 2–3 (b), and 4–5 (c) pentads relative to all flash droughts. The linear annual trends in the proportion of flash droughts at different onset times are estimated based on the Sen's slope estimator, and statistical significances in trends are determined based on the MK test for the entire study period (2000–2020). The shaded areas represent the range of uncertainty in three different datasets.

with an annual growth rate of 0.15% to 0.91% (Figure 11.3). The increasing trend is observed in most regions, especially in South Asia, Southeast Asia, and Central North America, where the proportion of flash droughts with 1-pentad onset has increased by more than 20%. This indicates that more flash drought events are developing within one week in these regions. Conversely, the proportions of flash droughts developing in 2–3 and 4–5 pentads show a statistically significant ($P < 0.05$) decrease from 56.08–66.25% and 14.98–28.32% in 2000 to 44.60–48.90% and 4.17–10.29% in 2020 (Figure 11.3), respectively, suggesting that flash droughts take less time to develop. Similar trends are observed in most sub-regions, with the largest decreasing trend seen in Southeast Asia, followed by South Africa, South Asia, and the Amazon Basin.

Overall, flash droughts are most likely to occur in humid and semi-humid regions. The onset development of flash droughts is becoming faster, and approximately 33.64–46.18% of flash droughts occur within 5 days for the period 2000–2020, which poses a significant challenge for drought monitoring, and thus emphasizes the need for effective mitigation and adaptation strategies to reduce the negative impacts on agriculture, water resources, and ecosystems.

UNDERLYING MECHANISM OF FLASH DROUGHT

Investigating the underlying mechanism of flash droughts is crucial as it can help us better understand the factors that trigger these extreme events, facilitating the development of effective early warning systems and mitigation strategies. Previous studies suggest that the occurrence of flash droughts is associated with moisture imbalance that is related to precipitation (P) and temperature (T) (Pendergrass et al., 2020). The increase (decrease) in P can lead to the corresponding increase (decrease) in SM, and the increase (decrease) in T also contributes to the decrease (increase) in SM. Thus, P and T are the key factors influencing the occurrence of SM decline. Furthermore, the simultaneous occurrence of extreme high T and P deficit may exacerbate the depletion of SM to develop a flash drought event (Nguyen et al., 2019). On the other hand, land–atmosphere interactions also play an important role in accelerating soil drying. P reduces the input of water into the soil, leading to insufficient SM (Zhou et al., 2019; Liu et al., 2020). This may cause high near-surface air T and vapor pressure deficit (VPD) through land–atmosphere coupling, eventually causing the occurrence of flash droughts. Thus, there is still a significant gap in the understanding of the mechanisms driving the occurrence of flash droughts from the perspective of land–atmosphere coupling, highlighting the need for further research improving our understanding of flash drought development.

It is widely accepted that there is a strong negative correlation between SM and VPD, and the low SM is always accompanied by the high VPD. In turn, the high VPD enhances the atmospheric evaporative demand, which can exacerbate the decline in SM through land–atmosphere interactions (Zhou et al., 2019). As expected, SM is strongly correlated with VPD: high VPD and low SM, and low VPD and high SM (Figure 11.4). Certain regions with a high value of Probability multiplication factor (PMF: PMF is defined as the ratio of the joint probability of extreme low SM and high VPD and the probability if they are assumed independent, and a PMF of 1 therefore represents no increase in the probability of co-occurrence) are consistent with those experiencing a high frequency of flash drought events. Even if the thresholds of SM and VPD change, similar patterns can be obtained. A positive correlation ($R = 0.63$, the points of Tibet and Sahara are excluded from correlation analysis because the PMF values of almost all points are less than 1, suggesting that there is no increase in the co-occurrence probability) between the frequency of flash droughts occurring in 19 regions and PMF is found (Figure 11.5(a)), especially for the Southeast Asia, Amazon Basin, and Eastern North America where the joint probability of bivariate extremes is approximately

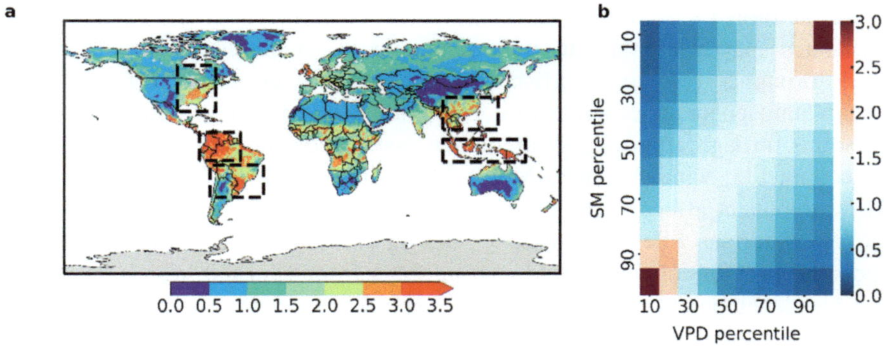

FIGURE 11.4 (a) Mean PMF of concurrent extreme SM (below the 30th percentile) and VPD (above the 90th percentile) for the period of 2000–2020 based on three datasets. (b) Mean probability (%) of percentile bins of SM and VPD across all land grid points.

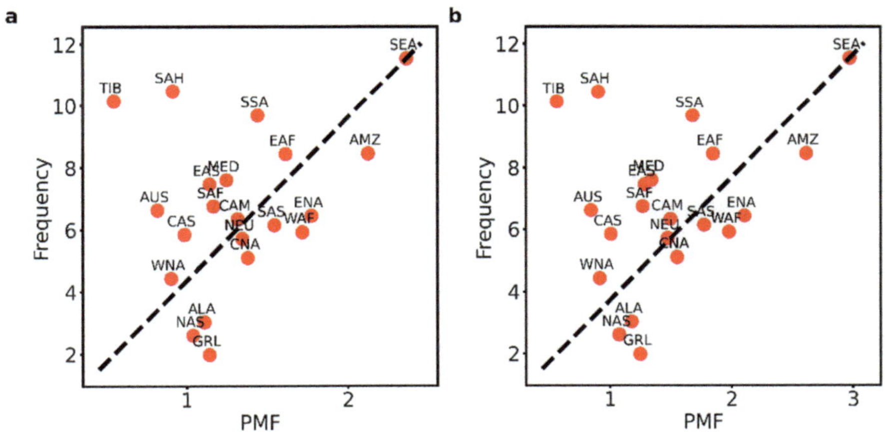

FIGURE 11.5 (a) Spatial relationship between the frequency of occurrence of flash droughts and the PMF of SM and VPD extremes (below the 30th percentile SM and above the 90th percentile VPD). (b) Spatial relationship between the frequency of occurrence of flash droughts and the PMF of SM and VPD extremes (below the 20th percentile SM and above the 90th percentile VPD). Each point in c and d is the mean of the results from three datasets.

1.77–2.36 times higher than the probability of the independent combination. This indicates that the strong coupling of SM and VPD through land–atmosphere feedback contributes to the rapid decline in SM. Moreover, the positive correlation between the frequency of flash droughts occurring in 19 regions and PMF will increase to 0.65 when the threshold of extreme SM decreases to the 20th percentile, suggesting that the intensity of SM–VPD coupling will increase when there is a further decline in SM (Figure 11.5(b)). Therefore, the development of flash droughts can be accelerated when SM depletion co-occurs with atmospheric aridity, and the joint influence of SM depletion and atmospheric aridity further enhances the rapid onset of flash droughts.

In addition to the local land–atmosphere coupling, large-scale atmospheric dynamics may also play an important role in triggering the occurrence of flash droughts (Mahto & Mishra, 2020; Nguyen et al., 2021). Moreover, ocean-atmosphere teleconnections may influence the development of flash droughts through atmospheric circulation associated with a rapid decline in SM in conjunction with P deficit (Mo & Lettenmaier, 2016). Vegetation is also a key component affecting the rapid onset development of flash droughts because of its important role in mediating the transpiration. Crops can suffer from moisture stress more quickly by pumping water from deep soil, which may trigger a rapid development of flash droughts. Exploring mechanisms of flash droughts would be challenging but a promising direction for unraveling the mystery of flash drought onset.

DROUGHT-RELATED COMPOUND EXTREMES

Compound events have been receiving increasing attention from the scientific community in recent years. Particularly, the drought-related compound extremes have attracted widespread attention, which refer to the simultaneous occurrence of droughts and other extreme weather events such as heatwaves, wildfires, and intense rainfall. These events can exacerbate the impacts of droughts since they can cause soil moisture depletion, water scarcity, and vegetation stress. A heatwave can accelerate the evaporation of soil moisture, which can worsen the intensity and duration of a drought. For example, the 2019/2020 mega-fires in Australia were closely associated with record-breaking temperatures, and both extremes were induced, to a large extent, by widespread dryness (King et al., 2020; Boer et al., 2020; Collins et al., 2021). Similarly, intense rainfall can amplify the impact of drought because the water absorption capacity of the soil can decrease after drought, which is highly likely to cause severe flooding events. For example, there was a fast transition from drought to pluvial flooding within one week over South Carolina of United States in September 2015, which brought a once-in-a-thousand-years flood (He & Sheffield, 2020). Thus, understanding the evolution and underlying mechanism of drought-related compound extremes facilitates the development of effective strategies for compound event monitoring, forecasting, and management.

Particularly, many regions around the world have been affected by the abrupt transition between droughts and pluvials in recent years (Zscheischler et al., 2020; Chen et al., 2020, 2022, 2023). Admittedly, the changes in frequency and intensity of droughts and pluvials are inherently interconnected and governed by the same underlying hydrological processes and atmospheric dynamics, which may lead to higher hydroclimatic variability with response to a warming climate (He & Sheffield, 2020). Figure 11.6 shows the spatial pattern of the frequency of compound drought and pluvial events for a recent warming period of 1980–2020, corresponding to the three satellite-based datasets (Table 11.1). Compound drought and pluvial events are more frequent over eastern North America, Europe, East Asia, Southeast Asia, southern Australia, southern Africa, and southern South America (Figure 11.6). The interannual changes in the global mean frequency of

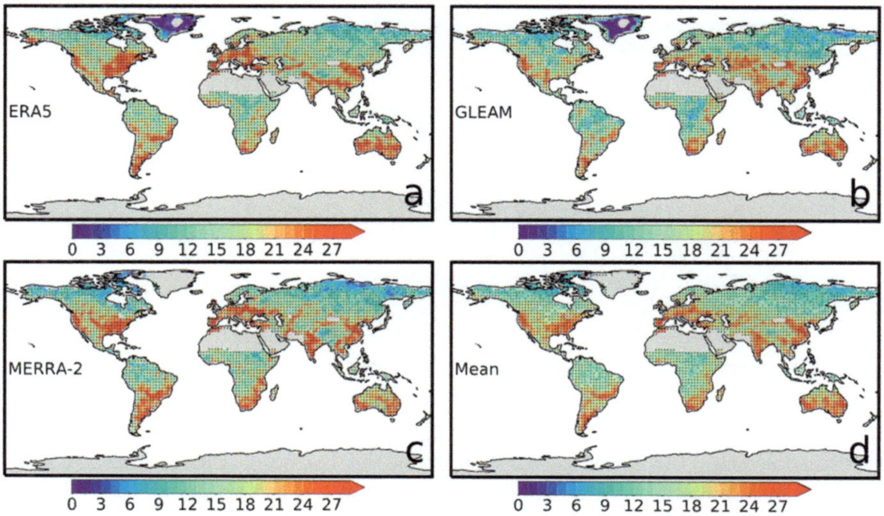

FIGURE 11.6 Spatial pattern of the frequency of DWTs during the period of 1980–2020. Spatial pattern of the total number of DWTs for each pixel during the period of 1980–2020 based on (a) ERA5, (b) GLEAM, and (c) MERRA-2 datasets (d) shows the Mean spatial pattern. Stippling denotes regions where the frequency is statistically significant at the 99% level based on 3,000 bootstrap tests.

compound drought and pluvial events estimated as the mean value of grid cells show a significant ($p < 0.01$) increasing trend during the period of 1980–2020. And the global mean frequency during the period of 2001–2020, is larger than that observed during the period of 1981–2000 (Figure 11.7). Thus, there is an increase in the frequency of compound droughts and pluvial events, leading to great potential risk associated with the transition between dry and wet extremes. The detection of spatiotemporal changes of compound drought and pluvial events provides meaningful insights into the risk assessment of compound extremes, facilitating policymakers and stakeholders to develop sound adaptation plans.

Exploring mechanisms of compound drought and pluvial events would be in a preliminary stage, and it is also a challenging direction for unraveling the physical processes behind the occurrences of compound drought and pluvial events. It is widely accepted that extreme events, especially for drought to pluvial, are associated with atmospheric phenomena, including large-scale teleconnections, blocking patterns, frontal weather systems, etc. In addition to the large-scale atmospheric dynamics, the local land–atmosphere feedback also plays an important role in the occurrence of compound drought and pluvial events. Land–atmosphere interactions can modify atmospheric dryness, cloud cover and moisture convergence and therefore the likelihood of cloud/precipitation generation, potentially influencing the occurrence of post-drought pluvial events. Thus, considering the influence of large-scale atmospheric dynamics on the land–atmosphere couplings in a more comprehensive manner can advance our understanding of the occurrence and evolution of abrupt shift from drought to pluvial.

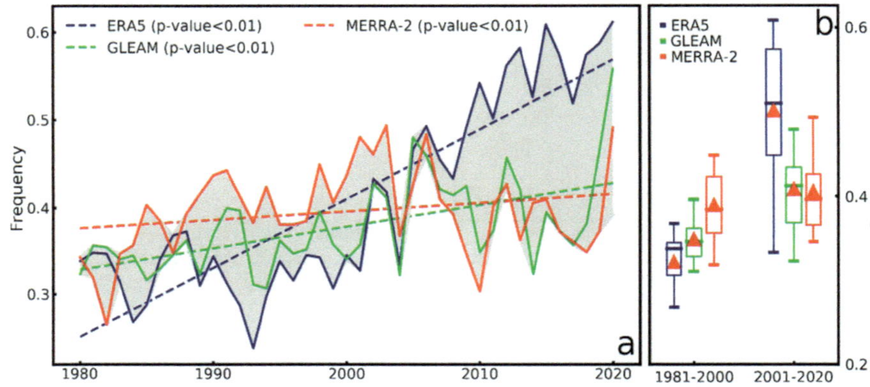

FIGURE 11.7 (a) Annual time series (solid lines) of globally averaged drought and pluvial compound events. The linear annual trends (dashed lines) for ERA5, MERRA-2, and GLEAM are estimated based on the Sen's slope estimator, and statistical significances in trends are determined based on the MK test for 1980–2020. The gray shading represents the uncertainty across three datasets. (b) Box plots of the total number of drought and pluvial compound events for each pixel during the periods of 1981–2000 and 2001–2020. The short horizontal line inside the box represents the 50th percentile, and the top and bottom of the box represent the 75th and 25th percentiles, respectively. The top and bottom of the line represent the 95th and 5th percentiles, respectively. The red triangles represent the mean values.

SUMMARY AND CONCLUSIONS

Geospatial techniques have proven to be an effective tool in understanding the emerging drought risk in a warming climate. The spatiotemporal evolution of flash drought has been examined in detail, revealing the rapid onset and severe impacts of the emerging type of droughts. Further, the onset development of flash drought is becoming faster, posing a significant challenge for drought monitoring. Although some of drought monitoring systems are available on a daily or weekly basis (e.g., the U.S. Drought Monitor, the Climate Prediction Center, the Drought Impact Reporter, etc.), but not for all countries. Thus, assessing the relatively fast onset development timescales of flash drought can provide valuable information for upgrading drought monitoring systems and indicators worldwide. Additionally, the underlying mechanisms of flash drought have been studied, shedding light on the complex interactions between land and atmospheric conditions. On the other hand, the study of drought-related compound extremes highlights the need for an integrated approach to drought management that considers the interactions between multiple stressors such as heatwaves, wildfires, and floods.

To sum up, geospatial analysis of emerging drought risk under a warming climate provides a valuable contribution to the understanding of complex dynamics of emerging droughts and their impacts. Our findings highlight the need for proactive and integrated management strategies that can help communities, economies, and ecosystems to adapt to the increasing frequency and severity of emerging droughts in a changing climate.

REFERENCES

Abbass, K., Qasim, M. Z., Song, H., Murshed, M., Mahmood, H., & Younis, I. (2022). A review of the global climate change impacts, adaptation, and sustainable mitigation measures. *Environmental Science and Pollution Research*, 29(28), 42539–42559.

AghaKouchak, A., et al. (2020). Climate extremes and compound hazards in a warming world. *Annual Review of Earth and Planetary Sciences*, 48, 519–548.

Al-Hemoud, A., Al-Enezi, A., Al-Dashti, H., Petrov, P., Misak, R., AlSaraf, M., & Malek, M. (2023). Hazard assessment and hazard mapping for Kuwait. *International Journal of Disaster Risk Science*, 14(1), 143–161.

Arabameri, A., Chandra Pal, S., Santosh, M., Chakrabortty, R., Roy, P., & Moayedi, H. (2022). Drought risk assessment: integrating meteorological, hydrological, agricultural and socio-economic factors using ensemble models and geospatial techniques. *Geocarto International*, 37(21), 6087–6115.

Belal, A. A., El-Ramady, H. R., Mohamed, E. S., & Saleh, A. M. (2014). Drought risk assessment using remote sensing and GIS techniques. *Arabian Journal of Geosciences*, 7, 35–53.

Boer, M. M., Resco de Dios, V., & Bradstock, R. A. (2020). Unprecedented burn area of Australian mega forest fires. *Nature Climate Change*, 10(3), 171–172.

Chen, H., & Wang, S. (2022). Accelerated transition between dry and wet periods in a warming climate. *Geophysical Research Letters*, 49(19), e2022GL099766.

Chen, H., Wang, S., Zhu, J., & Wang, D. (2023). Projected changes in the pattern of spatially compounding drought and pluvial events over Eastern China under a warming climate. *Earth's Future*, 11, e2022EF003397.

Chen, H., Wang, S., Zhu, J., & Zhang, B. (2020). Projected changes in abrupt shifts between dry and wet extremes over China through an ensemble of regional climate model simulations. *Journal of Geophysical Research: Atmospheres*, 125, e2020JD033894.

Collins, L., Bradstock, R. A., Clarke, H., Clarke, M. F., Nolan, R. H., & Penman, T. D. (2021). The 2019/2020 mega-fires exposed Australian ecosystems to an unprecedented extent of high-severity fire. *Environmental Research Letters*, 16, 044029.

Crausbay, S. D. et al. (2017). Defining ecological drought for the twenty-first century. *Bulletin of the American Meteorological Society*, 98, 2543–2550.

Dunne, A., & Kuleshov, Y. (2023). Drought risk assessment and mapping for the Murray–Darling Basin, Australia. *Natural Hazards*, 115(1), 839–863.

Ebi, K. L., et al. (2021). Extreme weather and climate change: population health and health system implications. *Annual Review of Public Health*, 42(1), 293–315.

Gerken, T., Bromley, G. T., Ruddell, B. L., Williams, S. & Stoy, P. C. (2018). Convective suppression before and during the United States Northern Great Plains flash drought of 2017. *Hydrology and Earth System Sciences*, 22(8).

He, X. & Sheffield, J. (2020). Lagged compound occurrence of droughts and pluvials globally over the past seven decades. *Geophysical Research Letters*, 47(14), e2020GL087924.

Hoerling, M. et al. (2014). Causes and Predictability of the 2012 Great Plains Drought. *Bulletin of the American Meteorological Society*, 95, 269–282.

Hoque, M. A. A., Pradhan, B., & Ahmed, N. (2020). Assessing drought vulnerability using geospatial techniques in northwestern part of Bangladesh. *Science of The Total Environment*, 705, 135957.

Jeyaseelan, A. T. (2003). Droughts & floods assessment and monitoring using remote sensing and GIS. *Satellite remote sensing and GIS applications in agricultural meteorology*, 291.

King, A. D., Pitman, A. J., Henley, B. J., Ukkola, A. M., & Brown, J. R. (2020). The role of climate variability in Australian drought. *Nature Climate Change*, 10, 177–179.

Liou, Y. A., & Mulualem, G. M. (2019). Spatio–temporal assessment of drought in Ethiopia and the impact of recent intense droughts. *Remote Sensing*, 11(15), 1828.

Liu, L. et al. (2020). Soil moisture dominates dryness stress on ecosystem production globally. *Nature Communications*, 11, 4892.

Mahto, S. S., & Mishra, V. (2020). Dominance of summer monsoon flash droughts in India. *Environmental Research Letters*, 15(10), 104061.

Mo, K. C. & Lettenmaier, D. P. (2016). Precipitation deficit flash droughts over the United States. *Journal of Hydrometeorology*, 17(4), 1169–1184.

Mullapudi, A., Vibhute, A. D., Mali, S., & Patil, C. H. (2022). A review of agricultural drought assessment with remote sensing data: Methods, issues, challenges and opportunities. *Applied Geomatics*, 1–13.

Nguyen, H. et al. (2019). Using the evaporative stress index to monitor flash drought in Australia. *Environmental Research Letters*, 14(6) 064016.

Nguyen, H., Wheeler, M. C., Hendon, H. H., Lim, E. P., & Otkin, J. A. (2021). The 2019 flash droughts in subtropical eastern Australia and their association with large-scale climate drivers. *Weather and Climate Extremes*, 32, 100321.

Otkin, J. A., Anderson, M. C., Hain, C. & Svoboda, M. (2015). Using temporal changes in drought indices to generate probabilistic drought intensification forecasts. *Journal of Hydrometeorology*, 16(1), 88–105.

Otkin, J. A. et al. (2018). Flash droughts: a review and assessment of the challenges imposed by rapid-onset droughts in the United States. *Bulletin of the American Meteorological Society*, 99, 911–919.

Pendergrass, A. G. et al. (2020). Flash droughts present a new challenge for subseasonal-to-seasonal prediction. *Nature Climate Change*, 10, 191–199.

Qing, Y., Wang, S., Ancell, B., & Yang, Z-L. (2022) Accelerating flash droughts induced by the joint influence of soil moisture depletion and atmospheric aridity. *Nature Communications*, 13, 1139.

Svoboda, M. et al. (2002). The drought monitor. *Bulletin of the American Meteorological Society*, 83, 1181–1190.

Wang, L., Yuan, X., Xie, Z., Wu, P. & Li, Y. (2016). Increasing flash droughts over China during the recent global warming hiatus. *Scientific Reports*, 6, 30571.

You, J. & Wang, S. (2021). Higher probability of occurrence of hotter and shorter heat waves followed by heavy rainfall. *Geophysical Research Letters*, 48, e2021GL094831.

Yuan, X., Wang, L., Wu, P., Ji, P., Sheffield, J., & Zhang, M. (2019). Anthropogenic shift towards higher risk of flash drought over China. *Nature Communications*, 10(1), 1–8.

Yuan, X., Wang, L., & Wood, E. F. (2018). Anthropogenic Intensification of Southern African Flash Droughts as Exemplified by the 2015/16 Season. *Bulletin of the American Meteorological Society*, 99(1), S86–S90.

Yuan, X., Wang, Y., Ji, P., Wu, P., Sheffield, J., & Otkin, J. A. (2023). A global transition to flash droughts under climate change. *Science*, 380(6641), 187–191.

Zhou, S. et al. (2019). Land–atmosphere feedbacks exacerbate concurrent soil drought and atmospheric aridity. *Proceedings of the National Academy of Sciences of the United States of America*, 116(38), 18848–18853.

Zscheischler, J., Martius, O., Westra, S., Bevacqua, E., Raymond, C., Horton, R. M., et al. (2020). A typology of compound weather and climate events. *Nature Reviews Earth & Environment*, 1(7), 333–347.

12 Urban Summer Air Temperature Forecasting through a Fusion of a Numerical Weather Prediction Model with Machine Learning

Dongjin Cho, Siwoo Lee, and Jungho Im

INTRODUCTION

Due to global warming, the intensity, frequency, and duration of heat waves and tropical nights have recently increased worldwide (Coffel et al., 2017; Perkins et al., 2012). Extreme air temperatures in summer can cause socio-economic problems such as thermal casualties and excessive electricity consumption (Baniassadi et al., 2018; Hifumi et al., 2018; Klinenberg, 2015; Russo et al., 2019). Half of the population lives in cities with higher temperatures than surrounding rural areas resulting from their highly heterogeneous landscape, complex infrastructure, and various anthropogenic activities (Schulze and Langenberg, 2014; Stewart, 2011). Reliable summer air temperature forecasting for urban areas is essential to mitigate the damage from thermal disasters effectively.

Numerical Weather Prediction (NWP) models have been used to predict air temperatures for operational forecasting. NWP models are based on atmospheric dynamics of physical parameters (Bauer et al., 2015; Hong et al., 2018; Kalnay, 2003), and there has been a steady improvement in NWP models over the past decades. Nonetheless, NWP models have systematic and random biases resulting from coarse spatial resolution, inaccurate boundary conditions, and imperfect parameterization of physical processes. To reduce such biases, post-processing techniques using statistical approaches have been developed based on Kalman filter and model output statistics (MOS), resulting in improvement of prediction accuracy (Whan et al., 2021; de Carvalho et al., 2011; Cho et al., 2022).

DOI: 10.1201/9781003244561-15

MOS is widely used among the post-processing approaches, which uses the statistical relationship between forecasts of NWP model and observations at weather stations (Glahn and Lowry, 1972; Jacks et al., 1990). Multi-linear regression (MLR) has been adopted for operational use because of its simpleness, but several limitations remain. First, an MLR-based post-processing model is required to build a model by stations because it cannot consider the nonlinearity of the data. Second, it is sensitive to the bias change of updating NWP model. NWP models are continuously updated to improve the accuracy of NWP models themselves. This update makes inconsistent NWP models, which affects the prediction accuracy of the MLR-based post-processing model.

Thanks to increasing computational power over decades, numerous previous studies have applied machine learning approaches, a subset of artificial intelligence (AI), such as artificial neural network (ANN), random forest (RF), and support vector regression (SVR) in MOS, because they can simulate the nonlinear relationships between predictors and predictands (Cho et al., 2020; Shin and Yi, 2019; Yi et al., 2018; Zhang et al., 2022). These machine learning approaches have the advantage of being able to produce a spatial distribution of bias-corrected air temperature forecasts through the development of a post-processing model that works for a multitude of stations using spatially continuous input variables. For example, Yi et al (2018) and Shin and Yi (2019) produced the bias-corrected spatial distribution maps of next-day maximum air temperature forecast from the Local Data Assimilation and Prediction System (LDAPS) model using SVR in Seoul, South Korea. Zhang et al. (2022) applied geographically weighted regression to produce the hourly air temperature forecast with a high spatial resolution from the Weather and Research Forecasting (WRF) model in Zhangjiakou city, Hebei Province, China.

Although various machine learning-based post-processing approaches have been suggested, they generally do not consider the use of temporal and spatial information. In recent years, deep learning approaches such as recurrent neural work (RNN) and convolutional neural network (CNN) have been applied in weather forecasting fields (Chen et al., 2020; Cho et al., 2022; Park et al., 2019). RNN extracts important features from time-series pattern of weather components, whereas CNN extract important spatial information from gridded data. Park et al. (2019) utilized a long short-term model (LSTM) to predict the temperature in South Korea, showing higher accuracy than the LDAPS for 24 h temperature predictions. Chen et al. (2020) compared MLR, RF, and 3D fully CNN for the post-processing of the temperature forecast from the European Centre for Medium-Range Weather Forecasts (ECMWF) over Tianjin, China. They indicated that 3D fully CNN outperformed MLR and RF.

In this chapter, we introduce two recently proposed AI-based post-processing case studies for improving the NWP model's next-day summer air temperature forecast in Seoul, South Korea (Cho et al., 2020; Cho et al., 2022). These two studies have different purposes to solve the limitations of the MLR-based post-processing approach. Cho et al. (2020) developed a multi-model ensemble (MME) aggregating RF, SVR, and ANN to produce reliable spatial distribution maps of bias-corrected maximum and minimum air temperatures (Tmax and Tmin, respectively) on the next day in summer. On the other hand, Cho et al. (2022)

evaluated the robustness of post-processing models including deep learning models and MME for updating the NWP model at the station level. The chapter will be organized as follow: AI algorithms mainly used for weather forecasting will be introduced first in Section 12.2. For two case studies, study area and NWP model will be described in Section 12.3. And then, two case studies will be presented in Sections 12.4 and 12.5. Finally, the final section will present the drawn conclusions. This chapter can help readers to know how machine learning can further improve the NWP model for urban air temperature forecasting.

MACHINE LEARNING ALGORITHMS

MACHINE LEARNING

RF is an ensemble machine learning approach by constructing multiple Classification and Regression Trees (CARTs) to predict a target variable (Breiman, 2001). RF grows a multitude of CARTs using a bootstrapping strategy that uses a subset of training samples and predictors at each node of the tree. And then, RF obtains final predictions by aggregating the output of each tree (majority for classification and simple average for regression). The use of both bootstrapping and aggregation is called bagging. With the bagging strategy, RF alleviates the disadvantages of the CARTs such as sensitivity and overfitting to training samples. The typical hyperparameters of RF are the number of random predictors used when growing the trees (Mtry) and the number of trees (Ntrees).

SVR is a supervised machine learning approach aiming to find optimal decision boundaries (hyperplane) that effectively separate training samples (Drucker et al., 1996). The optimal hyperplane is the linear boundary that maximizes the margin, the sum of the distances from the closest samples in the two domains to the linear boundary, and minimal empirical errors. To effectively find the optimal hyperplane, SVR generally extends training data from the original dimension to a higher dimension using a kernel function. Typically, three kernels are used in SVR: linear, polynomial, and radial basis functions (RBFs). There are two user-defined parameters in SVR: C that controls regression error and the margin and gamma that determine how much curvature used in a decision boundary.

ANN is an interconnected structure based on biological neural networks that operate and connect biological neurons in the human brain (Jain et al., 1996). Among the various neural network forms, multi-layer perceptron (MLP) is the most popular, which structure consists of layers for an input, more than one hidden, and an output with a back-propagation algorithm. The input layer gathers predictors for the neural network, while the output layer predicts the result for given predictors. The hidden layer is the intermediate layer between the input and output layer, which generates output by adjusting the weights of input variables using a nonlinear function (activation function). The performance of ANN depends greatly on its architecture. A complicated network with too many hidden layers and neurons can cause overfitting and poor generalization, while a basic network with few hidden layers and neurons may limit learning potential. As an activation function and a model optimization algorithm, a rectified linear unit (ReLU) and Adam optimizer are extensively used, respectively (Kingma and Ba, 2014; Nair and Hinton, 2010).

Deep Learning

In recent years, the accuracy of NWP temperature prediction has been further improved by applying deep learning techniques such as RNN and CNN (Cho et al., 2022; Gong et al., 2022; Zhang et al., 2020). First, RNN, has been known as a suitable ANN for dealing with time-sequential tasks with iterative learning in the model. RNN operates based on the function to store the output of previous steps and reintroduce them as input, allowing them to predict the output of a layer. The RNN architecture includes a mechanism to incorporate the output value of the hidden layer from the previous time point (t − 1) as the input value for the hidden layer at the next time point (t). This structure represents a repeating pattern in an ANN, where the outcome at the current time (t) influences the outcome at the next time (t + 1). Theoretically, RNN possess the capability to manage and learn long-term dependencies. However, it should be noted that empirical results often encounter the challenge of vanishing gradients as the input sequence length grows (Hochreiter et al., 1998). To address the issue of vanishing gradients, a variant of RNN called LSTM and gated recurrent units (GRU), which incorporate explicit memory, have emerged and attracted considerable attention.

LSTM, first introduced by Hochreiter and Schmidhuber (1997), is a more sophisticated form of RNN specifically designed to prevent the issues of diminishing and magnifying gradient problems and learn long-term dependencies. In contrast to the conventional RNN architecture, the LSTM stands out with three gates (i.e., forget, input, and output gates) and its interconnected network of four layers that actively engage in communication instead of relying on a single tanh layer. On the other hand, GRU is a simplified version of the traditional LSTM introduced by Cho et al. (2014), consisting of only two gates with an update gate and a reset gate. This approach not only resolves the problem effectively but also offers the advantage of improved time efficiency. The update gate regulates whether to discard or incorporate state information from the previous and current steps, whereas the reset gate governs retaining or disregarding the state information obtained from the preceding step.

In contrast to RNN, CNN has gained recognition due to its ability to extract essential spatial features from input data without prior knowledge (LeCun et al., 2015). This significance extends to weather forecasting tasks as well. CNN consists of convolutional layers, pooling layers, and fully connected layers. Convolutional layers generate feature maps by applying moving filters (or kernels) and activation functions involving dot product calculations to gridded input data. Pooling layers reduce the dimensions (width and height) of the feature map, which aids in mitigating overfitting, while returning the representative values, such as mean or maximum, using the applied moving filter. Combining convolutional and pooling layers plays a significant role in spatial feature extraction but can lead to excessive image size reduction. To prevent this, padding techniques using extra dummies (mainly zeros) are generally adopted to adjust the spatial size of the output feature maps. Finally, the fully connected layers convert lastly extracted two-dimensional feature maps to one-dimensional vectors. The flattened data are fed into the densely connected neural networks to generate the final output.

OVERVIEW OF STUDY AREA AND NUMERICAL WEATHER PREDICTION MODEL FOR CASE STUDIES

STUDY AREA: SEOUL

Our study area is the capital of South Korea, located at approximately 37.5°N and 127.0°E (Figure 12.1). Seoul is a mega-city with an area of 605 km^2 and a population of over 10 million. Geographically, Seoul is bounded by four distinct mountains, and the Han River separates it into its northern and southern halves. Due to rapid urbanization and dense infrastructures, Seoul has had a significant urban heat island (UHI) effect with a daily maximum UHI intensity of 4.5 °C on days without precipitation (Lee and Baik, 2010). Since Seoul is affected by the East Asian monsoon, it is hot and humid in summer with lots of days with hot weather over 30 °C and concentrated precipitation (i.e., seasonal rainfall was 892.1 mm in summer and 67.3 mm in winter during 1981–2010).

LOCAL DATA ASSIMILATION AND PREDICTION SYSTEM MODEL

The KMA is currently operating several NWP models based on the UK Met Office in May 2008. Among them, the Global Data Assimilation and Prediction

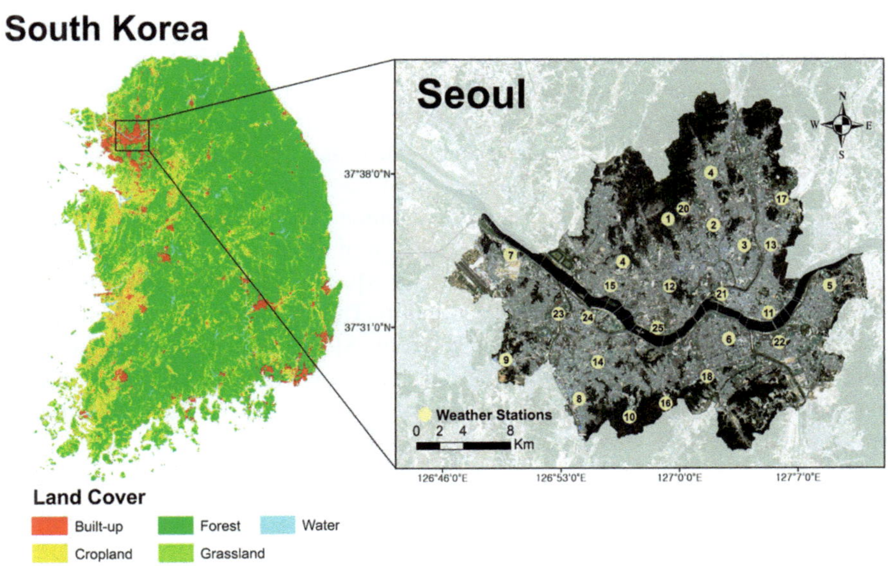

FIGURE 12.1 The study area of Seoul. The location of automatic weather stations (AWSs) used in Section 12.4 was presented.

System (GDAPS) model has been operating since May 2010, having a spatial resolution of 10–25 km over the global domain. However, due to its coarse spatial resolution, it is hard to use the GDAPS to capture extreme weather phenomena in small areas, especially in urban areas. To overcome such a limitation, the KMA is also operating the LDAPS. The LDAPS is designed to predict mesoscale weather phenomena over the Korean peninsula and the surrounding seas, performing dynamic downscaling by taking the initial condition information from the GDAPS. The LDAPS uses non-hydrostatic dynamics with semi-Lagrangian advection and semi-implicit time stepping (Orr et al., 2014). The LDAPS has a relatively high spatial resolution of 1.5 km and 70 pressure levels up to 40 km. It predicts the weather for up to 36–48 h with a time interval of 1 h every 6 h (i.e., 0, 6, 12, 18 UTC).

CASE STUDY 1: MACHINE LEARNING-BASED SPATIAL DISTRIBUTION PRODUCTION OF BIAS-CORRECTED URBAN AIR TEMPERATURE FORECAST

Data

This study focused on the hot summer season (July and August) from 2013 to 2017 since LDAPS data are available. The next-day LDAPS forecast data on the surface layer produced at 12 UTC (21 local time) were used: Tmax and Tmin, maximum and minimum relative humidity (RH), average wind speed (WS) and latent heat flux (LH), and 6-hour split average (e.g., each average of 0–5, 6–11, 12–17, and 18–23 h) cloud cover and precipitation forecast data on the next day.

The next-day Tmax and Tmin observations were acquired from 25 Automatic Weather Stations (AWSs) operated by KMA in Seoul (Figure 12.1). In addition, the present-day hourly air temperatures were also obtained to calculate the present-day Tmax and Tmin. The present-day Tmax and Tmin were employed as the maximum and minimum air temperature values between 0 and 21 local time to match the production time of the LDAPS (21 local time).

The elevation, slope, solar radiation, latitude, and longitude were used as auxiliary variables to reflect the spatial variation of air temperatures. The elevation was derived from Shuttle Radar Topography Mission (SRTM) Digital Elevation Model (DEM) with a spatial resolution of 30 m. The slope and solar radiation were obtained using the ArcGIS Spatial Analyst tool 'Slope' and 'Area Solar radiation' with DEM and day of year (DOY). The solar radiation was made on the assumption that incoming solar radiation follows the same temporal pattern by year. Lastly, latitude and longitude were extracted from the LDAPS grid.

Method

Figure 12.2 summarizes the entire procedure of the first case study. The next-day LDAPS forecast data, present-day in-situ Tmax and Tmin, and five auxiliary variables were used as input variables for forecasting next-day

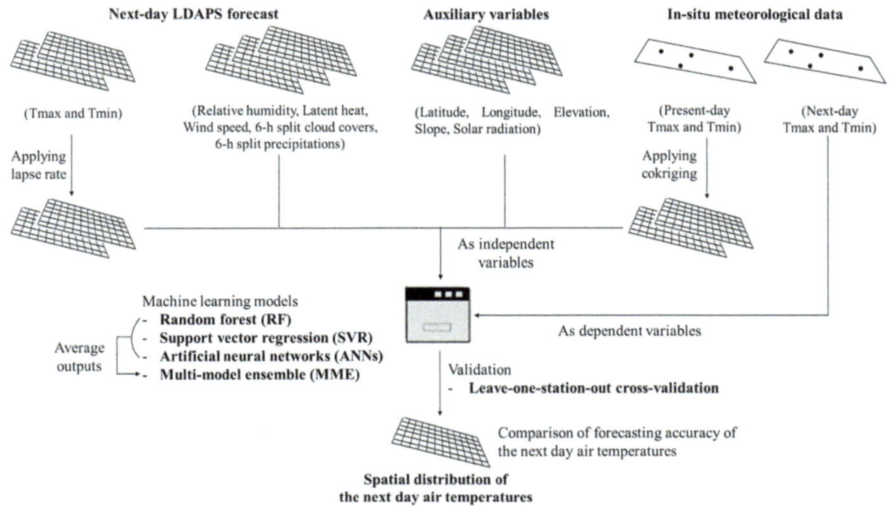

FIGURE 12.2 Flowchart of multi-model ensemble aggregating RF, SVR, and ANN proposed for producing spatial distribution maps of bias-corrected next-day Tmax and Tmin forecasts.

Tmax and Tmin. It should be noted that NWP models use the spatially smoothed elevation information as an initial condition to reduce grid-scale numerical noise. To use more reliable temperature information of the LDAPS, lapse rate equations with elevation developed by Yun et al. (2001) were adopted. To rasterize the point-scale in-situ Tmax and Tmin, the ordinary cokriging with elevation was used, which is well known to be better interpolation performance than other traditional approaches (e.g., Kriging and Inverse Distance Weighted (IDW)) for air temperatures (Ishida and Kawashima, 1993). RF, SVR, and ANN models were developed, and an MME combining RF, SVR, and ANN was then developed.

It noted that reliable and robust forecasting performance is essential for operational forecasting. A single statistical model is challenging to have universally better performance than others. With the bias-variance trade-off in statistical model, MME can improve the reliability and generalization of predictive models (Adeodato et al., 2011; Gorissen et al., 2009; Zhang et al., 2018). This study developed an MME model based on the simple average ensemble that can determine the weights of each ensemble member without additional training and is simple to construct.

To evaluate the generalization ability of spatial bias correction, hindcast-based leave-one-station-out cross validation (LOSOCV) was performed during July and August from 2015 to 2017. For example, samples at all stations except for station A from the first day of the study period to July 31, 2016 were trained to predict air temperature at station A on August 1, 2016. The coefficient of determination (R^2), bias (Equation (12.1)), and root mean square error (RMSE) were used for accuracy assessment.

$$Bias = \sum_{1}^{n} \frac{(\hat{y}_i - y_i)}{n}, \tag{12.1}$$

where y_i and $\hat{y}_i$ are the measured and predicted values, respectively, and n is the number of samples. To summarize the degree of improvement of machine learning models compared to the LDAPS, the skill score (SS; Equation (12.2)) based on RMSE was also used (Libonati et al., 2008; Wilks, 2011).

$$SS = \frac{(RMSE_{LDAPS} - RMSE_{BC})}{RMSE_{LDAPS}} \times 100\% \tag{12.2}$$

where $RMSE_{LDAPS}$ and $RMSE_{BC}$ are the RMSEs of LDAPS and bias correction models. Positive (negative) SS values indicate the improved (worsen) performance by bias correction models.

Results and Discussion

Table 12.1 shows the performance of LDAPS for next-day Tmax and Tmin forecasts by 25 stations in Seoul. The LDAPS had generally high temporal R^2 values ranging from 0.69 to 0.77 and 0.77 to 0.87 for forecasting next-day Tmax and Tmin, respectively. However, despite having a high R^2, there were stations that had a high RMSE than the average with a higher absolute bias than one degree. In particular, the RMSE and absolute bias of the LDAPS were significantly high at stations 4 and 20, where the differences between the topographic height in the LDAPS and the SRTM elevation were greater than 200 m. This means that differences in surface elevation data used as initial conditions cause large systematic errors in temperature forecasts (Zheng et al., 2017; Li et al., 2018; Zhang et al., 2009). In addition, the bias tendency of the LDAPS was different for the next-day Tmax and Tmin forecasts. Except for stations 4 and 20, the LDAPS had a cold bias at most stations such as stations 7 and 19 for the next-day Tmax forecast, whereas it had a warm bias for the next-day Tmin forecast. It implies that the LDAPS underestimates the next-day daily temperature range.

Figure 12.3 depicts the LDAPS performance and hindcast-based LOSOCV results of all bias-corrected models. The LDAPS had an R^2 of 0.69 (0.77), a bias of −0.85 °C (0.51 °C), and an RMSE of 2.08 °C (1.43 °C) for the next-day Tmax (Tmin) forecast, whereas all bias-corrected models resulted in various improvement levels, with R^2 ranging from 0.73 to 0.77 (0.83 to 0.85), bias from −0.15 to −0.07 °C (−0.05 to 0.02 °C) and RMSEs from 1.62 to 1.71 °C (1.04 to 1.13 °C). Among bias-corrected models, the RF had the highest RMSE with the lowest R^2 for forecasting both next-day Tmax and Tmin. SVR and ANN models showed similar results, but the SVR had a lower RMSE than the ANN from LOSOCV. These results imply that the SVR had a better spatial generalization ability of bias correction among the single models. On the other hand, the MME showed a slightly lower RMSE when compared to single models, which indicates that the MME improved the spatially generalized ability of bias correction (Adeodato et al., 2011; Zhang et al., 2018).

TABLE 12.1

Performance of the LDAPS by 25 Stations During July and August from 2013 to 2017 for Next-Day Tmax and Tmin Forecasts, Respectively

Station Number	Next-Day Tmax Forecast			Next-Day Tmin Forecast			ΔElev (m)
	R^2	Bias (°C)	RMSE (°C)	R^2	Bias (°C)	RMSE (°C)	
1	0.76	0.10	1.44	0.87	1.01	1.29	41
2	0.77	−0.37	1.51	0.85	1.35	1.64	52
3	0.74	0.54	1.68	0.86	0.87	1.24	3
4	0.75	−3.46	3.79	0.83	−2.14	2.35	247
5	0.71	−0.85	1.83	0.85	0.37	0.96	21
6	0.77	−0.23	1.47	0.84	0.33	0.98	−32
7	0.73	−1.58	2.18	0.81	−0.07	1.02	6
8	0.76	−0.30	1.51	0.84	0.42	1.04	−35
9	0.71	−1.51	2.23	0.82	0.40	1.06	−18
10	0.75	−0.29	1.50	0.77	1.48	1.88	−97
11	0.74	−0.04	1.58	0.83	0.69	1.15	−10
12	0.69	0.23	1.86	0.86	1.91	2.11	−64
13	0.74	0.15	1.62	0.87	0.30	0.91	−40
14	0.73	0.08	1.55	0.81	1.09	1.48	−20
15	0.69	−0.06	1.71	0.86	1.22	1.51	−5
16	0.73	−0.06	1.50	0.82	1.59	1.86	−14
17	0.74	−0.33	1.57	0.84	1.51	1.84	−14
18	0.71	−1.94	2.63	0.81	−0.13	1.01	−24
19	0.73	−0.24	1.63	0.85	1.48	1.76	−26
20	0.76	−3.57	3.86	0.84	−1.75	1.97	230
21	0.75	0.58	1.63	0.83	0.45	1.07	20
22	0.73	−0.80	1.78	0.82	0.39	1.09	−6
23	0.73	−1.71	2.32	0.82	0.11	1.00	7
24	0.75	−0.88	1.73	0.82	0.20	1.02	−1
25	0.75	−0.14	1.57	0.84	0.60	1.14	−7
Average	0.74	−0.67	1.91	0.83	0.55	1.38	9

Note: ΔElev Represents the Difference between Topography Height in the LDAPS and SRTM Elevation Aggregated to 1.5 km.

The SS values of each bias correction model were analyzed by 25 stations (Figure 12.4). The SS values were highly different by stations, but all bias-corrected models had SS values between 20% and 60% at stations where the LDAPS had a remarkable systematic bias (e.g., stations 4, 7, 20, and 23 (stations 4, 10, 12, and 20) for the next-day Tmax (Tmin) forecast). From this result, it can be judged that machine learning models can correct the systematic bias of the LDAPS where stations are not provided in Seoul. For the single models, the SS value by stations differed from model to model. Contrastively,

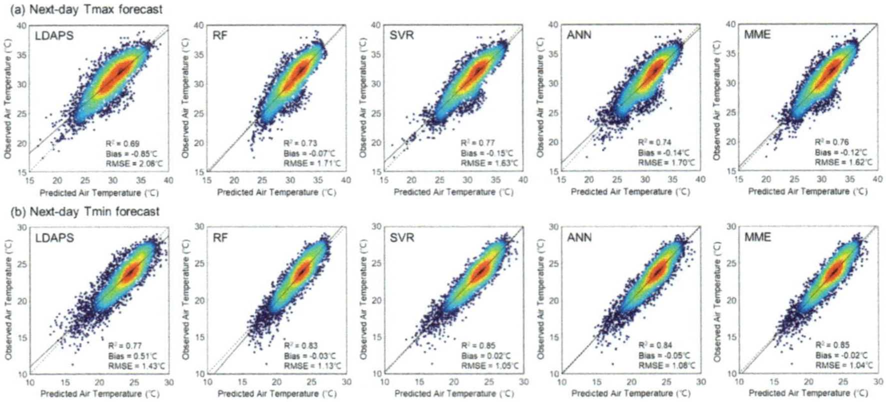

FIGURE 12.3 Results of the LDAPS and hindcast-based LOSOCV from RF, SVR, ANN, and MME during July and August from 2015 to 2017 for forecasting next-day Tmax and Tmin.

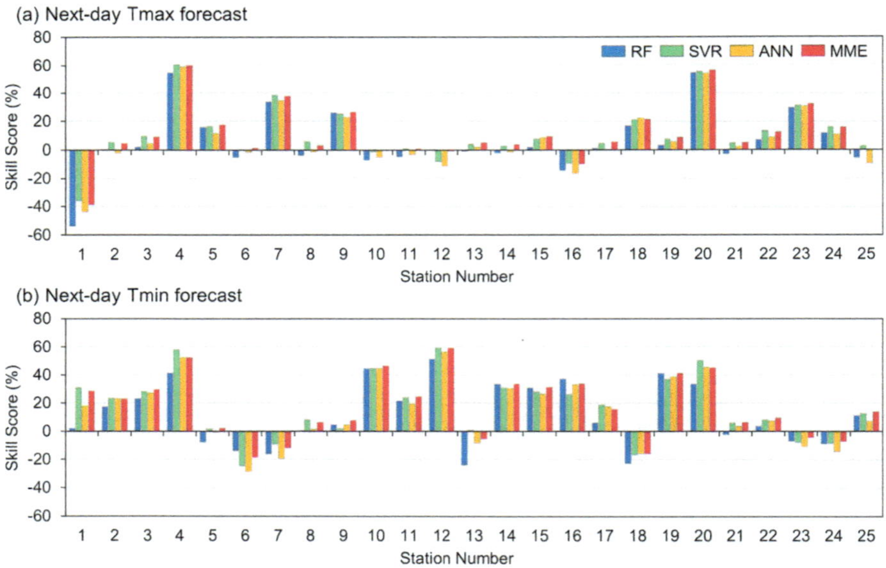

FIGURE 12.4 Skill score values from the hindcast-based LOSOCV results from RF, SVR, ANN, and MME by 25 stations.

the MME had similar values to the highest SS among the three single machine learning models at almost stations. This means the MME improved the robustness of single machine learning models. However, all bias-corrected models including the MME had negative SS values at some stations (e.g., stations 1 and 16 (stations 6 and 7) for the next-day Tmax (Tmin) forecast), where the LDAPS had no systematic bias close to zero and lower RMSE than that at other stations.

Figures 12.5 and 12.6 show the average spatial distribution maps of the LDAPS and all bias-corrected models for next-day Tmax and Tmin forecasts during 2015 to 2017. The LDAPS resulted in smoothed and clustered distribution that did not simulate the complex urban environment, showing low temperatures in the northern region around stations 4 and 20 where the initial elevation of the LDAPS was significantly different from the actual elevation. All bias-corrected models were similar distribution for forecasting both next-day Tmax and Tmin. The corrected temperature distributions were generally higher (lower) than the temperature distribution of the LDAPS for the next-day Tmax (Tmin) forecast. This implies that all models successfully corrected cold and warm biases of the LDAPS for

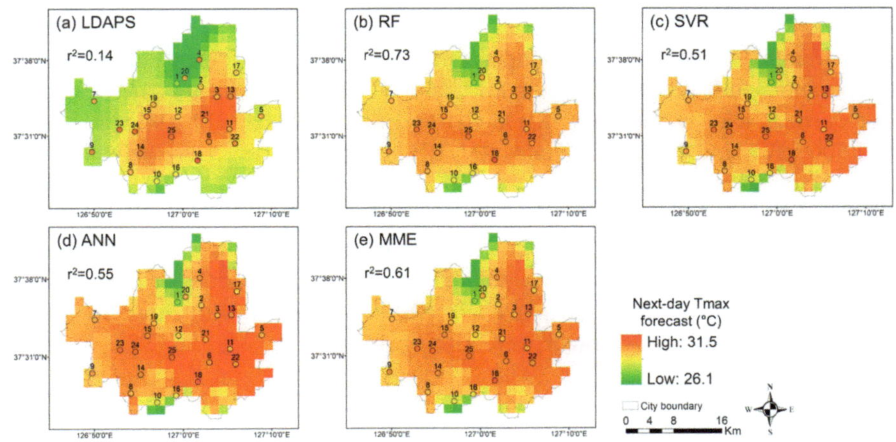

FIGURE 12.5 Spatial distribution maps of the next-day Tmax forecast from LDAPS, RF, SVR, ANN, and MME averaged during July and August from 2015 to 2017.

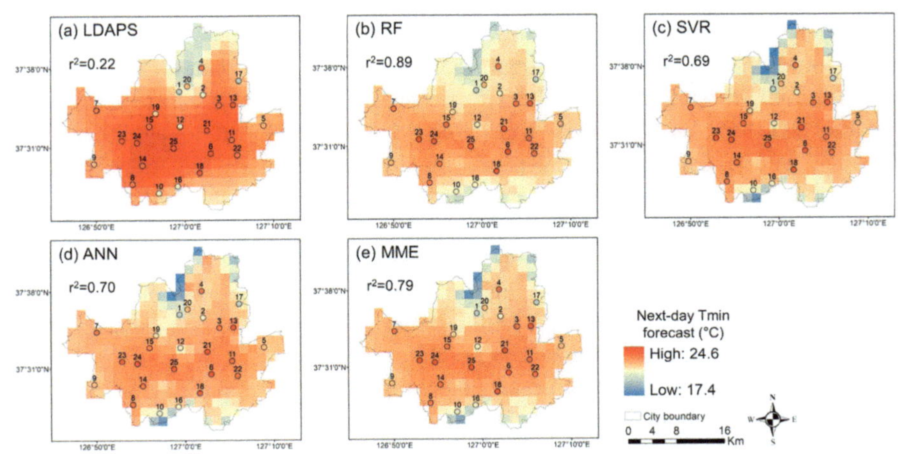

FIGURE 12.6 Spatial distribution maps of the next-day Tmin forecast from LDAPS, RF, SVR, ANN, and MME averaged during July and August from 2015 to 2017.

next-day Tmax and Tmin forecasts, respectively. Interestingly, the RF had narrower temperature ranges than the other models, but it had the highest spatial correlation values (R^2 of 0.73 (0.89) for next-day Tmax (Tmin) forecast). The MME showed the results of integrating the advantages of RF, which has high spatial correlation, and SVR and ANN, which have higher temperature ranges than the RF.

CASE STUDY 2: DEEP LEARNING-BASED POST-PROCESSING OF URBAN AIR TEMPERATURE FORECAST

DATA

The present-day three hourly analysis data and the next-day hourly forecast data for isobaric surface and grid surface data were used in this study. In total, 70 variables related to air temperature were used as potential predictors: temperature, RH, geopotential height, U and V wind components, and horizontal and vertical wind velocity at four different pressure levels (i.e., 925, 850, 700, and 500 hPa); and Tmax and Tmin, RH, LH, 10-m wind direction and velocity at the surface layer.

The study period was on the summer season (June-August) from 2013 to 2020. The observed next-day Tmax was obtained from 24 AWSs in Seoul, which satisfied the following two conditions. First, stations where the installation location changed during the study period were excluded. Second, stations with more than 10% missing values for each summer were also excluded.

METHOD

This study developed and compared four different single post-processing approaches with their ensemble model (MME). MLR, a method used for operational forecasting, and SVR, which performed well in case study 1, were selected. In addition, GRU and CNN were selected among the deep learning approaches. MLR and SVR used daily meteorological data at the weather stations as input. Meanwhile, the GRU used sequential hourly meteorological data as input to learn their diurnal cycle trend, and the CNN learned spatial information around the weather stations. Lastly, MME based on SS was developed to aggregate their outputs. Before developing post-processing models, feature selection based on Pearson's correlation coefficient was performed by station to find the optimal combination of features (Chandrashekar and Sahin, 2014; Hall, 1999). All meteorological variables, except for Tmax and Tmin, were converted into daily mean data. And then, the final input variables having significant correlation coefficient values (≥ 0.3 or ≤ -0.3) with a target variable during the training period were selected among the 70 potential predictors by stations.

The GRU learned diurnal time-series data of present-day three hourly analysis data and next-day hourly forecast data. It noted that analysis and forecast data of LDAPS has different sequence sizes. So, this study developed the GRU composing of two branches. These two branches consist of eight and 24 consecutive data for analysis and forecast data, respectively (Figure 12.7(a)), having the same structure

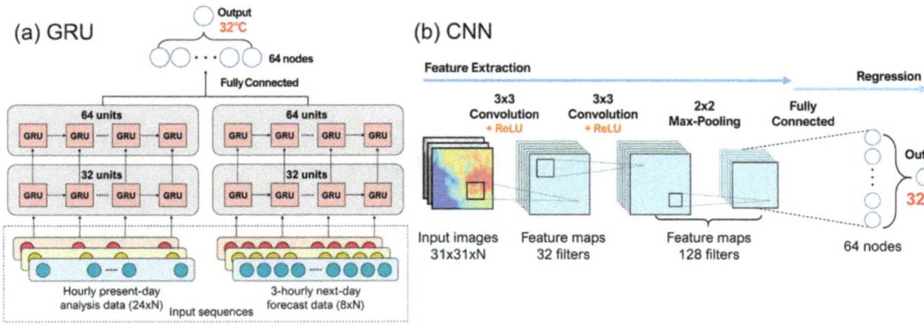

FIGURE 12.7 The architectures of (a) GRU and (b) CNN used in Case Study 2.

consisting of two stacked GRU layers with 32 and 64 units in the first and second layers, respectively. Lastly, the merge and fully connected layers with 64 nodes and the ReLU activation function were constructed to combine the outputs of the two branches and produce the final prediction value, respectively.

Meanwhile, the CNN learned 31 × 31 spatial features of meteorological variables from the LDAPS surrounding each station. The connected structure of two 3 × 3 convolutional layers with 32 filters and 128 filters, and one 2 × 2 max-polling layer was selected (Figure 12.7(b)). After the max-polling layer, a fully connected layer with 64 nodes and the ReLU activation function was used. For training GRU and CNN, the Mean Absolute Error (MAE) and Adam optimizer were used as loss function and optimizer, respectively. Among the training samples, 90% and 10% were randomly divided for training and validation to find the optimal weight of parameters. The batch size and epochs were set to 64 and 300 for the GRU and 32 and 700 for the CNN.

Unlike the producing bias-corrected spatial distribution performed in case study 1, the weighted average ensemble (WAE) can be an effective MME approach in developing post-processing at each seen (known) station reflecting the relative significance of the ensemble members (Adhikari and Agrawal, 2012; Fathi et al., 2019). This study developed an MME model that calculates the weight of ensemble members based on SS (Equation (12.3)) and used LDAPS and four different single post-processing models (i.e., MLR, SVR, GRU, and CNN) as ensemble members.

$$W_k = \frac{SS_k - \min(SS_k)}{\Sigma(SS_k - \min(SS_k))} \quad (12.3)$$

where k is the ensemble member. It noted that SS indicates the accuracy improvement by the post-processing model compared to the LDAPS, so the SS value of the LDAPS is zero. Thus, if all post-processing models outperform the LDAPS, then LDAPS does not contribute to the final prediction. However, if at least one post-processing model is worse than the LDAPS, then the corresponding post-processing model does not contribute to the final prediction. Performance over the 30 days prior to the forecast day was used to calculate the SS.

For model evaluation, hindcast validation was conducted from 2019 to 2020. This study the difference in SS (dSS), which indicates the accuracy improvement of machine learning compared to MLR.

$$dss = SS_{ML} - SS_{MLR} \quad (12.4)$$

where SS_{ML} and SS_{MLR} are the SS values of machine learning and MLR models, respectively.

Results and Discussion

As a hindcast validation result for the summer of 2019–2020 (Figure 12.8), the SVR showed comparable performance to MLR, showing insignificant dSS values at

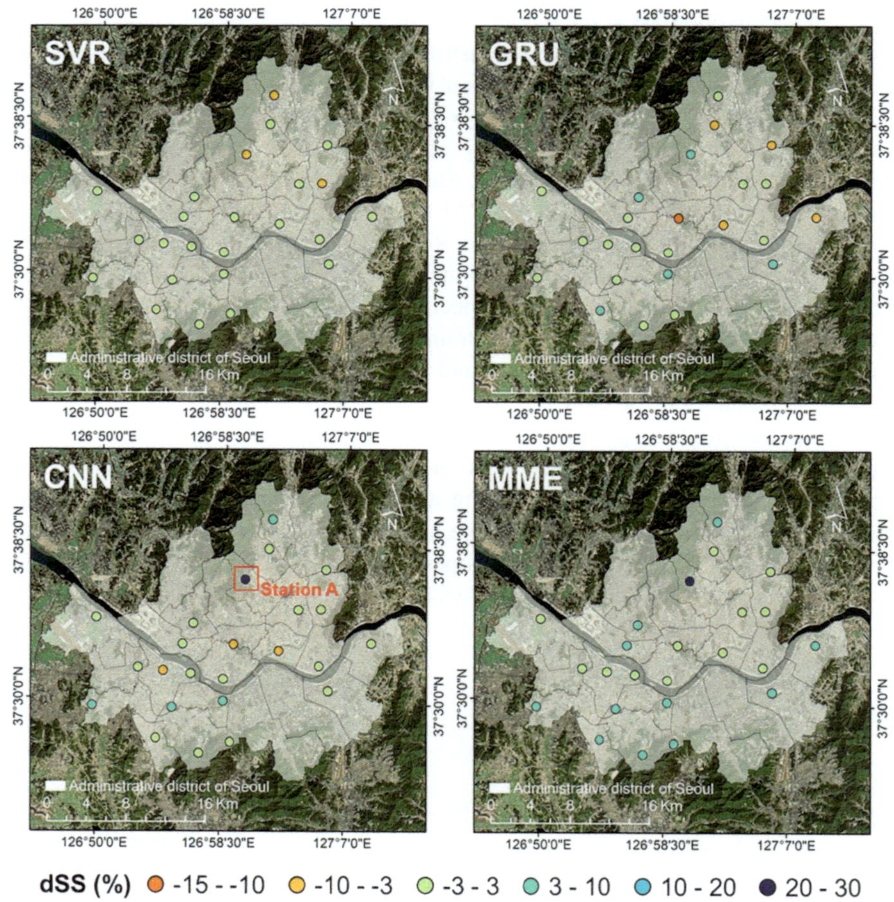

FIGURE 12.8 The dSS values from the hindcast validation results based on SVR, GRU, and CNN by 24 stations. The dSS values are displayed by the colors, and station A is used for further analysis.

most stations. In the case of GRU and CNN, dSS values were different for stations, and they showed dSS values of ≤ −3% and ≥3% at some stations. On the other hand, the MME had a more reliable and robust performance than the others, with dSS values of ≥3% at most stations. Interestingly, CNN and MME significantly improved performance with dSS values of ≥20% at station A.

The temporal pattern of the next-day Tmax forecast based on LDAPS and all post-processing models and in-situ observations was analyzed at station A during the validation period (Figure 12.9(a)). As a result, LDAPS tended to underestimate the next-day Tmax at the beginning of the validation period, while all post-processing models showed similar patterns to in-situ observations by reducing such a tendency of LDAPS. However, after the middle of the validation period, the LDAPS had more accurate results than the previous, not showing an under-estimation tendency. Despite this LDAPS model's bias change, MLR and SVR still produced higher values than the raw forecast of LDAPS, resulting in over-estimation. On the other hand, CNN and MME were robust to such bias change of the LDAPS, showing similar results to the in-situ observations.

When analyzing the yearly bias change of LDAPS (Figure 12.9(b)), it was insignificant at the average of 24 stations, while it was significant at station A, especially between 2019 and 2020. The change in the initial conditions around station A can be a primary reason for the bias change of LDAPS, but further studies are needed to analyze it in the future. As a result of analyzing the correlation between the bias change of LDAPS at 24 stations and the dSS of the CNN (Figure 12.9(c)), it showed a very high correlation with an r of 0.9. These results indicate that the CNN is robust to the bias change of LDAPS.

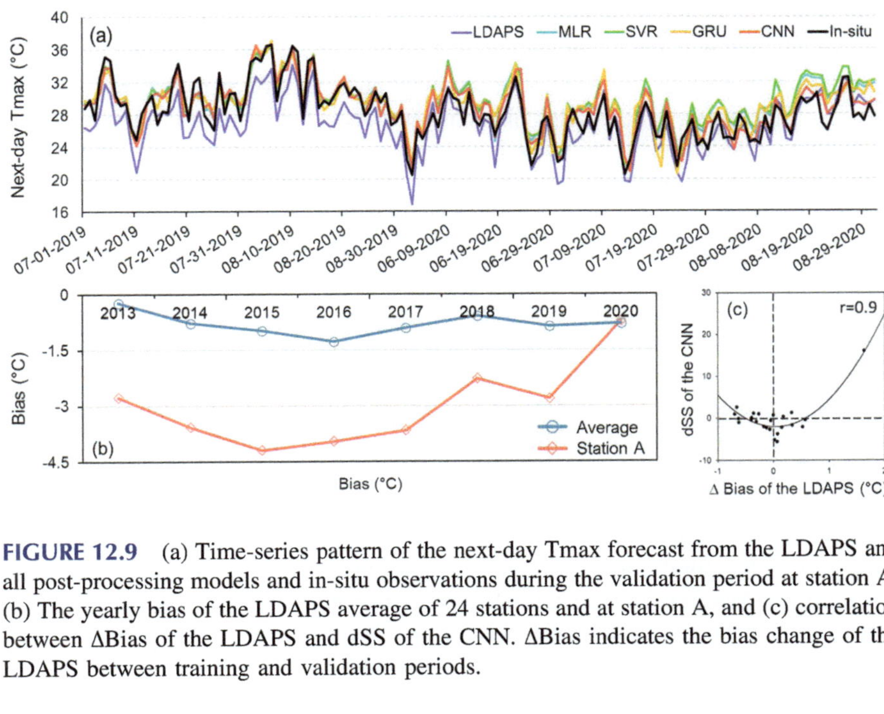

FIGURE 12.9 (a) Time-series pattern of the next-day Tmax forecast from the LDAPS and all post-processing models and in-situ observations during the validation period at station A. (b) The yearly bias of the LDAPS average of 24 stations and at station A, and (c) correlation between ΔBias of the LDAPS and dSS of the CNN. ΔBias indicates the bias change of the LDAPS between training and validation periods.

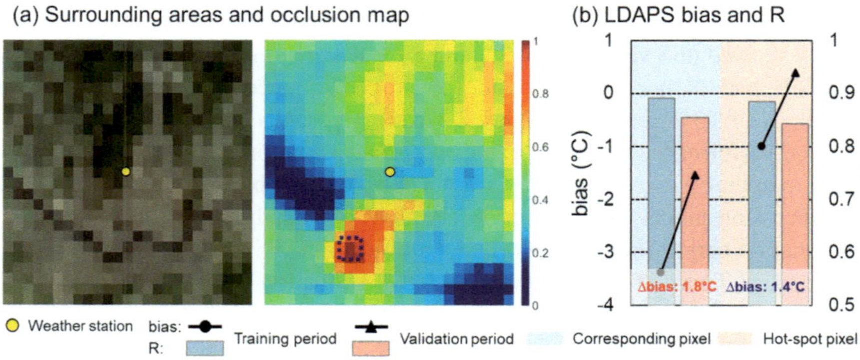

FIGURE 12.10 (a) Surrounding area at station A and occlusion map for the CNN. The yellow dot indicates the location of station A (corresponding pixel). The blue dashed box represents the location with the highest spatial contribution (hot-spot pixel). (b) The bias and R between the next-day Tmax forecasts of the LDAPS at corresponding and hot-spot pixels and in-situ observations at station A. ΔBias indicates the bias change of the LDAPS between training and validation periods.

To investigate why CNN can be robust to the bias change of LDAPS, the sensitivity analysis on CNN was further performed by measuring the degree of validation accuracy decrease when each pixel is converted to meaningless (out-of-range) values. As a result, a hot-spot pixel that has the greatest influence on the CNN performance was different from the weather station location (corresponding pixel) (Figure 12.10(a)). The next-day Tmax forecasts of LDAPS at the hot-spot and corresponding pixels were compared with the observations at station A (Figure 12.10(b)). LDAPS forecast at those two pixels showed a high temporal correlation with observations at both training and validation periods. However, the hot-spot pixel showed a smaller bias change between the training and validation periods than the corresponding pixel. This result means that CNN can alleviate the bias change of LDAPS by using spatial information with a high temporal correlation while having a much lower bias change.

CONCLUSION

This chapter produces two case studies on improving the forecast accuracy of an urban summer air temperature of the NWP model using AI. The first study successfully produced the spatial distribution of bias-corrected urban air temperature forecasts using AI. In addition, it improved the generalization ability through the MME model with an ensemble of three different machine learning models (i.e., RF, SVR, and ANN). The second study confirmed that CNN can be robust to the bias change of the operational NWP model by finding and using spatial information with a high temporal correlation and a small bias change of the NWP model. In addition, SS-based ensemble learning showed more reliable and robust performance when compared to single models, reflecting the relative significance of single models by stations. These two suggested machine learning-based approaches can be adopted in the operational forecasting of urban air temperature in summer. In addition, the findings of these two studies suggest that image-to-image translation (e.g., U-Net) can be more effective in producing spatial distribution maps of urban air temperature forecasts with improved accuracy for operational NWP models.

REFERENCES

Adeodato, P. J., Arnaud, A. L., Vasconcelos, G. C., Cunha, R. C., & Monteiro, D. S. (2011). MLP ensembles improve long term prediction accuracy over single networks. *International Journal of Forecasting*, 27(3), 661–671.

Adhikari, R., & Agrawal, R. K. (2012). A novel weighted ensemble technique for time series forecasting. In *Advances in Knowledge Discovery and Data Mining: 16th Pacific-Asia Conference, PAKDD 2012, Kuala Lumpur, Malaysia, May 29-June 1, 2012, Proceedings, Part I 16* (pp. 38–49). Springer Berlin Heidelberg.

Baniassadi, A., Heusinger, J., & Sailor, D. J. (2018). Energy efficiency vs resiliency to extreme heat and power outages: The role of evolving building energy codes. *Building and Environment*, 139, 86–94.

Bauer, P., Thorpe, A., & Brunet, G. (2015). The quiet revolution of numerical weather prediction. *Nature*, 525(7567), 47–55.

Breiman, L. (2001). Random forests. *Machine Learning*, 45, 5–32.

Chandrashekar, G., & Sahin, F. (2014). A survey on feature selection methods. *Computers & Electrical Engineering*, 40(1), 16–28.

Chen, K., Wang, P., Yang, X., Zhang, N., & Wang, D. (2020). A model output deep learning method for grid temperature forecasts in Tianjin area. *Applied Sciences*, 10(17), 5808.

Cho, D., Yoo, C., Im, J., & Cha, D. H. (2020). Comparative assessment of various machine learning-based bias correction methods for numerical weather prediction model forecasts of extreme air temperatures in urban areas. *Earth and Space Science*, 7(4), e2019EA000740.

Cho, D., Yoo, C., Son, B., Im, J., Yoon, D., & Cha, D. H. (2022). A novel ensemble learning for post-processing of NWP Model's next-day maximum air temperature forecast in summer using deep learning and statistical approaches. *Weather and Climate Extremes*, 35, 100410.

Cho, K., Van Merriënboer, B., Gulcehre, C., Bahdanau, D., Bougares, F., Schwenk, H., & Bengio, Y. (2014). Learning phrase representations using RNN encoder-decoder for statistical machine translation. *arXiv preprint arXiv:1406.1078*.

Coffel, E. D., Horton, R. M., & De Sherbinin, A. (2017). Temperature and humidity based projections of a rapid rise in global heat stress exposure during the 21st century. *Environmental Research Letters*, 13(1), 014001.

de Carvalho, J. R. P., Assad, E. D., & Pinto, H. S. (2011). Kalman filter and correction of the temperatures estimated by PRECIS model. *Atmospheric Research*, 102(1–2), 218–226.

Drucker, H., Burges, C. J., Kaufman, L., Smola, A., & Vapnik, V. (1996). Support vector regression machines. *Advances in Neural Information Processing Systems*, 9.

Fathi, M., Azadi, M., Kamali, G., & Meshkatee, A. H. (2019). Improving precipitation forecasts over Iran using a weighted average ensemble technique. *Journal of Earth System Science*, 128, 1–17.

Glahn, H. R., & Lowry, D. A. (1972). The use of model output statistics (MOS) in objective weather forecasting. *Journal of Applied Meteorology and Climatology*, 11(8), 1203–1211.

Gong, B., Langguth, M., Ji, Y., Mozaffari, A., Stadtler, S., Mache, K., & Schultz, M. G. (2022). Temperature forecasting by deep learning methods. *Geoscientific Model Development*, 15(23), 8931–8956.

Gorissen, D., Dhaene, T., & De Turck, F. (2009). Evolutionary model type selection for global surrogate modeling. *Journal of Machine Learning Research*, 10, 2039–2078.

Hall, M. A. (1999). *Correlation-based feature selection for machine learning* (Doctoral dissertation, The University of Waikato).

Hifumi, T., Kondo, Y., Shimizu, K., & Miyake, Y. (2018). Heat stroke. *Journal of Intensive Care*, 6(1), 1–8.

Hochreiter, S. (1998). The vanishing gradient problem during learning recurrent neural nets and problem solutions. *International Journal of Uncertainty, Fuzziness and Knowledge-Based Systems*, 6(02), 107–116.

Hochreiter, S., & Schmidhuber, J. (1997). Long short-term memory. *Neural Computation*, 9(8), 1735–1780.

Hong, S. Y., Kwon, Y. C., Kim, T. H., Esther Kim, J. E., Choi, S. J., Kwon, I. H., … & Kim, D. I. (2018). The Korean Integrated Model (KIM) system for global weather forecasting. *Asia-Pacific Journal of Atmospheric Sciences*, 54, 267–292.

Ishida, T., & Kawashima, S. (1993). Use of cokriging to estimate surface air temperature from elevation. *Theoretical and Applied Climatology*, 47, 147–157.

Jacks, E., Bower, J. B., Dagostaro, V. J., Dallavalle, J. P., Erickson, M. C., & Su, J. C. (1990). New NGM-based MOS guidance for maximum/minimum temperature, probability of precipitation, cloud amount, and surface wind. *Weather and Forecasting*, 5(1), 128–138.

Jain, A. K., Mao, J., & Mohiuddin, K. M. (1996). Artificial neural networks: A tutorial. *Computer*, 29(3), 31–44.

Kalnay, E. (2003). *Atmospheric modeling, data assimilation and predictability*. Cambridge University Press.

Kingma, D. P., & Ba, J. (2014). *Adam: A method for stochastic optimization*. arXiv preprint arXiv:1412.6980.

Klinenberg, E. (2015). *Heat wave: A social autopsy of disaster in Chicago*. University of Chicago Press.

LeCun, Y., Bengio, Y., & Hinton, G. (2015). Deep learning. N*ature*, 521(7553), 436–444.

Lee, S. H., & Baik, J. J. (2010). Statistical and dynamical characteristics of the urban heat island intensity in Seoul. *Theoretical and Applied Climatology*, 100, 227–237.

Li, Y., Zhao, C., Zhang, T., Wang, W., Duan, H., Liu, Y., … & Pu, Z. (2018). Impacts of land-use data on the simulation of surface air temperature in Northwest China. *Journal of Meteorological Research*, 32(6), 896–908.

Libonati, R., Trigo, I., & DaCamara, C. C. (2008). Correction of 2 m-temperature forecasts using Kalman filtering technique. *Atmospheric Research*, 87(2), 183–197.

Nair, V., & Hinton, G. E. (2010). Rectified linear units improve restricted boltzmann machines. In Proceedings of the 27th international conference on machine learning (ICML-10) (pp. 807–814).

Orr, A., Phillips, T., Webster, S., Elvidge, A., Weeks, M., Hosking, S., & Turner, J. (2014). Met Office Unified Model high-resolution simulations of a strong wind event in Antarctica. *Quarterly Journal of the Royal Meteorological Society*, 140(684), 2287–2297.

Park, I., Kim, H. S., Lee, J., Kim, J. H., Song, C. H., & Kim, H. K. (2019). Temperature prediction using the missing data refinement model based on a long short-term memory neural network. *Atmosphere*, 10(11), 718.

Perkins, S. E., Alexander, L. V., & Nairn, J. R. (2012). Increasing frequency, intensity and duration of observed global heatwaves and warm spells. *Geophysical Research Letters*, 39(20).

Russo, S., Sillmann, J., Sippel, S., Barcikowska, M. J., Ghisetti, C., Smid, M., & O'Neill, B. (2019). Half a degree and rapid socioeconomic development matter for heatwave risk. *Nature communications*, 10(1), 136.

Schulze, H., & Langenberg, H. (2014). Urban heat. *Nature Geoscience*, 7(8), 553–553.

Shin, Y., & Yi, C. (2019). Statistical downscaling of urban-scale air temperatures using an analog model output statistics technique. *Atmosphere*, 10(8), 427.

Stewart, I. D. (2011). A systematic review and scientific critique of methodology in modern urban heat island literature. *International Journal of Climatology*, 31(2), 200–217.

Whan, K., Zscheischler, J., Jordan, A. I., & Ziegel, J. F. (2021). Novel multivariate quantile mapping methods for ensemble post-processing of medium-range forecasts. *Weather and climate extremes*, 32, 100310.

Wilks, D. S. (2011). *Statistical methods in the atmospheric sciences* (Vol. 100). Academic Press.

Yi, C., Shin, Y., & Roh, J. W. (2018). Development of an urban high-resolution air temperature forecast system for local weather information services based on statistical downscaling. *Atmosphere*, 9(5), 164.

Yun, J. I., Choi, J. Y., & Ahn, J. H. (2001). Seasonal trend of elevation effect on daily air temperature in Korea. *Korean Journal of Agricultural and Forest Meteorology*, 3(2), 96–104.

Zhang, C., Smith, M., & Fang, C. (2018). Evaluation of Goddard's LiDAR, hyperspectral, and thermal data products for mapping urban land-cover types. *GI Science & Remote Sensing*, 55(1), 90–109.

Zhang, G., Zhu, S., Zhang, N., Zhang, G., & Xu, Y. (2022). Downscaling hourly air temperature of WRF simulations over complex topography: A case study of Chongli District in Hebei Province, China. *Journal of Geophysical Research: Atmospheres*, 127(3), e2021JD035542.

Zhang, J., Cha, D. H., & Lee, D. K. (2009). Investigating the role of MODIS leaf area index and vegetation-climate interaction in regional climate simulations over Asia. *Terrestrial, Atmospheric & Oceanic Sciences*, 20(2).

Zhang, Z., & Dong, Y. (2020). Temperature forecasting via convolutional recurrent neural networks based on time-series data. *Complexity*, 2020, 1–8.

Zheng, W., Ek, M., Mitchell, K., Wei, H., & Meng, J. (2017). Improving the stable surface layer in the NCEP global forecast system. *Monthly Weather Review*, 145(10), 3969–3987.

Part IV

Digital Cities

13 GIS-Based Modeling for Estimating Urban Carbon Emissions

Cai Meng and Ren Chao

INTRODUCTION

More than half of the world's population lives in cities, which account for about 64% of global primary energy use and produce over 70% of carbon emissions from energy activities (International Energy Agency, 2016; IPCC, 2014). Thus, cities are the main targets for reducing carbon emissions and mitigating climate change. The United Nations estimates that the major increase in the global population will happen in cities from 2012 to 2050 (UN DESA, 2018). World carbon emissions are expected to increase due to continuous urbanization, with most of this growth occurring in developing countries.

Urban carbon emissions estimation is the basis of carbon emission reduction and low-carbon development. The spatial modeling of urban carbon emissions can facilitate the development of spatially distributed emission inventories and reveal the spatial patterns within an administrative unit (Ou et al., 2015). These inventories can provide a spatial visualization of the carbon emissions from both production and consumption. Therefore, it enables pragmatic and realistic assessment of emission mitigation, such as identifying the carbon emission hotspots, energy activities responsible for high emissions, etc. (Kanemoto et al., 2016), and lays the foundation for strategies and actions to mitigate climate change (Rong et al., 2020; Zhao et al., 2019). They can also be integrated with other spatial data to facilitate interdisciplinary cooperation to reduce carbon emissions. A group of scientists working on climate change issues have further appealed for the priority given to high-quality and high-resolution emission inventories and to understand the interactions between cities and climate for climate change mitigation (Bai et al., 2018). Thus, there is an urgent need for an in-depth examination of the spatial heterogeneity of urban CO_2 emissions for cities.

The recent development of geospatial techniques can serve as a cutting-edge tool to model the spatial distribution of urban carbon emissions. GIS-based modeling of urban carbon emissions has emerged as a powerful tool for estimating and mapping carbon emissions from urban areas, providing valuable insights for policymakers and urban planners. In order to assess carbon emissions and facilitate practical mitigation strategies, diverse GIS-based methodologies have been developed to

DOI: 10.1201/9781003244561-17

model the spatial variations of carbon emissions, including top-down analyzes that assign the emissions from a coarse spatial unit to a finer resolution, bottom-up models that aggregate the fine emission data such as point source emissions to the desired spatial grid, and hybrid methods involving both bottom-up and top-down approach. In order to further develop our knowledge in controlling and reducing carbon emissions, this chapter will explore the use of GIS-based modeling for estimating urban carbon emissions, highlighting key methods and applications in the field.

DATA SOURCES AND METHODOLOGIES FOR GIS-BASED EMISSIONS MODELING

Top-Down Method Using Nighttime Lights

Top-down methods allocate the carbon emissions within a large spatial unit to a high spatial resolution by adopting some algorithms or proxy data such as land use (Chuai & Feng, 2019), road length (Song et al., 2020). With the development of remote sensing techniques, the nighttime light (NTL) data have been the most frequently used proxy to distribute the statistical carbon emissions at the jurisdiction level. It can potentially reflect the socioeconomic conditions and human activities that are relevant to energy demand. The Defense Meteorological Satellite Program-Operational Linescan System (DMSP-OLS) and Visible Infrared Imaging Radiometer Suite onboard the Suomi National Polar-Orbiting Partnership (NPP-VIIRS) are the two most adopted NTL datasets.

There are generally two main procedures for downscaling statistical carbon emissions at the jurisdiction level. The first procedure is data preparation, including the calculation of statistical carbon emissions at the administrative level from energy reports, the calibration of the NTL data, and the extraction of urban areas. Secondly, by establishing the relationship between statistical carbon emissions and NTL, an emission value is assigned to each pixel of NTL on this basis and integrated into the urban scale.

Ordinary Least Squares (OLS) regression methods can be utilized to build the linear relationship between NTL data and carbon emissions. The lack of data verification can weaken the reliability of the simple linear relationship between NTL and statistical emissions (Wang et al., 2014). Besides, only the spatial or temporal relationship can be explored by linear regression models, which may cause deviations in modeling carbon emissions in both space and time dimensions. Panel data analysis can simultaneously link the statistical emissions and NTL in spatiotemporal dimensions. Therefore, panel data models have already been used for carbon emissions estimation (Cui et al., 2019; Han et al., 2018; Zhang et al., 2021). Moreover, the geographical and temporally weighted regression (GTWR) model has also been used to build the relationship between statistical carbon emissions and NTL data due to its considerations of spatial and temporal heterogeneity of CO_2 emissions. Using the GTWR and NPP-VIIRS, Zhang et al. (2020) mapped CO_2 emissions from coal boilers, thermal power plants, and natural gas boilers in 15 northern provinces from 2012 to 2017 with a 5 km × 5 resolution.

Bottom-Up Method

Bottom-up methods to model urban carbon emissions generally integrate emissions at a point or sectoral level and then allocate the emissions into the designated spatial unit. There are some open bottom-up inventories at the facility level, such as the China Cement Emission database (Liu et al., 2021), China coal-fired Power plant Emissions Database (F. Liu et al., 2015), the Vulcan dataset of the United States (http://vulcan.project.asu.edu/research.php.). The Vulcan dataset was developed by a bottom-up approach using seven major datasets containing the road networks, spatial information of point emission sources, the floor and areas of buildings, etc. (Gurney et al., 2009).

There are also other studies to develop carbon emission maps using bottom-up approaches. Household and personal surveys have also been applied to plot the carbon emissions (Rong et al., 2020; Yang et al., 2015). Wu et al. (2018) established a database for energy use intensity (EUI) from the Shanghai building energy efficiency monitoring platform for each building function and mapped the emissions of Shanghai based on the EUI and the building function. Zhang et al. (2018) deployed a traffic allocation model to simulate traffic in road networks through a gasoline consumption function. With the advances in big data development, taxi Global Positioning System (GPS) trajectory data from taxi companies have been adopted to map high-resolution taxi emissions by daily travel (Wu et al., 2023; Xia et al., 2020; Zhao et al., 2017). Furthermore, Wen et al. (2022) predicted real-time link-level vehicle emissions by using high-density traffic monitoring data and land use random forest models.

Hybrid Method

Hybrid approaches involving both bottom-up and top-down methods have been implemented to retain the accuracy of the bottom-up method and the efficiency of the top-down method. Rayner et al. (2010) constructed global gridded fossil fuel carbon emission quantification through the Fossil Fuel Data Assimilation System (FFDAS) that combined some elements of downscaling, bottom-up information, and data assimilation within a model of fossil fuel carbon emissions to disaggregate national emissions to a 0.25° global grid optimally. It is calculated based on a modified Kaya identity which expresses emissions as a product of areal population density, per capita economic activity, energy intensity of the economy, and carbon intensity of energy. There is also an Open-source Data Inventory for Anthropogenic Carbon dioxide (ODIAC) developed by the National Institute for Environmental Studies in Japan (Oda & Maksyutov, 2015). The ODIAC generates fossil fuel CO_2 emissions on a 1 km grid based on satellite observations of nighttime lights and a geocoded estimation of power plant CO_2 emissions.

The China High Resolution Emission Database (CHRED) was created by the Chinese Academy for Environmental Planning (Cai et al., 2018). The CHRED has the finest spatial resolution so far (1 km) among all the available emission data sources in China. The dataset was created primarily based on point emission sources from the industrial sector and complementary socioeconomic information from

Mainland China. Industrial emissions were modeled by the bottom-up approach, taking advantage of the point emission sources at the facility level containing approximately 1.5 million enterprises from the First China Pollution Source Census (FCPSC) dataset. Emissions from the transport, services, and agricultural sectors of the CHRED were disaggregated from statistical data using social-economic and urban land use data using top-down approaches. In addition, J. Wang et al. (2014) constructed a high spatial-resolution (10 km) map of carbon emissions for China, with emissions from the industrial and residential sectors also generated from the FCPSC dataset.

Tsinghua University developed The Multi-resolution Emission Inventory for China (MEIC, http://meicmodel.org/) (Li et al., 2017; Zheng et al., 2018). The dataset contains the spatial distribution of ten air pollutants and CO_2 in mainland China with the finest grid of 0.25° × 0.25°. The carbon emissions from power, cement, and steel industries are generated from two bottom-up inventories, i.e., the China Cement Emission database (Liu et al., 2021) and the Global Power Emissions Database (Tong et al., 2018). Important proxy data in the top-down method for downscaling emissions include population, roads, and power plants.

Apart from using the standardized framework, localized hybrid methods have also been applied in many cities. Furthermore, Cai et al. (2020) adopted the hybrid method for developing the spatial inventory in Hong Kong with a resolution of 100 m. Industrial and airport emissions were modeled using the bottom-up method. In the top-down approach, urban form and traffic flow were used as proxies for building and traffic emissions.

Cross-Comparison of the Methods

The characteristics of the three methods are summarized and compared in this section (Table 13.1). The top-down method, which uses NTL data, is widely applied at global, national, and regional scales due to its simplicity, public availability, and efficiency, achieving a medium level of 70% to 90%. However, the method may not accurately predict human activities relevant to transport and industrial processes since NTL data reflect socioeconomic characteristics instead of fossil fuel combustion during the night (Ghosh et al., 2010). Additionally, this method cannot perform well in developing countries since the relationship between the NTL pixel value and carbon emissions is less significant in developing countries than in developed countries (Doll et al., 2000). This may affect the accuracy of the top-down method in modeling the carbon emissions in developing countries with rapid urbanization and industrialization like China.

Through requiring more detailed data, the bottom-up method usually provides the most accurate results. In particular, the bottom-up method is mainly implemented by government authorities because it is suitable for local-scale assessment and, therefore, favored by governors at the city, district, community, or household levels. There are also limitations of the method. The bottom-up method provides more accurate results but requires extensive resources and detailed data, limiting its applicability and comparability across different cities (Jing et al., 2018).

CASE STUDIES OF GIS-BASED MODELING FOR URBAN CARBON EMISSIONS ESTIMATION

Linking the urban contextual form with urban carbon emissions can provide an opportunity for urban planning and urban morphological design strategies toward carbon emission reduction. By incorporating high-quality urban form data, more accurate spatial patterns of urban carbon emissions with fine resolution can be modeled to support more targeted and feasible low-carbon strategies. Hence, in this section, we presented two cases of GIS-based approaches to model high-resolution carbon emission inventories from open urban form data and demonstrated the approaches through implementation in China. This section will focus on the two largest urban agglomerations in China, namely the Yangtze River Delta (YRD) and Pearl River Delta (PRD) regions.

Modeling Spatiotemporal Carbon Emissions for Two Mega-Urban Regions

Data

Carbon emissions generated from fossil energy consumption were calculated for all 30 cities (11 cities of the PRD region and 19 cities of the YRD region). The emission factors were retrieved from IPCC (2006) and Z. Liu et al. (2015). Data on energy consumption were taken from the energy balance table from China Energy Statistical Yearbook and city statistical yearbooks. Socioeconomic information, including GDP and population data for each city, was also retrieved from city statistical yearbooks.

This study chose VIIRS Stray Light Corrected Nighttime Day/Night Band Composites (https://lpdaac.usgs.gov/products/vnp09gav001/) as the primary proxy data to predict spatiotemporal carbon emissions. For each year, the final output of the NTL image (unit: nano-W cm−2 sr−1) was a collection of the mean DN value of the pixels among all the monthly products within the year.

The Local Climate Zone (LCZ) scheme provides a standardized way to characterize global cities based on their morphology and function (Stewart &

TABLE 13.1
Cross-Comparison of the Methods

Methods	Top-Down Method from Nightlights	Bottom-Up Method
Spatial coverage	Global coverage	Usually local coverage
Finest spatial resolution	500 m	Point
Temporal resolution	Monthly or annual	Usually a single year
Accessibility	Freely available	Usually requiring non-public data
Accuracy	Medium accuracy (70% to 90%)	Usually most accurate
Applications	Regional, national-scale studies	Local estimation at the city, district, community or household levels

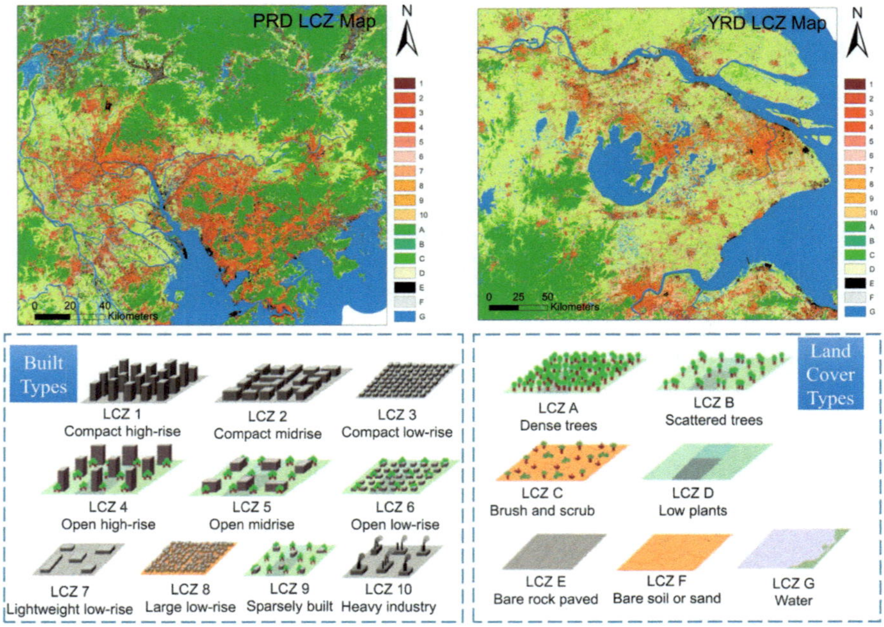

FIGURE 13.1 LCZ map of the PRD and YRD regions in 2016 (Cai et al., 2023).

Oke, 2012). Compared with previous land use/land cover products with a single urban class, it provides a detailed investigation of the built environments and characterized land surface structure and covers ten built types (LCZ 1–10) and seven natural types (LCZ A-G) (Figure 13.1). Moreover, LCZ links land cover types and urban morphology with corresponding thermal properties, which is closely related to energy consumption (Du et al., 2023). Furthermore, the LCZ scheme has demonstrated strong capability in characterizing the spatial distribution of air pollutants (Shi et al., 2019). Accounting for the urban form through LCZ classification can thereby provide a new opportunity to model energy consumption and the spatial variation of carbon emissions.

The LCZ maps were produced based on various remote sensing products such as Landsat 8, DEM, DSM, Sentinel-2, and a random forest classifier. The accuracy assessment shows their overall accuracy is about 76% (M. Cai et al., 2018; Chung et al., 2021; Wang et al., 2019).

Research Steps

According to the LCZ maps, LCZ 1–10 represented urban areas and the natural LCZ classes (LCZ A-G) were integrated into one class as natural land cover. Moreover, to focus on the impact of urban compaction, LCZ 1–6 were further combined into two categories, i.e., the compact urban forms (LCZ 1–3), and open urban forms (LCZ 4–6). Therefore, 13 LCZ classes (12 built classes of LCZ 1–10, compact LCZ, and open LCZ, as well as one natural class) in total were analyzed in this study.

The urban form of the study area was quantified by a series of metrics that can offer detailed spatial patterns of different land use/landscape types at both class-level and landscape-level based on the LCZ maps (Haines-Young & Chopping, 1996; Neel et al., 2004). The class-level landscape metrics include the percentage of landscape types (PLAND), Largest Patch Index (LPI), Aggregation Index (AI), Connectance Index (CONNECT), which can describe spatial patterns of classes within a predefined land lot area. The landscape-level metrics include the contagion index (CONTAG) and Shannon's Evenness Index (SEI), which can provide information about the diversity of the land cover/land use types. There are 52 class-level landscape metrics (13 LCZ classes for each class-level landscape metrics) and two landscape-level metrics urban form indicators that were deployed as urban form indicators. The 54 metrics were calculated at a 500 m grid level on the platform of Fragstats v 4.2.1(McGarigal et al., 2012). Furthermore, in order to focus on carbon emissions in urban areas, this study excludes grids where the natural LCZ is completely dominant, i.e., grids where the LPI of the natural LCZ is 100%

The NTL data and the 54 urban form factors were regarded as potential independent variables while the city-level carbon emissions are the dependent variable. The panel data is at a city-year level. The statistical model assumes a linear relationship between the predictors and CO_2 emissions at the city level, and this relationship can be applied to all grid cells ($500 \times 500\ m^2$) of the two regions.

This study considered a linear regression model with time-fixed effects to capture the possible time trends and involve the temporal heterogeneity for a more accurate and stable prediction of carbon emissions. The relationship between the predictors and the city-level carbon emissions was established by accounting for the time-fixed effects (Equation 13.1):

$$CE_{ij} = \alpha_1 Var_1 + \alpha_2 Var_2 + \ldots + \alpha_n Var_n + \gamma + \beta_j + \varepsilon_{ij} \tag{13.1}$$

where β_j denotes a year-specific adjustment to the intercept γ of the regression model. The model was further validated by F-test and a Hausman test to decide between fixed or random effects. Once the relationship was proved by the tests, it is valid to use the selected predictors as proxies to estimate CO_2 emissions at the urban scale via a top-down model. Such a statistical relationship was then applied to all predictors at the grid level (500 m) for each year to obtain the spatiotemporal carbon emissions.

Spatiotemporal Carbon Emissions

Figure 13.2 presents the year-on-year changes of gridded CO_2 emissions in the two regions based on the predictive panel data model. From 2012 to 2013, the PRD region witnessed significantly increased emissions in most cities, which is likely to be related to the continuous urban expansion during this period (Figure 13.2(a)). There are also dispersive declines in Guangzhou and Foshan. Similar to the PRD region, a large magnitude of increase in carbon emissions occurs in the YRD region from 2012 to 2013. Some scattered decreases are located in Nanjing, Suzhou, and Hangzhou.

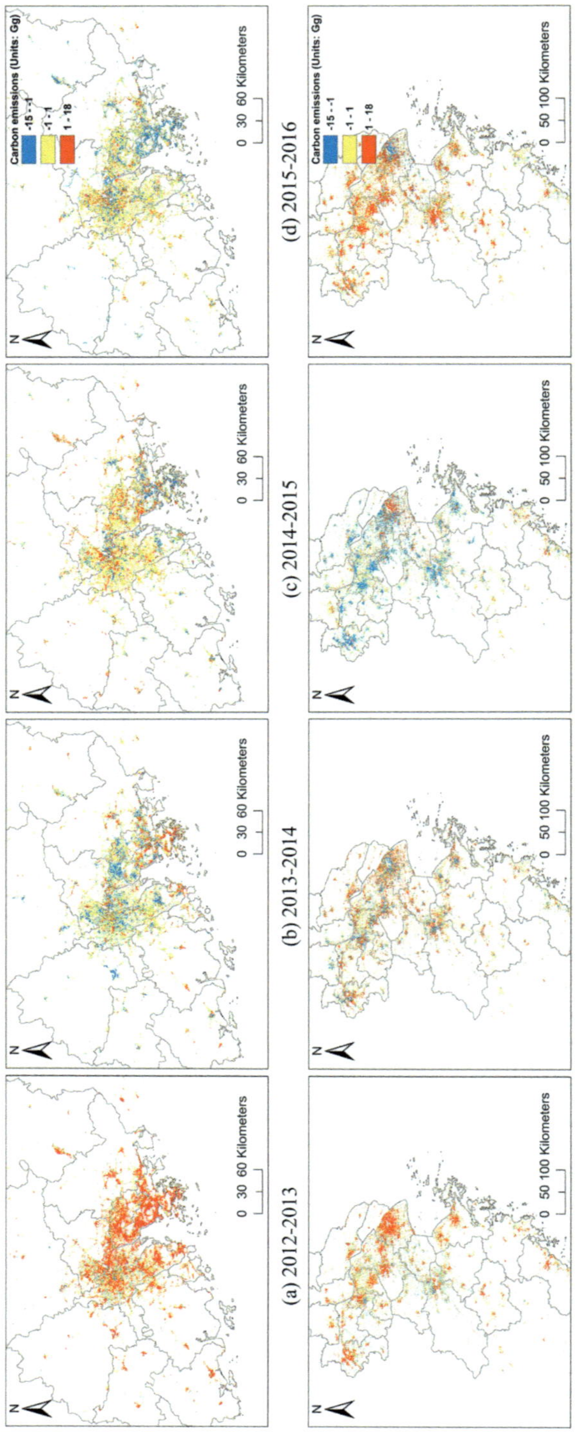

FIGURE 13.2 Year on year change of carbon emissions from 2012 to 2016 in the PRD region (a–d), and YRD region (e–h) (Cai et al., 2023).

From 2013 to 2014, a large expansion of emission decrease has been detected in the PRD region, covering most of the urban areas of Dongguan, Guangzhou, Foshan, Zhongshan, Zhaoqing and Yunfu. Some concentrated growth is located in Hong Kong and Shenzhen while some scattered increases are in other cities in the region. Meanwhile, the YRD region is concurrent with a more mixed pattern of growth and decline in carbon emissions (Figure 13.2(f)). Frequent blue pixels that represent declines are distributed in the urban cores of the region, especially in Shanghai, Hangzhou, Ningbo, Nanjing, and Suzhou-Wuxi-Changzhou. The increases are more often distributed in the urban fringes in the YRD region.

From 2014 to 2015, the PRD region has a large area of decline in emissions in most cities, especially in Hong Kong, Shenzhen, Guangzhou, Foshan and Zhongshan (Figure 13.2(g)). The decline hotspots have shifted from the southwest to the southeast of the region compared with the changes from 2013 to 2014. Some scattered increase is identified in Guangzhou, Shenzhen, and Foshan. The YRD region also shows prevailing decreases in emissions, with some increases in Shanghai and Suzhou (Figure 13.2(c)). the spatial patterns have fewer variations in the urban fringes of the two regions during the study period.

From 2015 to 2016, the PRD region exhibits a generally downward pattern (large area covered with blue and yellow color in Figure 13.2(d)), while some mixed changes are identified in the Guangzhou-Foshan area. Further declines in carbon emissions have been observed in major cities in the PRD region, including Guangzhou, Shenzhen, Foshan and Hong Kong. Differing from the downtrend in the PRD region, the YRD region has growing carbon emissions in the urban centers of most cities (Figure 13.2(h)). There is also concentrated reduction of carbon emissions in the urban centers of Shanghai.

In general, year-on-year changes in carbon emissions in the PRD region are more uniform and have an overall decreasing pattern, demonstrating the region has achieved integrated and coordinated development. However, the year-on-year changes in the YRD region are more diverse and mixed in different cities, indicating that the coordinated development has not yet been fully realized of the region, and the emission reduction measures and effects are not consistent across cities.

Modeling High-resolution of Carbon Emissions for Two Mega-Urban Regions

Model Description

In this study, a hybrid model involving both top-down and bottom-up methods was applied. Four sectors (transport, business, industry, and residence) will be accounted for in this model. Energy-related emissions, including direct emissions from fossil fuels as well as indirect emissions from heating and power generation, were accounted for each sector.

Firstly, the total emissions in each sector will be calculated following the IPCC guidelines (IPCC, 2006). The emissions that have specific spatial information and can be directly derived from openly available sources, such as civil aviation, marine, will be modeled using the bottom-up method that directly assigns the emissions to the spatial grid of the emission source.

For emissions without detailed spatial information or publicly available sources, the spatial distribution of the carbon emissions can be modeled by the top-down method that distributes the statistical data to the emission source using proxy data. For building emissions, including the business, industrial, and residential sectors, buildings are considered emission sources. Building attributes are identified as key energy performance indicators and form the basis of the building typology categorization (Christen et al., 2010). Building attributes relevant to energy consumption include building use, as well as morphological attributes such as building volume, window, wall, and roof areas (Christen et al., 2010). The building volume density (BVD) represents the building morphology and provides an estimate of the space of the building (Ren et al., 2019). The integration of BVD and building use can be used to assign the energy demand and thus is key in quantifying building carbon emissions. Therefore, building attributes, including the building use and the BVD, will be used as proxy data for allocating the emissions from buildings.

Building morphology was obtained by a simple and efficient approach using open data (Ren et al., 2019). The Maps Static API was used to extract the building footprints, and the building heights were generated from a free digital surface model named AW3-D30. A spatial join was then applied in GIS between the extracted points of building use and the extracted building to assign the building use attributes. Thereafter, the building footprints and building heights were used to calculate the BVD for each building use.

The total emissions from buildings (industrial, business, and residential sectors) were further refined as the emissions from each building use, based on the percentage of the energy consumption of each building use in each sector. Thereafter, the emissions from each building use were proportionally assigned to each building based on the BVD for each building use. Therefore, the building carbon emissions of a certain pixel p can be calculated as:

$$CE_p = CE_i \times \frac{E_j}{E_i} \times \frac{BVD_p}{\Sigma BVD_j} \tag{13.2}$$

where i and j represent the sector and the building use in the sector for the pixel p, respectively. E means energy consumption. BVD_p is the BVD value for the pixel p and the ΣBVD_j is the total BVD value for the building use j.

Transportation emissions from fossil fuels mainly consist of road traffic emissions, civil aviation emissions, and marine emissions. The civil aviation emissions and marine emissions were calculated based on the data from the civil aviation department and Hong Kong energy end-use report, respectively. Therefore, the emissions from civil aviation and marine were directly assigned to the geographical location of the airport and the ferries. The indirect emissions in the transport sector are the emissions from the MTR. So, the transportation emissions from electricity were equally assigned to the rasterized grid of the MTR route.

Emissions from the road network were calculated by excluding the marine, civil aviation and MTR emissions from the total transport emissions. The road network is regarded as the emission source. Traffic flow is the product of traffic density and velocity, so the traffic flow along the road network can reflect the patterns of traffic emissions (Xia & Shao, 2005). Traffic flow will be modeled and used as a proxy to calculate the transport emissions. The road network emissions were distributed to the spatial grid of the road network based on the proportion of the traffic flow. In this study, the Lagrangian model proposed by Xia and Shao (2005) was used to simulate traffic flow on a complex road network based on the Annual Average Daily Traffic. This simple method will be quite efficient if there is sufficient road network data.

Therefore, the model for creating the inventory contains six steps (Figure 13.3): 1) preparation of input data; 2) accounting total carbon emissions by sector; 3) extraction of proxy data, including building use, building morphology, and traffic flow; 4) disaggregation of the statistical data on carbon emissions to the level of emission sources; 5) visualization of results; 6) valuation of the results. The total urban carbon emissions were mapped out by aggregating the emissions from different sectors. The emissions were finally averaged in each Tertiary Planning Unit (TPU) of Hong Kong for further planning recommendations. TPU was designed for town planning purposes, and there are 291 TPUs in Hong Kong.

Spatial Distribution of Urban Carbon Emissions

The emissions from different sectors as well as the total emissions were mapped out on a 100 m-resolution raster grid (Figure 13.4). The areas and TPUs of high emissions can be identified. For the transport sector, high emissions were identified in the Cross-Harbour Tunnel and some major roads connecting the New Territories and Kowloon. The emissions from the business sector are the largest among all sectors. Commercial buildings are mainly located in Kowloon and Hong Kong Island, and the high emissions in this sector are concentrated in these areas. Emissions from the residential sector are relatively low but cover a large area of Hong Kong. There are few industries in Hong Kong so that the emissions from the industrial sector are not evident from the mapping results. The total emissions mapping can reflect that the high-density urban areas in central Hong Kong have higher emissions and should be taken seriously into account in the low-carbon development. Also, the results are capable of identifying the variations of carbon emissions across different buildings. The emissions of each TPU were summarized in Figure 13.4(f), and TPUs with high emissions, such as Yau Tsim Mong, Central & Western can be detected, therefore, the local planners and policymakers can target these TPUs for developing low-carbon strategies.

CHALLENGES AND OPPORTUNITIES IN GIS-BASED EMISSIONS MODELING

With the development of urban data science, several future directions of the development of the spatial inventory can be identified:

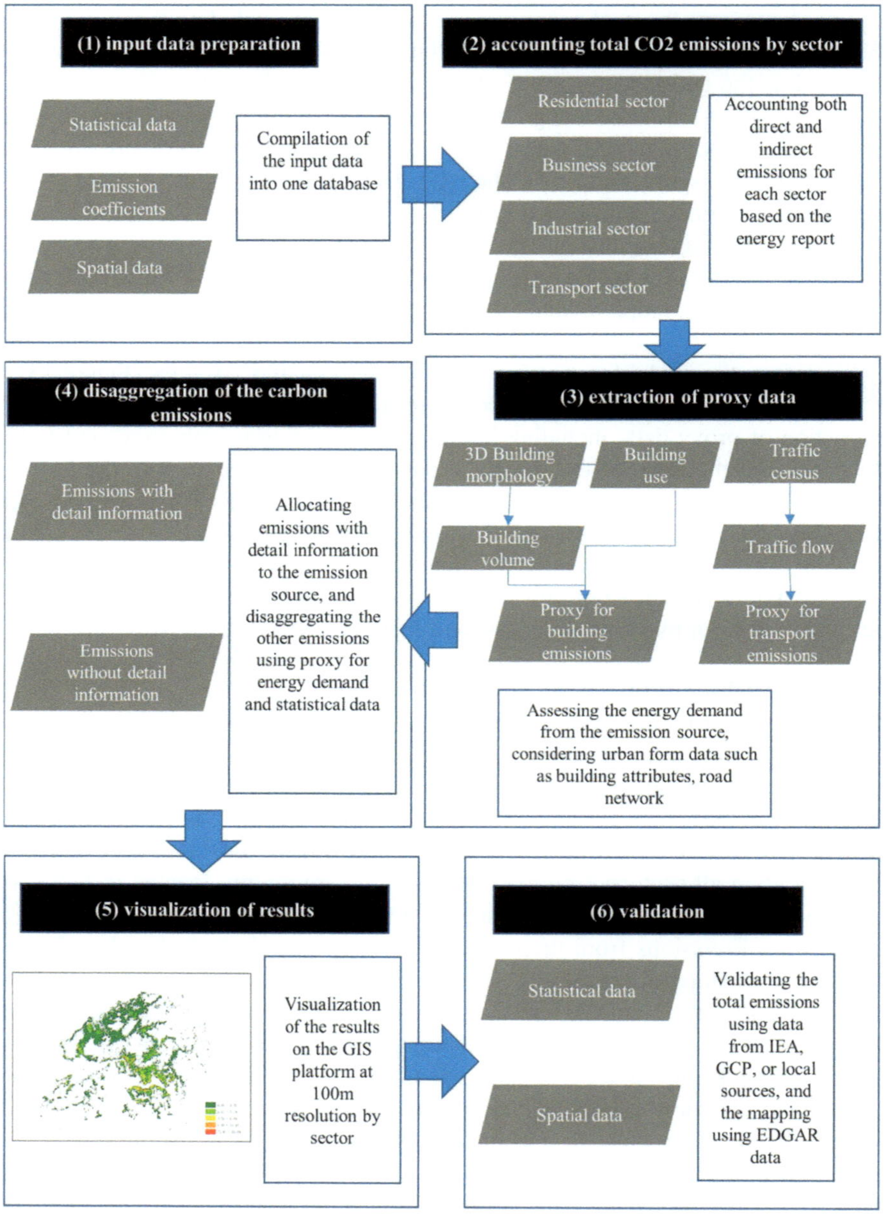

FIGURE 13.3 The procedure for developing a spatial carbon inventory (Cai et al., 2020).

Involving high-quality urban form data: With the development of urban form extraction techniques (Ren et al., 2019), further work involving high-quality urban form data in the modeling process is necessary to obtain more accurate spatial patterns of urban carbon emission for Chinese cities. Also, adopting urban form data in the modeling can support researchers, urban planners, and policymakers to

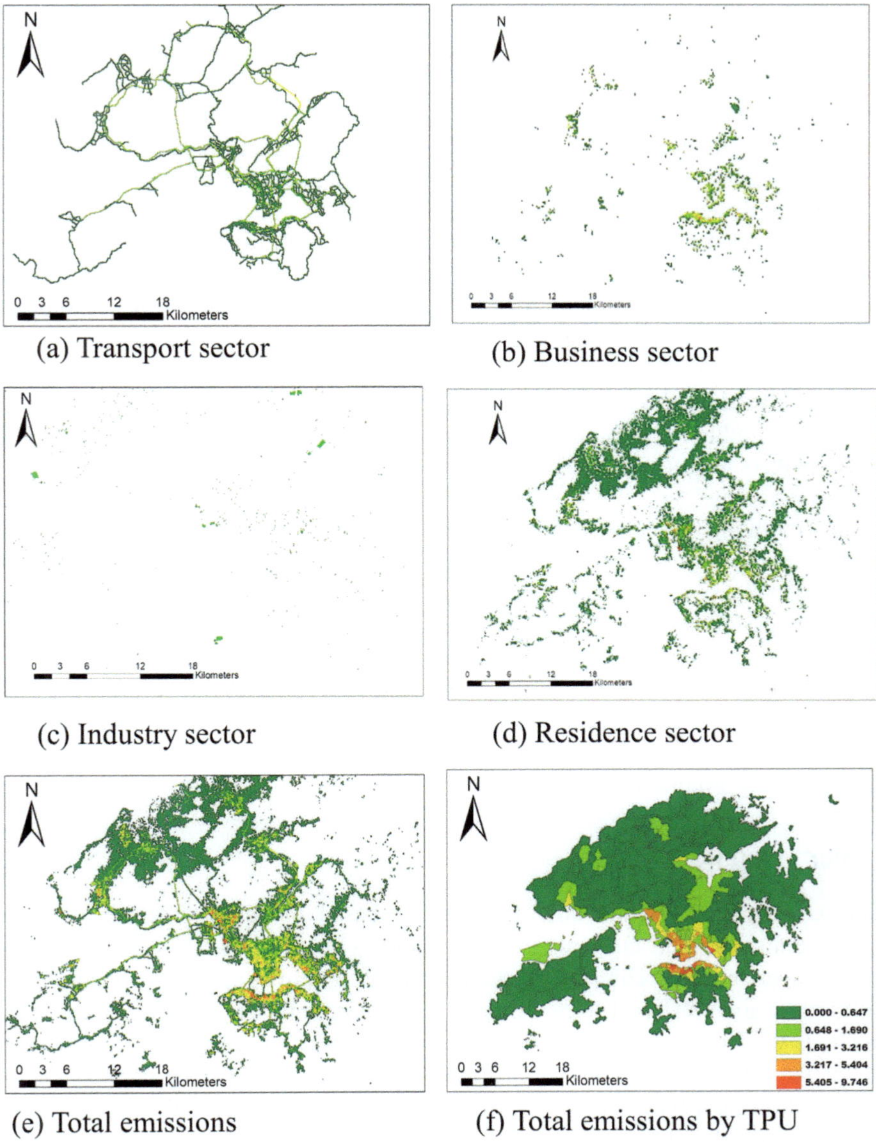

FIGURE 13.4 Carbon emissions in (a) transport sector, (b) business sector, (c) industry sector, (d) residence industrial sector, (e) total emissions and (f) average carbon emissions in each TPU (unit: Kt) (Cai et al., 2020).

have in-depth knowledge of the impact of the urban form and devise corresponding planning strategies.

Developing more accurate and comprehensive models: Future research could focus on improving the accuracy and comprehensiveness of GIS-based models of carbon emissions. This could involve developing more advanced machine learning algorithms, constructing sensor networks, incorporating more detailed data on energy consumption and emission sources, and improving the accuracy of remote sensing data.

Improving temporal resolution: Current GIS-based carbon emissions models typically provide annual or monthly estimates, which may not capture the dynamic nature of carbon emissions. Future research could focus on developing real-time monitoring systems using sensor networks and other sources to collect data on carbon emissions and other environmental factors with high temporal resolution. This could improve our understanding of the factors that drive carbon emissions and help policymakers and urban planners develop more effective interventions to reduce emissions in real time.

Evaluating the effectiveness of mitigation strategies: GIS-based models could be used to evaluate the effectiveness of different mitigation strategies, such as promoting public transportation and adjusting urban forms. This could help policymakers and urban planners identify the most effective strategies for reducing carbon emissions.

Incorporating climate change projections: Future research could incorporate climate change projections into GIS-based models of carbon emissions. This could help policymakers and urban planners anticipate the impacts of climate change on carbon emissions and develop more effective adaptation strategies.

Developing generic methods for different cities: Future research could focus on developing generic methods for conducting high-resolution urban carbon emission modeling for different cities. This could involve developing open data platforms and standardized methods for collecting and analyzing data, which researchers and policymakers around the world could use.

CONCLUSION

This chapter reviews and analyzes the GIS-based modeling methods of urban carbon emissions. The currently available datasets and methods for spatial modeling are summarized. The common methods include top-down approaches using NTL, bottom-up analyzes, and hybrid methods. The strengths and weaknesses of the methods were compared to explore the future needs and trends in the development of spatial models. The top-down method based on the NTL can be implemented to predict the spatial variations of the regional and national carbon emissions using openly available data sources. However, the accuracy of the product can be influenced by the data quality of the NTL and the underestimation of the emissions from the transport and industrial sectors. The bottom-up method has generally been conducted locally by government authorities or planning departments and has been able to secure accurate carbon emission data. However, the universal application of this method is limited by data availability.

This chapter presented two cases in mega-urban regions in China using GIS-based methods and open urban form data, which can more accurately model the spatial pattern of urban carbon emissions. The spatial inventory tool can help support policy decisions regarding urban energy resilience and low-carbon development for different cities. The potential opportunities in GIS-based modeling of carbon emissions include involving high-quality urban form data, developing more accurate and comprehensive models, improving temporal resolution, and developing generic methods for different cities. Newly developed urban morphology extraction technology and geospatial techniques enable the development of more accurate inventories of urban carbon emissions at higher spatiotemporal resolutions, contributing to the development of generic methods for high-resolution urban carbon emission modeling in global cities.

REFERENCES

Bai, X., Dawson, R. J., Ürge-Vorsatz, D., Delgado, G. C., Barau, A. S., Dhakal, S., Dodman, D., Leonardsen, L., Masson-Delmotte, V., & Roberts, D. C. (2018). *Six Research Priorities for Cities and Climate Change*. In: Nature Publishing Group.

Cai, B., Liang, S., Zhou, J., Wang, J., Cao, L., Qu, S., Xu, M., & Yang, Z. (2018). China high resolution emission database (CHRED) with point emission sources, gridded emission data, and supplementary socioeconomic data. *Resources, Conservation and Recycling*, 129, 232–239. 10.1016/j.resconrec.2017.10.036

Cai, M., Ren, C., Shi, Y., Chen, G., Xie, J., & Ng, E. (2023). Modeling spatiotemporal carbon emissions for two mega-urban regions in China using urban form and panel data analysis [Article]. *Science of the Total Environment*, 857, Article 159612. 10.1016/j.scitotenv.2022.159612

Cai, M., Ren, C., Xu, Y., Lau, K. K.-L., & Wang, R. (2018). Investigating the relationship between local climate zone and land surface temperature using an improved WUDAPT methodology–A case study of Yangtze River Delta, China. *Urban Climate*, 24, 485–502.

Cai, M., Shi, Y., & Ren, C. (2020). Developing a high-resolution emission inventory tool for low-carbon city management using hybrid method – A pilot test in high-density Hong Kong (under review). *Energy and Buildings*.

Christen, A., Coops, N. C., Kellett, R., Crawford, B., Heyman, E., Olchovski, I., Tooke, T. R., & van der Laan, M. T. (2010). A LiDAR-based urban metabolism approach to neighbourhood scale energy and carbon emissions modelling University of British Columbia].

Chuai, X., & Feng, J. (2019). High resolution carbon emissions simulation and spatial heterogeneity analysis based on big data in Nanjing City, China. *Science of the Total Environment*, 686, 828–837.

Chung, L. C. H., Xie, J., & Ren, C. (2021). Improved machine-learning mapping of local climate zones in metropolitan areas using composite Earth observation data in Google Earth Engine. *Building and Environment*, 199, 107879.

Cui, Y., Zhang, W., Wang, C., Streets, D. G., Xu, Y., Du, M., & Lin, J. (2019). Spatiotemporal dynamics of CO_2 emissions from central heating supply in the North China Plain over 2012–2016 due to natural gas usage. *Applied Energy*, 241, 245–256.

Doll, C. H., Muller, J.-P., & Elvidge, C. D. (2000). Nighttime imagery as a tool for global mapping of socioeconomic parameters and greenhouse gas emissions. *AMBIO: A Journal of the Human Environment*, 29(3), 157–162.

Du, R., Liu, C.-H., Li, X.-X., & Lin, C.-Y. (2023). Effect of local climate zone (LCZ) and building category (BC) classification on the simulation of urban climate and air-conditioning load in Hong Kong. *Energy*, 271, 127004. 10.1016/j.energy.2023.127004

Ghosh, T., Elvidge, C. D., Sutton, P. C., Baugh, K. E., Ziskin, D., & Tuttle, B. T. (2010). Creating a global grid of distributed fossil fuel CO_2 emissions from nighttime satellite imagery. *Energies*, 3(12), 1895–1913. 10.3390/en3121895

Gurney, K. R., Mendoza, D. L., Zhou, Y., Fischer, M. L., Miller, C. C., Geethakumar, S., & de la Rue du Can, S.(2009). High resolution fossil fuel combustion CO_2 emission fluxes for the United States. *Environ Sci Technol*, 43(14), 5535–5541.

Haines-Young, R., & Chopping, M. (1996). Quantifying landscape structure: A review of landscape indices and their application to forested landscapes. *Progress in physical geography*, 20(4), 418–445.

Han, J., Meng, X., Liang, H., Cao, Z., Dong, L., & Huang, C. (2018). An improved nightlight-based method for modeling urban CO_2 emissions. *Environmental Modelling & Software*, 107, 307–320. 10.1016/j.envsoft.2018.05.008

International Energy Agency. (2016). *Energy Technology Perspectives 2016: Towards Sustainable Urban Energy Systems*. OECD.

IPCC. (2006). *2006 IPCC Guidelines for National Greenhouse Gas Inventories*. Intergovernmental Panel on Climate Change.

IPCC. (2014). IPCC, 2014: climate change 2014: synthesis report. Contribution of Working Groups I II and III to the Fifth Assessment Report of the intergovernmental panel on Climate Change. IPCC, Geneva, Switzerland, 151.

Jing, Q., Bai, H., Luo, W., Cai, B., & Xu, H. (2018). A top-bottom method for city-scale energy-related CO_2 emissions estimation: A case study of 41 Chinese cities. *Journal of Cleaner Production*, 202, 444–455.

Kanemoto, K., Moran, D., & Hertwich, E. G. (2016). Mapping the carbon footprint of nations. *Environ Sci Technol*, 50(19), 10512–10517.

Li, M., Liu, H., Geng, G., Hong, C., Liu, F., Song, Y., Tong, D., Zheng, B., Cui, H., & Man, H. (2017). Anthropogenic emission inventories in China: a review. *National Science Review*, 4(6), 834–866.

Liu, F., Zhang, Q., Tong, D., Zheng, B., Li, M., Huo, H., & He, K. (2015). High-resolution inventory of technologies, activities, and emissions of coal-fired power plants in China from 1990 to 2010. *Atmospheric Chemistry and Physics*, 15(23), 13299–13317.

Liu, J., Tong, D., Zheng, Y., Cheng, J., Qin, X., Shi, Q., Yan, L., Lei, Y., & Zhang, Q. (2021). Carbon and air pollutant emissions from China's cement industry 1990–2015: trends, evolution of technologies, and drivers. *Atmospheric Chemistry and Physics*, 21(3), 1627–1647.

Liu, Z., Guan, D., Wei, W., Davis, S. J., Ciais, P., Bai, J., Peng, S., Zhang, Q., Hubacek, K., & Marland, G. (2015). Reduced carbon emission estimates from fossil fuel combustion and cement production in China. *Nature*, 524(7565), 335–338.

McGarigal, K., Cushman, S. A., & Ene, E. (2012). FRAGSTATS v4: spatial pattern analysis program for categorical and continuous maps. Computer software program produced by the authors at the University of Massachusetts, Amherst. Available at the following web site: http://www.umass.edu/landeco/research/fragstats/fragstats.html.

Neel, M. C., McGarigal, K., & Cushman, S. A. (2004). Behavior of class-level landscape metrics across gradients of class aggregation and area. *Landscape ecology*, 19(4), 435–455.

Oda, T., & Maksyutov, S. (2015). ODIAC fossil fuel CO_2 emissions dataset.

Ou, J., Liu, X., Li, X., & Shi, X. (2015). Mapping global fossil fuel combustion CO_2 emissions at high resolution by integrating nightlight, population density, and traffic network data. *IEEE Journal of Selected Topics in Applied Earth Observations and Remote Sensing*, 9(4), 1674–1684.

Rayner, P. J., Raupach, M. R., Paget, M., Peylin, P., & Koffi, E. (2010). A new global gridded data set of CO_2 emissions from fossil fuel combustion: Methodology and evaluation. *Journal of Geophysical Research*, 115(D19). 10.1029/2009jd013439

Ren, C., Cai, M., Li, X., Shi, Y., & See, L. (2019). Developing a rapid method for 3-dimensional urban morphology extraction using open-source data Sustainable Cities and Society.

Rong, P., Zhang, Y., Qin, Y., Liu, G., & Liu, R. (2020). Spatial patterns and driving factors of urban residential embedded carbon emissions: An empirical study in Kaifeng, China [Article]. *Journal of Environmental Management*, 271, Article 110895. 10.1016/j.jenvman.2020.110895

Shi, Y., Ren, C., Lau, K. K.-L., & Ng, E. (2019). Investigating the influence of urban land use and landscape pattern on PM2. 5 spatial variation using mobile monitoring and WUDAPT. *Landscape and Urban Planning*, 189, 15–26.

Song, X., Hao, Y., & Zhu, X. (2020). Vehicular emission inventory establishment and characteristics research in the Yangtze River Delta Urban Agglomeration [Article]. *Huanjing Kexue Xuebao/Acta Scientiae Circumstantiae*, 40(1), 90–101. 10.13671/j.hjkxxb.2019.0350

Stewart, I. D., & Oke, T. R. (2012). Local climate zones for urban temperature studies. *Bulletin of the American Meteorological Society*, 93(12), 1879–1900.

Tong, D., Zhang, Q., Davis, S. J., Liu, F., Zheng, B., Geng, G., Xue, T., Li, M., Hong, C., & Lu, Z. (2018). Targeted emission reductions from global super-polluting power plant units. *Nature Sustainability*, 1(1), 59–68.

UN DESA. (2018). World Urbanisation Prospects, 2018 Revision.

Wang, J., Cai, B., Zhang, L., Cao, D., Liu, L., Zhou, Y., Zhang, Z., & Xue, W. (2014). High resolution carbon dioxide emission gridded data for China derived from point sources [Research Support, Non-U.S. Gov't]. *Environ Sci Technol*, 48(12), 7085–7093. 10.1021/es405369r

Wang, R., Cai, M., Ren, C., Bechtel, B., Xu, Y., & Ng, E. (2019). Detecting multi-temporal land cover change and land surface temperature in Pearl River Delta by adopting local climate zone. *Urban Climate*, 28, 100455.

Wang, S., Fang, C., Guan, X., Pang, B., & Ma, H. (2014). Urbanisation, energy consumption, and carbon dioxide emissions in China: A panel data analysis of China's provinces. *Applied Energy*, 136, 738–749. 10.1016/j.apenergy.2014.09.059

Wen, Y., Wu, R., Zhou, Z., Zhang, S., Yang, S., Wallington, T. J., Shen, W., Tan, Q., Deng, Y., & Wu, Y. (2022). A data-driven method of traffic emissions mapping with land use random forest models. *Applied Energy*, 305, 117916. 10.1016/j.apenergy.2021.117916

Wu, J., Jia, P., Feng, T., Li, H., Kuang, H., & Zhang, J. (2023). Uncovering the spatiotemporal impacts of built environment on traffic carbon emissions using multi-source big data. *Land Use Policy*, 129, 106621. 10.1016/j.landusepol.2023.106621

Wu, Y., Sharifi, A., Yang, P., Borjigin, H., Murakami, D., & Yamagata, Y. (2018). Mapping building carbon emissions within local climate zones in Shanghai. *Energy Procedia*, 152, 815–822.

Xia, C., Xiang, M., Fang, K., Li, Y., Ye, Y., Shi, Z., & Liu, J. (2020). Spatial-temporal distribution of carbon emissions by daily travel and its response to urban form: A case study of Hangzhou, China [Article]. *Journal of Cleaner Production*, 257, Article 120797. 10.1016/j.jclepro.2020.120797

Xia, L., & Shao, Y. (2005). Modelling of traffic flow and air pollution emission with application to Hong Kong Island. *Environmental Modelling & Software*, 20(9), 1175–1188. 10.1016/j.envsoft.2004.08.003

Yang, S., Wang, C., Lo, K., Wang, M., & Liu, L. (2015). Quantifying and mapping spatial variability of Shanghai household carbon footprints. *Frontiers in Energy*, 9(1), 115–124. 10.1007/s11708-015-0348-8

Zhang, R., Matsushima, K., & Kobayashi, K. (2018). Can land use planning help mitigate transport-related carbon emissions? A case of Changzhou. *Land Use Policy*, 74, 32–40. 10.1016/j.landusepol.2017.04.025

Zhang, W., Cui, Y., Wang, J., Wang, C., & Streets, D. G. (2020). How does urbanization affect CO_2 emissions of central heating systems in China? An assessment of natural gas transition policy based on nighttime light data [Article]. *Journal of Cleaner Production*, 276, Article 123188. 10.1016/j.jclepro.2020.123188

Zhang, Y., Pan, J., Zhang, Y., & Xu, J. (2021). Spatial-temporal characteristics and decoupling effects of China's carbon footprint based on multi-source data [Article]. *Journal of Geographical Sciences*, 31(3), 327–349. 10.1007/s11442-021-1839-7

Zhao, J., Ji, G., Yue, Y., Lai, Z., Chen, Y., Yang, D., Yang, X., & Wang, Z. (2019). Spatio-temporal dynamics of urban residential CO_2 emissions and their driving forces in China using the integrated two nighttime light datasets. *Applied Energy*, 235, 612–624.

Zhao, P., Kwan, M.-P., & Qin, K. (2017). Uncovering the spatiotemporal patterns of CO_2 emissions by taxis based on Individuals' daily travel. *Journal of Transport Geography*, 62, 122–135.

Zheng, B., Tong, D., Li, M., Liu, F., Hong, C., Geng, G., Li, H., Li, X., Peng, L., & Qi, J. (2018). Trends in China's anthropogenic emissions since 2010 as the consequence of clean air actions. *Atmospheric Chemistry and Physics*, 18(19), 14095–14111.

14 The Relationship between Visual Space and Elders' Mental Health within 15-Minute Life Circles Using SVIs

A Case Study of Beijing

Jiang Ying

INTRODUCTION

Population ageing has become one of the most considerable demographic trends in the 21st century. The share of population aged 60 and above in China was 18.7%, or 264 million, while about 190 million people are aged 65 or above, accounting for 13.5%, by the end of 2020 (*Seventh National Census Communiqué*, 2021). In the context of rapid urban development, the elderly living in modern cities is more likely to be plagued by negative emotions, such as depression, anxiety and loneliness, compared with young people. Previous researches on environmental psychology believe that the environment with positive characteristics can positively affect on psychology and emotions of people(Kaplan, 1983). Moreover, the quality of the surrounding built environment of the residence community directly influences the living quality of the residents, which create a significant impact on the life satisfaction and psychological health conditions of the residents.

For the past few years, the notion of a 15-minute community living circle (15-minute CLC) has attracted wide attention. Wu et al. (2022)evaluated the spatial layout of public service facilities within the 15-minute CLC from the perspective of supply and demand balance. Based on the daily needs of residents, Qiu et al. (2022) also studied the coverage rate of facilities within 15-minute life circles of residential blocks and introduced the concept of development coordination to comprehensively evaluate the availability of public services by applying point of interest(POI) data. However, previous studies mainly concentrated on the analysis of the spatial layout and accessibility of public service facilities within 15-minute life circles of residence communities, while the visual perception of streets within the 15-minute CLC has received less attention.

DOI: 10.1201/9781003244561-18

Some noted map service providers, such as Google, Tencent, Baidu, etc., have provided the public with street view map services containing high-precision urban street view images, which can adequately describe the visual feelings of pedestrians in urban streets environment from the human perspective and can depict various elements contained in urban roads in detail, with extremely rich information. Deep learning (DL) technique provides technical support for image data analysis and understanding after computer science technologies has improved dramatically. Zhang et al. (2018) proposed a DL model to investigate the correlation between streetscape elements and six indicators of human perception, including security, vitality, beauty, depression, affluence and monotony and applied the model to predict human perception in SVIs, creating a city-wide distribution map of human perception in the two cities. Li and Ghosh (2018) used the green view index (GVI) obtained by Google Street View image (GSVIs) and the Walk Score to discover the relationship between the Body Mass Index (BMI) of residents of different genders and these two indicators in Cleveland. Wang, Liu, et al. (2019) evaluated six urban environmental perception indicators, including wealth, safety, liveliness, depression, boredom and beauty, through the results of semantic segmentation of Tencent SVIs (TSVIs) using the FCN-8s model, and assessed their relationship with the mind and body health of old people. However, previous studies on urban visual environment and analysis of the correlation between urban visual environment and health condition were primarily relied on the administrative division or 1 km buffer zones of residential communities, which may have deviated from the actual perception of residents, since the visual environment of the streets around the residential blocks that have a greater impact on the residents could not be systematically assessed in those researches based on administrative divisions and the 1 km buffer zone generated without the urban road network data might deviate from the actual travel range of people.

Based on previous studies, this study used a deep learning model to semantically segment street view data to study the visual environment within the 15-minute walking circles of residential areas in Haidian District, Beijing, combined with the notion of "15-minute CLC" and urban road network data and explores its relationship to the mental health condition of the aged.

STUDY AREA

Beijing is not only the capital of China but also the commercial and cultural center of China. By 2020, there are 16 administrative districts in Beijing, including Haidian District, Xicheng District, Dongcheng District, etc. Haidian District was chosen as the study area of this research, as shown in Figure 14.1, which is one of the most economically deprived urban areas in Beijing. In 2021, the economic aggregate and growth contribution ranked first in Beijing, while the regional GDP of the district reached 950.17 billion yuan and the per capita gross regional product was 47,100 dollars.

According to the seventh national census of Haidian District, by the end of 2020, the population aged 60 and over was 578,000, constituting 18.5% of the total population, of which 409,000 were aged 65 or over, accounting for 13.1%.

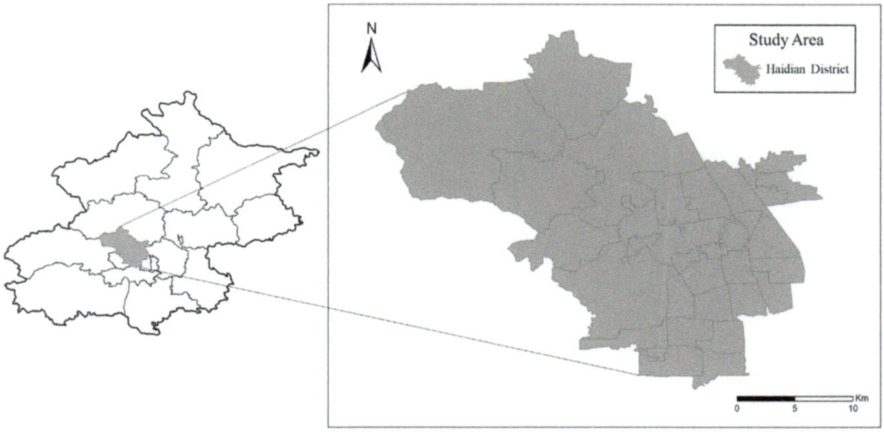

FIGURE 14.1 Location of Haidian District, Beijing.

Compared with the population structure in 2010, the number of people aged 60 and over increased by 206,000 people, representing a 7.2% growth, of which the population aged 65 and over increased by 134,000 people, accounting for a 4.7% increase. The ageing of the population in Haidian District has gradually deepening, therefore, Haidian District is an ideal research area to study the mental health of the elderly.

DATA

Mental Health Data

Data on the psychological health condition of old people, came from the depression and anxiety survey of the elderly conducted by researchers from the Renmin University of China in Haidian District, Beijing, in 2011, which can be downloaded from Chinese National Survey Data Archive(*The depression and anxiety survey of the elderly*). The survey used the short version of the Geriatric Depression Scale (GDS-15)(Sheikh & Yesavage, 1986) and Geriatric Anxiety Inventory (GAI) (Pachana et al., 2007) in order to accurately and effectively measure the level of depression and anxiety among the elderly in Beijing. The original form of the Geriatric Depression Scale (GDS) contains 30 questions, which could be time-consuming to fill out for the elderly and lead to fatigue, therefore, the survey measured the level of depression using a shortened version.

The respondents were the registered population over 60 years old in Haidian District, Beijing. They selected 45 neighborhood committees or villages from 13 streets (jiedao) in Haidian District by typical sampling. They systematically sampled individual respondents within the community, 30 from each community, and there were 1350 respondents in total. As the residential addresses of some respondents were missing in the original survey data, mental health data from 1231 respondents was finally obtained after data cleaning. They lived in 48 housing estates or villages in Haidian District, as seen in Figure 14.2.

FIGURE 14.2 Sampled housing estates or villages during the Depression and Anxiety Survey of the elderly in 2011.

Residence Communities Data in Haidian District

The residence communities used in the study on street visual perception in the 15-minute walking circles of residential blocks in Haidian District, Beijing, which can be seen in Figure 14.3, was obtained from Lianjia, which is one of the most widely used comprehensive real estate service platforms in China.

Street View Image Data

The street landscape assessment in the research was based on the reachable locations that people are able to walk to within 15 minutes. The walkability analysis in 15 minutes was combined with urban road network data, therefore, it would be more similar to the range that residents can walk to in daily life than the buffer analysis. Eight-hundred-eighty 15-minute walking life circles were finally obtained according to the community data obtained from Lianjia because the network data around eight communities were missing and could not meet the requirement of 15-minute walking, while 48 walking life circles were finally obtained based on the residential communities of respondents in the mental health questionnaire.

Baidu Street View Images (BSVIs) was the street view image data source of this research since GSV lacks the data of mainland China and the coverage of BSV is wider than TSV. The sampling points for gathering street view data were generated within the 15-minute life circles of communities, as shown in Figure 14.4.

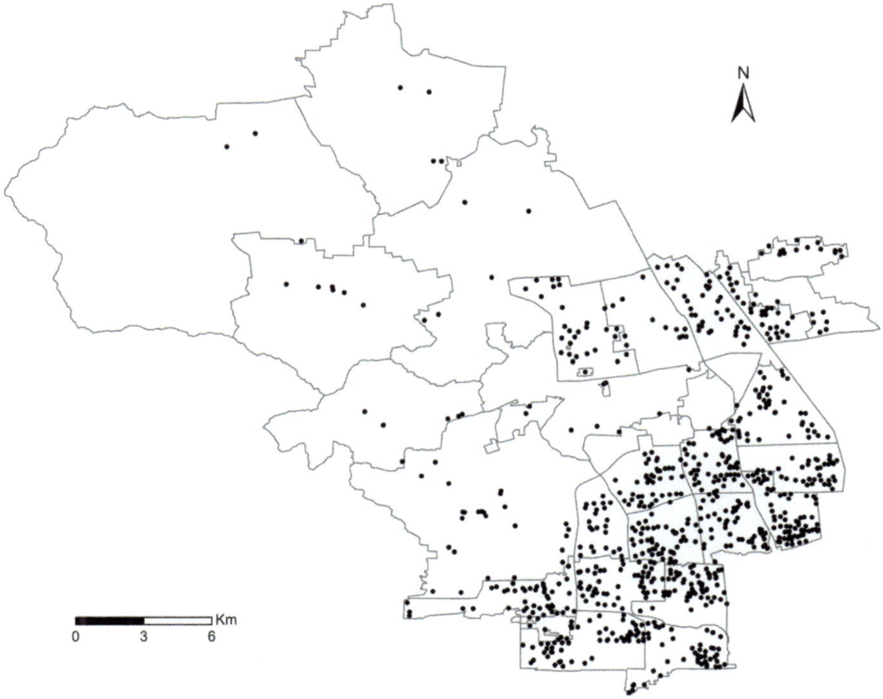

FIGURE 14.3 Residential blocks in Haidian District, Beijing.

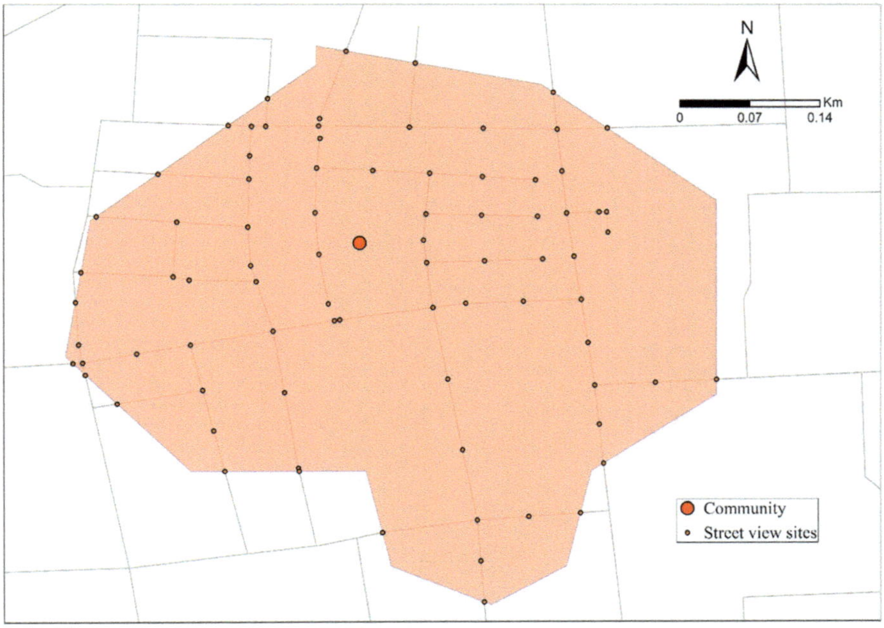

FIGURE 14.4 Example for street view image sites inside the 15-minute walking circle.

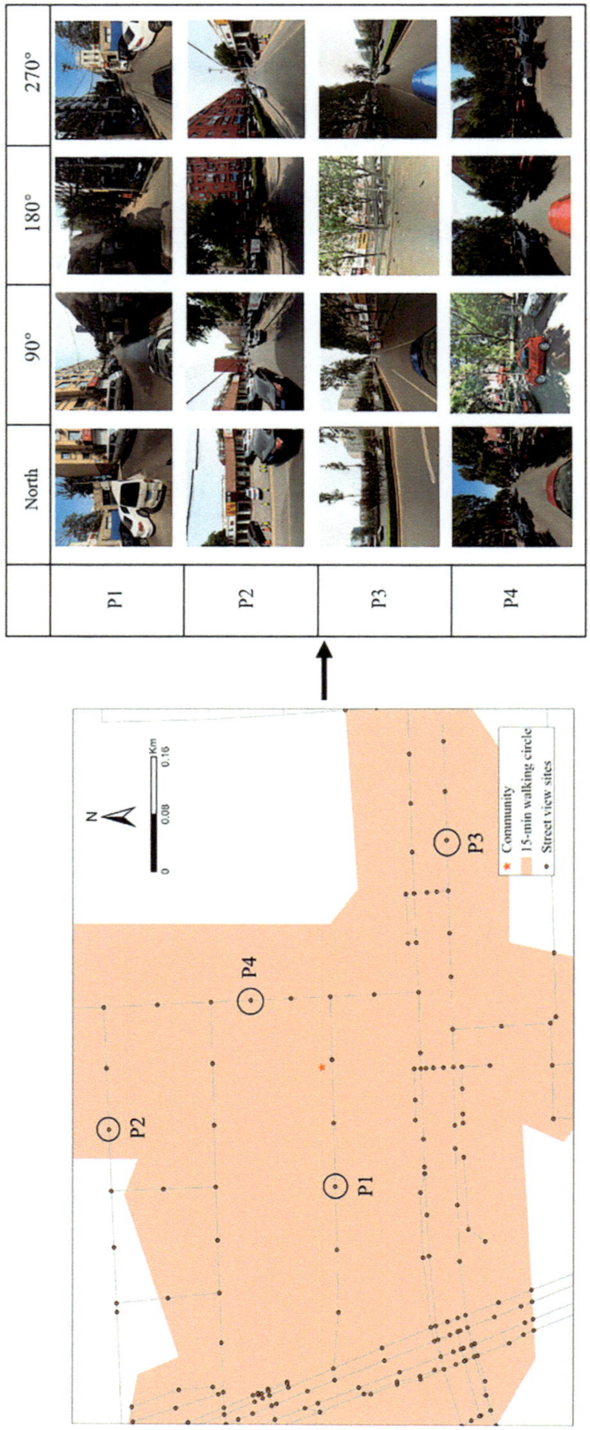

FIGURE 14.5 Example for the BSVIs collection.

TABLE 14.1
The Main Parameter Configuration of BSVIs in the Python Crawler

Parameter	Description	Value/Example
width	The width of a SVI, ranged from 10 to 1024	480 pixels
height	The height of an image, ranged from 10 to 512	480 pixels
panoid	The scene ID of an image, which is used to confirm the position of the image	09002200121706201623328918D
heading	The angle to the true north direction ranged from 0 to 360	0, 90, 180, 270 degrees
pitch	Pitch angle, which is the angle of the street view camera looking up or down, ranged from 0 to 90	0 degree
fov	Range in a horizontal direction	90

According to the 15-minute life circles for the residential blocks from Lianjia and the questionnaire, 48962 sampling points and 6421 sampling points were finally obtained respectively.

As the study would carry out the analysis of street greenness, SVIs used in the project were collected in the green season in order to reduce the analysis error of the study of greenery. The latest data on the green season in the BSV map was collected in 2017. In this research, street view data from 2017 and 2013 were used to analyze the visual perception of the streets within 15-minute walking circles of the residential blocks in the Haidian District and to analyze the relationship between mental health and street visual perception of the elderly.

After parsing the data by loading process of Baidu Panorama Map web pages, a python crawler was written to obtain four SVIs with four horizontal angles (0, 90, 180, 270 degrees) for each sampling point, which can be seen in Figure 14.5. Table 14.1 shows the main parameters of SVIs when collecting the data from BSV. As some roads are located in remote areas, the street views of those road sections were unable to be collected on the Baidu platform, resulting in empty data collected at some sampling points. Finally, 15,519 SVIs for 2013, and 110,933 SVIs for 2017, were obtained.

METHODOLOGY

Figure 14.6 shows the flowchart of the methodology in this research: 1) The pre-trained PSPNet model was applied to semantically segment SVIs from Baidu Map before calculating the pixel ratio of each stuff in each image. 2) Based on the pixel proportion of some kinds of stuff, including trees, grass, sky, building, etc, three visual indicators, covering greenness, openness and enclosure, would be chosen to calculate to get these visual indicators of each residential block within the

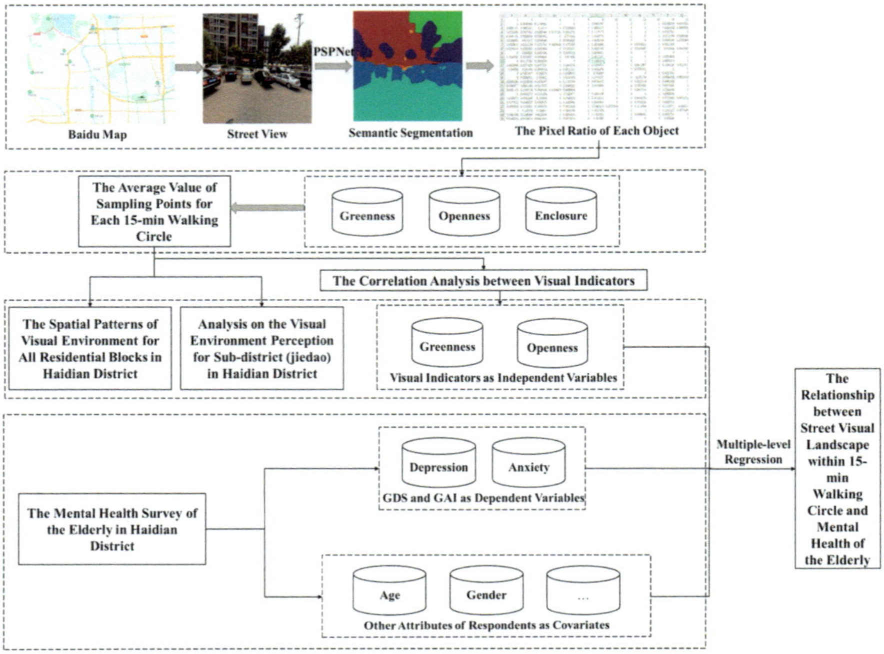

FIGURE 14.6 The flowchart showing the methodology in this research.

15-minute walking circle. 3) The assessment of street visual perception of pedestrians who lived in residential blocks in the Haidian District would be conducted, including the spatial patterns of each residential block and the visual environment with 15-minute CLC for sub-districts (jiedao). 4) After the correlation analysis between three visual indicators, greenness and openness would be selected as independent variables, and the multilevel regression analysis would be conducted to analyze the correlation between street visual perception factors and the mental health of the elderly, combining the GDS and GAI scores in the mental health questionnaire as dependent variables, and other attributes of respondents as covariates.

Semantic Segmentation for Street View Images

The PSPNet model was used to conduct image semantic segmentation of BSVIs in this study, which is based on the MXNet(*MXNet*) framework, an open-source high-performance modular ML library (Chen et al., 2015). The pre-trained model in the research was trained by the ADE20K dataset. Figure 14.7 shows the results of image semantic segmentation on BSVIs by the pre-trained PSPNet model. In the pictures of semantic segmentation result, different colors represent different categories of items. After semantic segmentation, the proportion of pixels of each item in each image was calculated for the assessment of the visual perception indicators of streets.

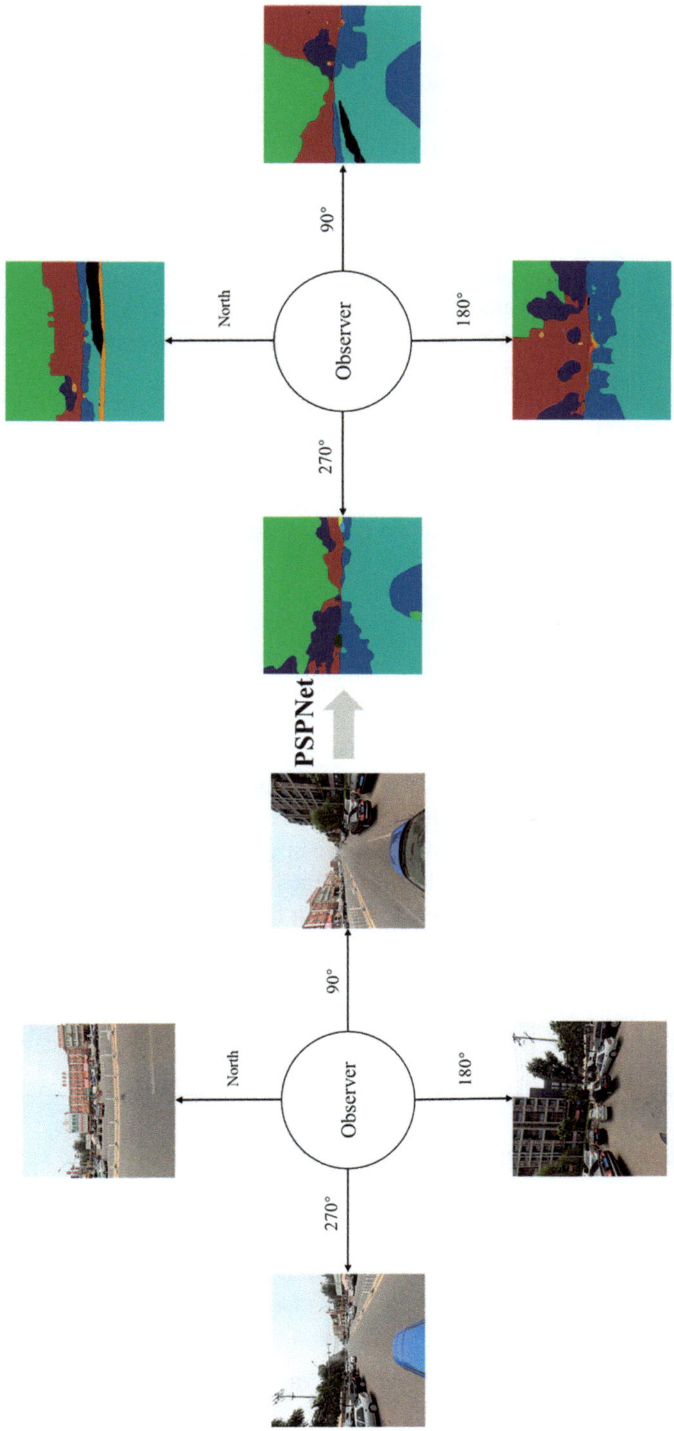

FIGURE 14.7 Image semantic segmentation on BSVIs by the PSPNet model.

Street Visual Perception Assessment

Various indexes have been discussed to evaluate visual perception of pedestrians on urban streets and to understand how the urban landscape influences the psychological well-being of citizens in previous studies, such as Dai et al. (2021), Helbich et al. (2019), Wang, Helbich, et al. (2019), Wang, Liu, et al. (2019), Wang, Lu, et al. (2019), etc. According to these studies, the study selected greenness, openness and enclosure as three visual indicators for the assessment of street visual perception of pedestrians. Table 14.2 shows the calculation formulas for these three indexes. Compared with NDVI, greenness is closer to the human perspective, meaning that it better captures how humans perceive green spaces and vegetation in their surroundings. The urban greenness index in this study was calculated based on the pixel ratio of trees, grasses and plants in SVIs.

Openness is the ratio of visibility of the sky from the perspective of the human eye (Dai et al., 2021). Assessing the level of openness can help planners to propose effective plans to enhance the appeal of the street landscape and increase the comfort degree of public events (Dzhambov et al., 2018; Ma et al., 2021). The openness index in this study was calculated based on the proportion of sky pixels in street view pictures.

Enclosure describes the level of encapsulation of streets for pedestrians (Dai et al., 2021). The information on this enclosure structure can measure the level of encapsulation of the building environment for pedestrians and connects it with people's perception of space limitations and closeness, and a place's livability and security (Ewing & Handy, 2009; Porta & Renne, 2005; Zhang et al., 2012). Referring to Yin and Wang (2016) and Wang, Lu, et al. (2019), the value of enclosure was calculated by the pixel proportion of buildings and trees in images.

In Table 14.2, N represents the number of images at each sampling point ($1 \leq N \leq 4$, some sampling points do not have street view images at some angles). The greenness, openness and enclosure of each sampling point are the average values of the visual indicators of the SVIs from four angles of the sampling point. The correlation analysis of pixel proportions of different angles of buildings, sky, trees, grass and plants showed that the pixel proportions of different angles of each type are highly correlated with each other at the significance level of 95%, as seen in Table 14.3. Therefore, it is an appropriate method to calculate the

TABLE 14.2
The Visual Perception Indicators in the Study

Indicators	Formula	Expression
Greenness	$G_i = \frac{1}{N}\sum_{j=1}^{N} T_j + \frac{1}{N}\sum_{j=1}^{N} Grass_j + \frac{1}{N}\sum_{j=1}^{N} P_j$	T_j represents the pixel ratio of trees, $Grass_j$ is the percentage of grass, P_j is the percentage of plants
Openness	$O_i = \frac{1}{N}\sum_{j=1}^{N} S_j$	S_j represents the pixel ratio of the sky
Enclosure	$E = \frac{1}{N}\sum_{j=1}^{N} B_j + \frac{1}{N}\sum_{j=1}^{N} T_j$	B_j represents the percentage of buildings, T_j represents the pixel ratio of trees

TABLE 14.3

The Correlation Analysis of the Pixel Proportions of Different Degrees of Building, Sky, Tree, Grass, and Plant

(a)	Building (0°)	Building (90°)	Building (180°)	Building (270°)
Building (0°)	1	0.268**	0.218**	0.276**
Building (90°)	0.268**	1	0.248**	0.317**
Building (180°)	0.218**	0.248**	1	0.27**
Building (270°)	0.276**	0.317**	0.27**	1
**p < 0.05				

(b)	Sky (0°)	Sky (90°)	Sky (180°)	Sky (270°)
Sky (0°)	1	0.345**	0.264**	0.292**
Sky (90°)	0.345**	1	0.266**	0.302**
Sky (180°)	0.264**	0.266**	1	0.302**
Sky (270°)	0.292**	0.302**	0.302**	1
**p < 0.05				

(c)	Tree (0°)	Tree (90°)	Tree (180°)	Tree (270°)
Tree (0°)	1	0.328**	0.24**	0.308**
Tree (90°)	0.328**	1	0.271**	0.336**
Tree (180°)	0.24**	0.271**	1	0.342**
Tree (270°)	0.308**	0.336**	0.342**	1
**p < 0.05				

(d)	Grass (0°)	Grass (90°)	Grass (180°)	Grass (270°)
Grass (0°)	1	0.247**	0.14**	0.162**
Grass (90°)	0.247**	1	0.244**	0.214**
Grass (180°)	0.14**	0.244**	1	0.276**
Grass (270°)	0.162**	0.214**	0.276**	1
**p < 0.05				

(e)	Plant (0°)	Plant (90°)	Plant (180°)	Plant (270°)
Plant (0°)	1	0.153**	0.109**	0.169**
Plant (90°)	0.153**	1	0.134**	0.115**
Plant (180°)	0.109**	0.134**	1	0.176**
Plant (270°)	0.169**	0.115**	0.176**	1
**p < 0.05				

greenness, openness, and enclosure of each sampling point by calculating the average ratio of the building, sky, tree, grass, and plant pixels of the four street view pictures from different angles. The values of each factor in residence communities were determined by averaging the outcomes of each residential block. To calculate these average values, the data from each sample point within the 15-minute walking circle was considered, the results were averaged to represent the overall visual characteristics of each residential block. And the values of visual perception indicators in different sub-districts (jiedao) were generated by the average values of residential blocks in the sub-districts (jiedao).

The Spatial Clustering of Visual Indicators for Residential Communities

The Getis-Ord Gi* model(Manepalli et al., 2011) was applied in this study to discover the clustering characteristics of high and low values of greenness, openness and enclosure in the 15-minute walking life circles of residential communities in the Haidian District. The model can be described by the following formula (Manepalli et al., 2011):

$$G_i^* = \frac{\sum_{j=1}^{n} w_{i,j}x_j - \bar{X}\sum_{j=1}^{n} w_{i,j}}{S\sqrt{\frac{[n\sum_{j=1}^{n} w_{i,j}^2 - (\sum_{j=1}^{n} w_{i,j})^2]}{n-1}}} \tag{14.1}$$

where G_i^* is the z-score, x_j represents the value for the community j, $w_{i,j}$ represents the space weight between community i and j, n is the number of communities in total, and:

$$\bar{X} = \frac{\sum_{j=1}^{n} x_j}{n} \tag{14.2}$$

$$S = \sqrt{\frac{\sum_{j=1}^{n} x_j^2}{n} - (\bar{X})^2} \tag{14.3}$$

If a statistically significant z-score is positive, a higher z-score means a tighter cluster of high values (hot spots). And if a statistically significant z-score is negative, a lower z-score means a tighter cluster of low values (cold spots).

Regression Analysis between Visual Perception of Urban Environment and Mental Health of the Elderly

Depression and anxiety were included as dependent variables in the analysis model. The GDS score in the questionnaire was used to quantitatively assess the degree of depression in respondents, and the GAI score was applied to measure the anxiety level of respondents.

In the regression analysis model, the street visual perception indicators were the independent variables. After calculating three visual indicators of all sampling points, including greenery, openness, and enclosure, the correlation analysis of these three indexes was carried out in order to assess the correlation between them and the results are shown in Table 14.4. The results illustrated that the absolute

TABLE 14.4
The Correlation between Three Visual Perception Indicators

	Greenness	Openness	Enclosure
Greenness	1	–0.452	0.552
Openness	–0.429	1	–0.938
Enclosure	0.552	–0.938	1

value of the correlation coefficient for openness and enclosure is 0.938. Then, the multicollinearity analysis between the three indices was also performed. The variance inflation factors (VIF) value for the enclosure is 10.913 (VIF > 10), the VIF value for openness is 9.516 (5 < VIF < 10), and the VIF value for greenness is 1.608 (VIF < 5). As there was a multicollinearity of openness and enclosure, the enclosure was excluded from the analysis of mental health and street visual perception. After excluding the enclosure variable, the results of the multicollinearity analysis showed that the VIF value for openness is 1.357 (VIF < 5), and the VIF value for greenness is 1.26 (VIF < 5), therefore, the variables for greenness and openness were selected as independent variables in the regression analysis to avoid collinearity between independent variables.

A series of individual variables were controlled in the multilevel regression analysis, including gender, age, ethnicity, education level, marital status, household registration, number of chronic diseases, living status, self-care ability, and social status, as seen in Table 14.5. Referring to previous studies (Wang, Liu, et al., 2019), some of the variables in the questionnaire were reclassified for better regression analysis: 1) The gender variable was divided into male and female groups (man = 1, woman = 0). 2) The ethnicity variable was divided into Han and ethnic minority groups (Han = 1, ethnic minorities = 0). 3) The variable of marital status was to reclassify the three categories of spouses living together, having spouses living together, widowed/divorced/unmarried in the original questionnaire into the with-spouse and without-spouse groups, of which people who are widowed or divorced, or unmarried belong to the non-spouse class and others are classified as the with-spouse class (with-spouse = 1, without-spouse = 0). 4) The Hukou variable was divided into two groups, including non-agricultural and agricultural account(hukou) (non-agricultural account (hukou) = 1, agricultural account (hukou) = 0). 5) The variable of living status was to reclassify three classes in the original questionnaire, including the living-only-with-spouse group, the living-with-children group, and the living-alone group, into the living-alone and non-living-alone class, among which people who live with spouse and children belong to the non-living-alone group, and others are classified to the living-alone group (non-living-alone = 1, living-alone = 0). 6) The variable of self-care ability includes the limited group and the unlimited group, which was classified according to the answers to the series of questions of the ability to live in daily life, and those people are considered limited to their ability to live as long as they have at least one limited ability, otherwise they are considered to be unlimited (limited = 1, unlimited = 0). 7) The level of education was reclassified

TABLE 14.5
Explanation of Variables

Variables	Description
Age	Continuous variable
Ethnic Groups	Binary variable: Han = 1, ethnic minorities = 0
Gender	Binary variable: man = 1, woman = 0
Hukou	Binary variable: non-agricultural account (hukou) = 1, agricultural account (hukou) = 0
Education	Nominal variable: college/university and above = 3, secondary school = 2, junior school = 1, illiteracy = 0
Marriage status	Binary variable: with-spouse = 1, without-spouse = 0
Living status	Binary variable: non-living-alone = 1, living-alone = 0
Health status	Nominal variable: good = 2, fair = 1, poor = 0
Functional ability	Binary variable: limited = 1, unlimited = 0
Social relationship quantity	Continuous variable: The number of people with whom they can see and meet.
Social relationship quality	Continuous variable: The number of people who are able to talk with them and can them when they need.
Chronic diseases	Continuous variable: The number of chronic diseases the respondents have.

in seven categories in the original questionnaire into four categories. (illiteracy = 0, junior school = 1, secondary school = 2, college/university and above = 3). 8) The variable for the health status in the original questionnaire was reclassified into the poor, fair and good groups, in which the very poor and poor groups in the original questionnaire were classified as the poor class, while the good and very good groups were classified as the good class (poor = 0, fair = 1, good = 2). 9) The variable for social status was social relationship quantity and quality, of which the value for social relationship quantity refers to the number of people with whom they are able to see and meet at least once a month, and social relationship quality is the number of relatives and friends who can talk with them and can them when they need. Among these variables, gender, ethnicity, marital status, living status, hukou and functional ability are binary variables, and education level and health status are nominal variables, while age, the number of chronic diseases, and the quantity and quality of social interaction are continuous variables.

The multilevel regression model(Luke, 2004) was selected to study the correlation between the urban visual environment of the street within 15-minute walking circles of residential communities and the psychological health outcomes of older adults since variables can be divided at both individual and community levels. The model can be described as follows:

Individual level:

$$R_{jm} = \beta_{0j} + \sum_{1}^{i} \beta_{ij} X_{ijm} + e_{jm} \tag{14.4}$$

$$e_{jm} \sim N(0, \sigma_e^2) \tag{14.5}$$

Community level:

$$\beta_{0j} = \gamma_{00} + \sum_1^n \gamma_{0n} Y_{nj} + u_{0j} \tag{14.6}$$

$$u_{0j} \sim N(0, \sigma_{u0}^2) \tag{14.7}$$

$$\beta_{ij} = \gamma_{i0} \tag{14.8}$$

where R_{jm} is the degree of depression or anxiety of each individual m who lives in residential block j, β_{0j} is the average value for GDS or GAI of residential community j, X_{ijm} represents individual factors for each respondent m, β_{ij} represents partial regression coefficients of individual factors, and Y_{nj} represents visual perception variants for residential block j, and γ_{0n} describes partial regression coefficients of visual perception variants, e_{jm} and u_{0j} are errors at individual and community level, respectively.

Several models were included in the study to evaluate the associations between the different indicators and the mental health outcomes of the elderly. Model 0 analyzed the association between GDS, GAI and control variables, respectively, excluding the street visual indicators. Models 1 and 2 analyzed the correlation between the visual perception factors and GDS and GAI scores by adding the greenness and openness variables, respectively. Models 3–5 were applied to test the robustness of Models 1–2. Referring to previous studies(Wang, Lu, et al., 2019), models 3a and 3b removed sample data from respondents who suffer from eye diseases or disorders that can impair visual perception, such as cataract, glaucoma, Alzheimer's disease, or neurological disease, which negatively affected their perception of street view landscape, resulting in a total of 833 samples. In Models 4a and 4b, the sample data from respondents with difficulty in walking were deleted, because they were less likely to walk for 15 minutes outside the residential blocks, and the street visual landscape had less impact on their mental health, and finally, 919 samples were left. Models 5a and 5b excluded those respondents who are older than 85 years old and conducted the analysis on the remaining 1201 sample data, as they might have cognitive biases in the visual perception of streetscape. The multilevel regression analysis was conducted in SPSS.

RESULTS

Street Visual Perception within 15-Minute Walking Circles of Residence Communities

The spatial distribution of estimated street visual perception of 15-minute life circles for residential blocks in the study area is shown in Figure 14.8. In terms of

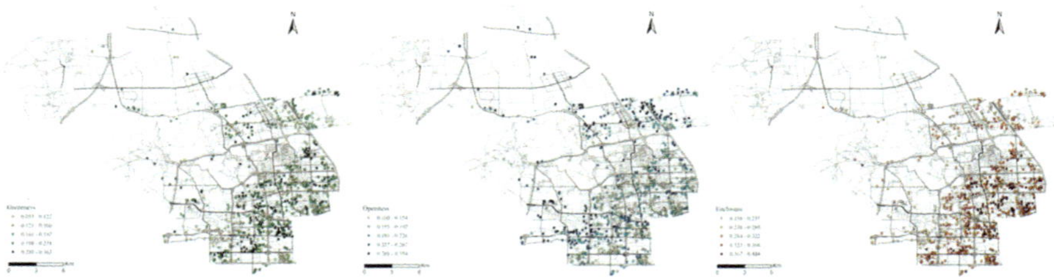

(a) Psychological greenery (b) Outdoor openness (c) Outdoor enclosure

FIGURE 14.8 The spatial distribution of street visual indices: (a) Psycologicalgreenery, (b) outdoor openness, and (c) outdoor enclosure.

overall distribution trends for street visual indicators, the changing trend for greenery from the urban area to the suburb was not apparent, however, there were aggregations for the communities with high and low greenness in 15-minute living circles inside the Third-Ring Road (Third-RR), the Fourth-Ring Road (Fourth-RR), the Fifth-Ring Road (Fifth-RR) and outside the Fifth-RR, while the values for openness and enclosure of residential communities extremely varied with the distance from the communities in the study area to the center of Beijing.

The communities with high and low greenness within the ranges of walking accessibility in 15 minutes show mixed distribution, which can be seen in Figure 14.8(a). However, the results of hot spots analysis in Figure 14.9(a) illustrate that the residential areas which have greater and poorer street greenness are clustered in the study area. There are three dense clusters of residential blocks with high greenness within 15-minute walking circles with 99% confidence, including the south, east and northeast of Haidian District. The residential community cluster with high value of the green view index (GVI) was the most significant in the area near Yuyuantan Park and Zizhuyuan Park inside the Fourth-RR in the south of Haidian District, where many universities, such as Capital Normal University (CNU), Beijing Business University (BBU), Beijing University of Chemical Technology (BUCT), and some primary and secondary schools are located. In the intersection of the eastern part of Haidian District and Chaoyang district, there was a cluster of residential communities with high-value greenness in the Xueyuan sub-district (jiedao). And there are many well-known universities in the administrative area of the Xueyuan sub-district (jiedao), such as Beijing Forestry University (BFU), China Agricultural University (CAU), Beijing University of Science and Technology (BUST), Beijing Language and Culture University (BLCU), etc.

The aggregation of residential blocks with the high or low value of greenness is not significant in other zones and other communities with high greenness show a sporadic distribution trend. However, even in these scattered locations, they still tended to be situated around urban parks, country parks and other major green areas of the city, such as Xiangshan Park and Xiangshan International Golf Club in the southwest of Haidian District. Residential communities with low greenness within 15-minute walking circles were mainly concentrated in the north of Beijing West Railway Station, at the boundary of the southwest corner of Haidian District and Shijingshan District, both sides of the Beijing-Tibet Expressway in the northeast of

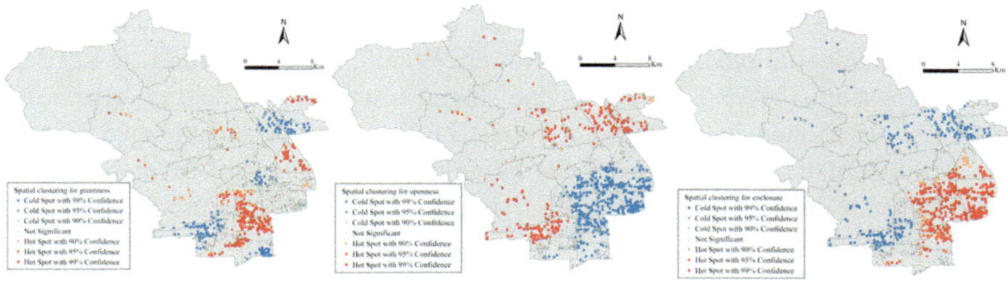

(a) Greenness (b) Openness (c) Enclosure

FIGURE 14.9 The aggregation for residence communities with high and low values for visual perception indicators: (a) Greenness, (b) openness, and (c) enclosure.

Haidian District, and near Haidian Huangzhuang that is a famous commercial district in Beijing.

In the terms of openness and enclosure, the overall spatial patterns of the two indexes show opposite trends, as shown in Figure 14.9(b) and (c). Overall, the 15-minute living circles of the residential areas outside the Fourth-Ring (Fourth-R) exhibited a notably higher level of openness compared to those situated within the Fourth-R. Conversely, the visual closure within 15-minute life circles for those communities was significantly lower than that of the residential areas within the Fourth-R. This is mainly caused by advanced urban construction and high land utilization rate within the Fourth-RR, and the developed economy in an urban area with dense buildings and more high-rise buildings. However, some communities located in the area between the North Fourth-R and North Fifth-R also had lower values for openness and higher degrees of enclosure, which might because the residential blocks are closed to Wudaokou and Liudaokou. These two zones are highly dynamic emerging business districts in the north of Beijing with many newly-built office buildings.

The sub-districts (jiedao) with high values for greenness and others with poor greenness present mixed distribution, which can be seen in Figure 14.10(a). It indicates that there is no considerable aggregation in the urban green construction in the streets around residential areas at the sub-district (jiedao) level in Haidian District.

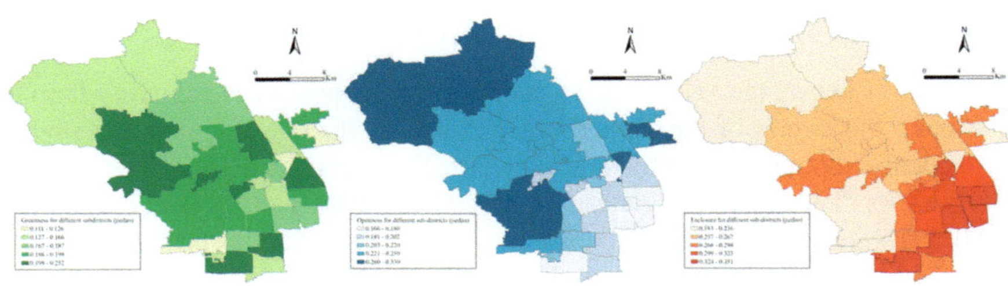

(a) Psychological greenery (b) Outdoor openness (c) Outdoor enclosure

FIGURE 14.10 The visual perception of different sub-districts (jiedao) in the level of residential blocks: (a) Psychological greenery, (b) outdoor openness, and (c) outdoor enclosure.

The top three sub-districts (jiedao) for greenness in the 15-minute walking circles of residential blocks in the study were Wanliu (25.2%), Wenquan (22.3%), and Xiangshan (22.2%). Wanliu, the highest-ranked sub-district (jiedao)for greenness, is located in the middle of the Haidian District and the two parts of this sub-district are on both sides of the Summer Palace. The land development and utilization rate in this area are not high and the land use types are mainly urban parks, mountain parks and other green spaces, including Haidian Park, Yuquan Mountain, Liangshan Park, etc., for which there are a large number of urban green spaces around the residential communities and the street green rate is relatively high in this area. Wenquan and Xiangshan sub-districts (jiedao), which ranked second and third in the value of greenness, are located in the west of Haidian District, and are famous leisure resorts in Beijing, far from the center of Beijing. There are Beijing Xiangshan Park, Xishan National Forest Park, National Botanical Garden and other tourist attractions in these two sub-districts, and Wenquan park, Baijiatuan Park and other urban parks have been built close to residential areas in these two sub-districts. The sub-districts (jiedao) with poor greenness were Tiancunlu (11.1%) and Dongsheng (12.6%), which may be due to the wide roads without street trees in the area and the late planting time of trees that are the core greening of roads around the communities, resulting in the short growth time and the small canopy for these trees.

In terms of the spatial distribution of openness for sub-districts (jiedao) in Figure 14.10(b), the openness degree of different sub-districts (jiedao) shows an increasing trend from the Fourth-R to the outside of the Fifth-R, while the value for Sujiatuo and Shangzhuang sub-districts (jiedao), which are farthest from the center of Beijing in Haidian District, both exceeded 28%. In Figure 14.10(c), the values for enclosure in the study area show a different spatial pattern to that of openness, while the values for enclosure of sub-districts (jiedao) which are closer to the downtown of Beijing were higher and other sub-districts (jiedao) farther from the inner city had lower values for the enclosure. The sub-districts (jiedao) with high and low values for openness and enclosure within the Fourth-RR were mixed distributed, which cause by the might difference in urban development and land use of the sub-districts (jiedao) located inside the Four-R.

Relationship between Visual Perception within 15-Minute Walking Circle of Residence Communities and Mental Health Outcomes

Figure 14.11 shows the values for greenness and openness of the streets within the 15-minute life circles of the 48 residential blocks in the questionnaire. Among the 48 residence communities in the questionnaire, the average values of greenness and openness for these housing estates were 0.203 and 0.183, respectively, and 0.079 and 0.062 were the standard deviations of these two indicators, which can be seen in Table 14.6. The community, namely Jiancaicheng Dongli, had the highest value for greenness, 0.389, while the residential block with the lowest value was Ximulou with 0.052 for greenness of the streets within the 15-minute CLC. The residential area with the highest degree of openness, 0.393, was Niegezhuang Village, and Shuangyushu Dongli had the lowest degree, 0.07.

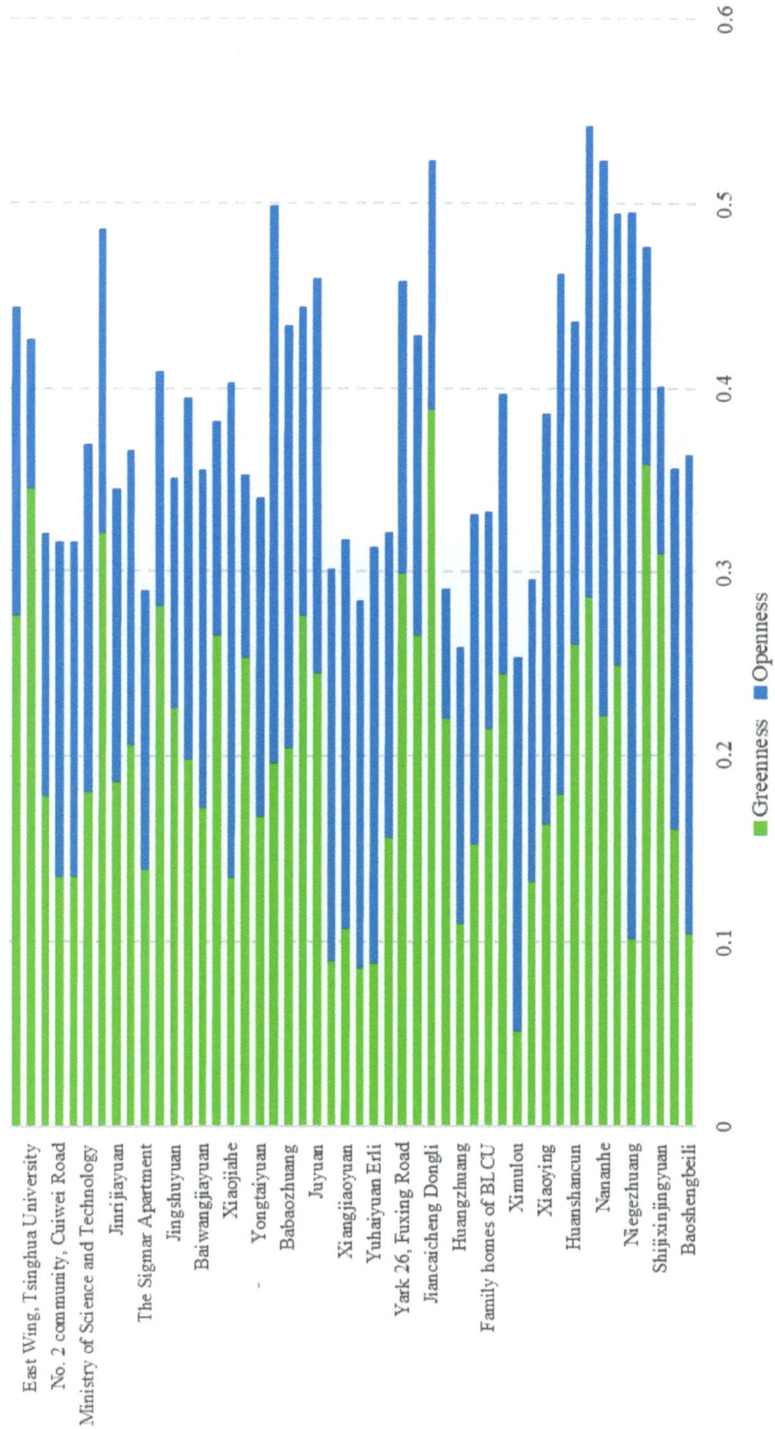

FIGURE 14.11 The values for greenness and openness of residential areas in the questionnaire.

TABLE 14.6
Description of Variables

Independent Variates	Average Value (SD)
Greenness	0.203(0.079)
Openness	0.183(0.062)

(a) Independent variates

Dependent Variates	Average Value (SD)
Geriatric Depression Scale (GDS-15)	3.155(2.538)
Geriatric Anxiety Inventory (GAI)	2.265(4.266)

(b) Dependent variates

Control Variates	Percentage/Average Value (SD)
Age	70.74(7.03)
Hukou (%)	
Non-agricultural	81.15%
Agricultural	18.85%
Ethnic Groups (%)	
Han	96.26%
Ethnic Minorities	3.74%
Education (%)	
Illiteracy	11.13%
Junior school	20.36%
Secondary school	43.95%
College/university and above	24.56%
Functional ability (%)	
Limited	47.28%
Unlimited	52.72%
Chronic diseases	2.31(2.11)
Social ties indicators	
Social ties quantity	9.4(8.96)
Social ties quality	13.81(13.2)
Gender (%)	
Man	40.29%
Woman	59.71%
Marital status (%)	
With a spouse	77.66%
Without a spouse	22.34%
Health status (%)	
Poor	11.54%
General	44.27%
Good	44.19%
Living status (%)	
Living alone	8.53%
None living alone	91.47%

Table 14.6. statistically describes the depression and anxiety of the respondents in the mental health survey, as well as their characteristics, such as gender, age, education level, marital status, etc. Among 1231 respondents, the mean GDS was 3.155 and the standard deviation was 2.538. The mean of GAI was 2.265, and the standard deviation was 4.266. The mean age of the respondents was 70.74, and the standard deviation was 7.03. The respondents with non-agricultural hukou accounted for 81.15%, and 18.85% of the respondents had agricultural hukou. Han people accounted for 96.26% and 3.74% of the respondents were ethnic minorities. In terms of education level, 11.13% of the respondents were untaught, 20.36% of them reached primary school level, 43.95% of them achieved middle school level, and 24.56% of them had studied at university. The proportion of the elderly without functional limitations was 52.72%, and 47.28% of the respondents were functionally limited. The mean number of chronic diseases was 2.31, and the standard deviation was 2.11. The mean of social relationship quantity was 9.4, and the standard deviation for social relationship quantity was 8.96, while the mean of social relationship quality was 13.81, and the standard deviation for social relationship quality was 13.2. Male respondents accounted for 40.29%, while 59.71% of respondents were female. Respondents who were married account for 77.66 per cent and those without a spouse account for 22.34 per cent. The respondents in good physical condition accounted for 44.19%, and 44.27% of the elderly were in general physical condition, while 11.54% of the respondents believed that their health was poor. The elderly who did not live alone accounted for 91.47%, and 8.53% of the elderly lived alone.

Table 14.7 shows the results of several multilevel regression models. Model 0a and model 0b illustrate the correlation between the personal attributes of respondents and depression index and anxiety index, respectively, while Models 1 and 2 show the effects of greenness and openness on depression and anxiety in old people, controlling for covariates. Among control variables, the variables, namely age, marital status, health status, functional ability, social quality and the number of chronic diseases, had significant effects on the elderly's depression degree, while the variables, including age, physical health, functional ability, quality and quantity of social interaction, and chronic diseases had considerable effects on anxiety. The elderly's anxiety and depression degrees were significantly decrease with the increase of age, especially anxiety level, while the absolute value of the correlation coefficient between age and GAI (|−4.642|) was larger than that on GDS (|−3.225|). Depression and anxiety scores increased significantly for older adults who were illiterate compared to those with college degrees, but not for those with elementary and secondary school degrees, which might be because the illiterate old people who cannot read would encounter many difficulties in their daily life. The elderly with a spouse had significantly lower levels of depression than those without a spouse, referring to the correlation coefficient (2.201), while the marriage status had no considerable impact on GAI scores. The self-perceived health status could considerably influence the degree of depression and anxiety. Compared with the elderly in good health, the elderly in general health had a relatively high degree of depression and the GAI scores of these people also considerably increased, while the elderly in poor health had a much higher degree of depression and anxiety,

TABLE 14.7
The Results of Correlation Analysis Models

Variables	Model 0a (DV: GDS)	Model 0b (DV: GAI)	Model 1 (DV: GDS)	Model 2 (DV: GAI)
Age	−3.132**	−4.642***	−3.225**	−4.586***
Race (ref: Han)	0.133	1.500	0.024	1.411
Gender (ref: man)	−0.17	−1.745*	−0.089	−1.800*
Hukou (ref: non-agricultural)	0.98	−0.574	0.53	−1.68
Education (ref: college/university and about)				
Illiteracy	2.300**	2.416**	2.072**	2.118**
Junior school	1.541*	1.434*	1.209*	0.961*
Secondary school	2.406*	1.823*	2.145*	1.616*
Marriage status (ref: with a spouse)	2.283**	1.647*	2.201**	1.69*
Living status (ref: living-alone)	0.069	0.358	−0.073	−0.272
Health status (ref: good)				
Poor	9.020***	6.92***	8.852***	6.668***
General	3.619***	1.569***	3.554***	1.37***
Functional ability (ref: limited)	−3.404***	−2.089**	−3.279***	−1.901**
Social quantity	0.548	2.268**	0.369	2.087**
Social quality	−3.654***	−3.736***	−3.467***	−3.600***
The number of chronic diseases	6.303***	4.543***	6.325***	4.653***
Greenness			−1.857**	−1.18**
Openness			1.053**	2.554**
Constant	5.538***	4.701***	5.303***	3.754***
Random effects				
Number of residential communities	48	48	48	48
Number of residents	1231	1231	1231	1231
AIC	5570.052	6914.827	5559.865	6895.565

$*p < 0.10$. $**p < 0.05$. $***p < 0.01$.

according to the correlation coefficients on the group (CO[GDS] = 8.852 and CO [GAI] = 6.668). If elderly people could take care of themselves, they also had lower depression and anxiety scores than those who had functional limitations. More friends who could talk and help older people had positive impacts on the decreases in GDS and GAI scores, suggesting that close relatives and friends could alleviate the elderly' s depression and anxiety. Surprisingly, the number of friends who could meet and connect significantly increased the anxiety of the elderly, possibly because too many non-intimate friends or relatives may bring social pressure on the elderly, which may affect the health of the elderly with limited energy. An increase in the number of chronic diseases also had a great relationship with the rises on depression and anxiety among older adults (CO[GDS] = 6.325 and CO[GAI] = 4.653).

TABLE 14.8
The Robustness Test Results for the Correlation Analysis

Variables	Model 3a	Model 3b	Model 4a	Model 4b	Model 5a	Model 5b
Greenness	−1.254**	−1.219**	−1.169**	−1.161**	−1.690**	−1.730**
Openness	1.149**	2.740**	1.096**	2.788**	1.606**	2.927**

*$p < 0.10$. **$p < 0.05$. ***$p < 0.01$.

In Model 1, the greenness variable had a negative effect on depression scores at the 95% significance level (−1.857), while the openness variable would positively affect depression scores at the 95% significance level (1.053), which indicates that the GDS score of older adults had growth with the decreased green view index and increased visible sky index. And in Model 2, an increase in GVI could negatively affect anxiety scores at the 95% significance level (−1.18), whereas an increase in the degree of openness had positive relationship with a rise in GAI score at the 95% significance level (2.554). For the correlation coefficient of street greenness and openness in Models 1–2, the increase of street openness had a greater effect on GAI than on GDS, while the growth in street greenness had a nearly similar effect on depression and anxiety levels.

Models 3–5 were applied to test the robustness of the original models and their outcomes can be seen in Table 14.8. The results of Models 3–5 show that an increase in GVI would negatively influence GDS and GAI scores, and a rise in the value for openness would positively affect GDS and GAI scores.

DISCUSSION

Urban Planning and the Visual Environment around Residential Blocks

The study found that there were both residential communities with lower greenness in urban and suburban districts. And the values for the greenness of the communities located in the inner city were not significantly lower than that of the residential blocks in the suburban areas. These conclusions can also be found in previous studies (Li et al., 2015; Ye et al., 2019). The reason might be that green spaces in downtown areas are mainly street plants that can be more easily perceived by pedestrians, compared with rural green spaces (Ye et al., 2019). The values of visual openness and enclosure within the 15-minute walking circles of residence communities varied significantly with the level of urban economic growth and the degree of urbanization. Ma et al. (2021) also summarized a similar finding in Shenzhen. The residential communities with low visual openness and high visual enclosure in the 15-minute walking circles were aggregated in the central area of the urban spatial structure possibly due to a high land utilization rate and a high level of urbanization with a mass of high-rise office buildings in the area. The results of urban visual environment assessment provide valuable scientific insights for urban planning. Given the complex characteristics of the current urban visual environment, policies on urban planning and construction should be adapted to local conditions. For villages and towns far from the city center and areas with low land utilization rates, urban planners should emphasize the enhancement of urban street greening and the stimulation of block vitality, while highly developed zones should prioritize the diversity of cityscapes and architectural characteristics.

Interpretation of the Impacts of Urban Street Visual Environment on Mental Health of the Elderly

The finding in this research that the improvement of street greenery in the 15-minute living circles of communities had a negative correlation with the deeper depression and anxiety of the elderly is consistent with the previous researches on restorative environment theory, including attention recovery theory and stress relief theory, which has confirmed the positive impacts of a green environment on mitigating people's mental health problems (Kaplan & Talbot, 1983). Street greening can deaden the noise interference from people and vehicles, restore people's attention, and recover pedestrians' stress (Markevych et al., 2017), which could help the elderly population relax their mood and stabilize their thoughts when walking in such a street environment. Therefore, their depression and anxiety could also be more effectively regulated.

According to the regression analysis results in this study, the increase in the sky view ratio within the 15-minute walking circles around residential blocks might negatively affect the psychological health of the elderly and aggravate the degree of depression and anxiety in the elderly. Wang, Lu, et al. (2019) also found that people would have higher degrees of depression and anxiety when they are exposed to higher sky visibility. This might be due to open spaces with little shade from trees or buildings, or the absence of clear barriers between driveways and sidewalks, which can decrease the feeling of security of pedestrians, since a sense of security as a basic need second only to physiology is the premise of realizing the higher-level emotional needs of human beings, referring to Maslow's Hierarchy of Needs Theory (McLeod, 2007). The high value of sky openness could also negatively affect the walkability of the streets, which has been proved by Yin and Wang (2016). It possibly because the high sky visibility could reduce the willingness of the elderly to walk in the streets within the 15-minute life circles of residential communities by influencing the urban street microclimate and street thermal comfort. Less physical activities might cause an increase in the risk of chronic diseases, such as diabetes, cardiovascular and circulatory diseases, and the growing rate of physical function degradation with rising age, both of which are important factors in raising levels of depression and anxiety in older adults.

Strengths and Limitations

Several strengths can be emphasized in this study. First, the street visual landscape research within the 15-minute CLC applied the road network data, which means that the results would be more consistent with the travel paths in residents' daily life and it improves the reliability of the research results. Second, the study pays attention to the visual environment within 15-minute walking circles instead of public service facilities, which provides innovative ideas to planners for the improvement of the urban environment. Third, the method using SVIs and a deep learning model in this study can help reduce the reliance on subjective estimations and experiences in the preliminary research of policy formulation of urban planning, enhancing the effectiveness of planning schemes in the improvement of the urban environment. Fourth, SVIs from the green season were applied in this research, which enhances the reliability of the assessment results of urban street greening. Fifth, the correlation analysis between visual perception and mental health of the elderly population corroborates the influences of the street visual environment around residential communities on the mental health outcomes of the elderly with several robust results.

However, there are still numerous limitations in this study. First, SVIs used to analyze the visual environment of the 15-minute walking circles of residential communities in Haidian District were taken about five years ago. However, the street trees in most urban areas have matured and the inter-annual change in greenness would not change significantly. Second, in the process of obtaining street view data through sampling points, there were some missing data of sampling points, but the absence of data has little impact on the results since the visual indicators within a 15-minute life circle were represented by the average values. Third, since only one of the many street view data sources was selected in this study, it is still unknown whether the data source would cause bias in the results. Fourth, as SVIs are taken on motorways, there might be a difference between these images and the view of pedestrians on the sidewalk. Fifth, Haidian District in Beijing, one of the most deprived cities in China, was chosen as the study area in the correlation analysis between visual indicators and mental health outcomes of the elderly, therefore, there might be a deviation between the results of this research and the relationship to the elderly in underdeveloped areas or rural areas.

Future studies can apply a large amount of street view data which are obtained on sidewalks to increase the credibility of visual indicators in describing pedestrians' perceptions of the urban visual environment. Future researches could also select other less developed regions in China to figure out if there are differences in outcomes between deprived and undeveloped regions. And there are still many other metrics that can assess the urban visual environment and might have influences on mental health, such as traffic flow, street obstacles, etc. Other indicators could be considered in future researches to increase the diversity and effectiveness of the findings.

CONCLUSIONS

The research applied SVIs and advanced DL methods to study the street landscape within the 15-minute walking circles of the residential blocks in Haidian District, including greenness, openness and enclosure, and analyzed the correlation between it and the mental health outcomes of the elderly, which provides a quantitative method to evaluate the street visual environment around residential communities to help alleviate the depression and anxiety of the elderly. For the visual environment of 15-minute walking circles of residential communities in Haidian District, this research found that the residential communities with high GVI within the 15-minute life circles were mainly those areas close to schools, universities, urban parks and leisure resorts. The changing trend of openness and sense of enclosure within the 15-minute life circles of residential blocks was opposite. The residential areas with low openness and high visual enclosure were mainly clustered in the central area of spatial planning in Beijing. Combined with the elderly mental health survey data, the correlation analysis between the visual perception within 15-minute life circles of residential areas and the psychological health outcomes of old people found that the improvement of street greenery within the 15-minute living circles in the communities was significantly correlated with the reduction of depression and anxiety in the elderly, while an increase of sky openness had a high positive correlation with the deeper feeling of depression and anxiety.

The quantitative analysis of the urban visual environment from the perspectives of urban planning and spatial geography can validly identify the residential areas with poor visual environments and help planners to propose specific modification schemes for these areas. In the process of the goal achievement of healthy ageing, not only the implementation of pension policy and the improvement of medical treatment should be concerned, but also the impacts of the urban visual environment of 15-minute life circles of residential communities on the elderly's daily life should be considered in spatial planning and urban design to improve their mental health.

REFERENCES

Chen, T., Li, M., Li, Y., Lin, M., Wang, N., Wang, M., Xiao, T., Xu, B., Zhang, C., & Zhang, Z. (2015). Mxnet: A flexible and efficient machine learning library for heterogeneous distributed systems. arXiv preprint arXiv:1512.01274.

Dai, L., Zheng, C., Dong, Z., Yao, Y., Wang, R., Zhang, X., Ren, S., Zhang, J., Song, X., & Guan, Q. (2021). Analyzing the correlation between visual space and residents' psychology in Wuhan, China using street-view images and deep-learning technique. *City and Environment Interactions*, 11, 100069. 10.1016/j.cacint.2021.100069

Dzhambov, A. M., Markevych, I., Hartig, T., Tilov, B., Arabadzhiev, Z., Stoyanov, D., Gatseva, P., & Dimitrova, D. D. (2018). Multiple pathways link urban green-and bluespace to mental health in young adults. *Environmental Research*, 166, 223–233.

Ewing, R., & Handy, S. (2009). Measuring the unmeasurable: Urban design qualities related to walkability. *Journal of Urban Design*, 14(1), 65–84.

Helbich, M., Yao, Y., Liu, Y., Zhang, J., Liu, P., & Wang, R. (2019). Using deep learning to examine street view green and blue spaces and their associations with geriatric depression in Beijing, China. *Environment International*, 126, 107–117.

Kaplan, S. (1983). A model of person-environment compatibility. *Environment and Behavior*, 15(3), 311–332.

Kaplan, S., & Talbot, J. F. (1983). Psychological benefits of a wilderness experience. In *Behavior and the Natural Environment* (pp. 163–203). Springer.

Li, X., & Ghosh, D. (2018). Associations between body mass index and urban "green" streetscape in Cleveland, Ohio, USA. *International Journal of Environmental Research and Public Health*, 15(10), 2186.

Li, X., Zhang, C., Li, W., Kuzovkina, Y. A., & Weiner, D. (2015). Who lives in greener neighborhoods? The distribution of street greenery and its association with residents' socioeconomic conditions in Hartford, Connecticut, USA. *Urban Forestry & Urban Greening*, 14(4), 751–759.

Luke, D. A. (2004). Basic multilevel modeling. *Multilevel Modeling*, 10–53.

Ma, X., Ma, C., Wu, C., Xi, Y., Yang, R., Peng, N., Zhang, C., & Ren, F. (2021). Measuring human perceptions of streetscapes to better inform urban renewal: A perspective of scene semantic parsing. *Cities*, 110, 103086.

Manepalli, U. R., Bham, G. H., & Kandada, S. (2011). Evaluation of hotspots identification using kernel density estimation (K) and Getis-Ord (Gi*) on I-630. 3rd International Conference on Road Safety and Simulation,

Markevych, I., Schoierer, J., Hartig, T., Chudnovsky, A., Hystad, P., Dzhambov, A. M., De Vries, S., Triguero-Mas, M., Brauer, M., & Nieuwenhuijsen, M. J. (2017). Exploring pathways linking greenspace to health: Theoretical and methodological guidance. *Environmental Research*, 158, 301–317.

McLeod, S. (2007). Maslow's hierarchy of needs. *Simply Psychology*, 1(1-18).

MXNet. https://mxnet.apache.org/versions/1.9.1/

Pachana, N. A., Byrne, G. J., Siddle, H., Koloski, N., Harley, E., & Arnold, E. (2007). Development and validation of the Geriatric Anxiety Inventory. *International Psychogeriatrics*, 19(1), 103–114.

Porta, S., & Renne, J. L. (2005). Linking urban design to sustainability: formal indicators of social urban sustainability field research in Perth, Western Australia. *Urban Design International*, 10(1), 51–64.

Qiu, C., Wang, Z., & Liao, J. (2022). Research of the coverage rate of "15-minute living circle" based on walking accessibility. *Shanghai City Management*, 31, 77–85.

Seventh National Census Communiqué. (2021).

Sheikh, J. I., & Yesavage, J. A. (1986). Geriatric Depression Scale (GDS): recent evidence and development of a shorter version. *Clinical Gerontologist: The Journal of Aging and Mental Health.*

The depression and anxiety survey of the elderly. http://www.cnsda.org/index.php?r=projects/view&id=60493698

Wang, R., Helbich, M., Yao, Y., Zhang, J., Liu, P., Yuan, Y., & Liu, Y. (2019). Urban greenery and mental wellbeing in adults: Cross-sectional mediation analyses on multiple pathways across different greenery measures. *Environmental Research*, 176, 108535.

Wang, R., Liu, Y., Lu, Y., Zhang, J., Liu, P., Yao, Y., & Grekousis, G. (2019). Perceptions of built environment and health outcomes for older Chinese in Beijing: A big data approach with street view images and deep learning technique. *Computers, Environment and Urban Systems*, 78, 101386.

Wang, R., Lu, Y., Zhang, J., Liu, P., Yao, Y., & Liu, Y. (2019). The relationship between visual enclosure for neighbourhood street walkability and elders' mental health in China: using street view images. *Journal of Transport & Health*, 13, 90–102.

Wu, H., Sun, X., & Zhou, J. (2022). Construction of a performance evaluation method for public service facilities in a 15-minute life circle: An empirical study of Changsha city from the perspective of balanced supply and demand. *South Architecture* (6), 62–71. CNKI:SUN:NFJZ.0.2022-06-013.

Ye, Y., Richards, D., Lu, Y., Song, X., Zhuang, Y., Zeng, W., & Zhong, T. (2019). Measuring daily accessed street greenery: A human-scale approach for informing better urban planning practices. *Landscape and Urban Planning*, 191, 103434.

Yin, L., & Wang, Z. (2016). Measuring visual enclosure for street walkability: Using machine learning algorithms and Google Street View imagery. *Applied Geography*, 76, 147–153.

Zhang, F., Zhou, B., Liu, L., Liu, Y., Fung, H. H., Lin, H., & Ratti, C. (2018). Measuring human perceptions of a large-scale urban region using machine learning. *Landscape and Urban Planning*, 180, 148–160.

Zhang, J., Heng, C. K., Malone-Lee, L. C., Hii, D. J. C., Janssen, P., Leung, K. S., & Tan, B. K. (2012). Evaluating environmental implications of density: A comparative case study on the relationship between density, urban block typology and sky exposure. *Automation in Construction*, 22, 90–101.

15 Comparisons of Deep Learning Models for Dynamic Local Climate Zone Mapping

Haojie Chen, Cheolhee Yoo, and Qihao Weng

INTRODUCTION

The Local Climate Zone concept, which provides an objective protocol for measuring the magnitude of any city's climate and ecology, has been extensively used to address the limitation of using only urban and non-urban for urban climate and ecology issues (Xiang et al., 2023). The building structure and urban morphology were characterized to 10 built types based on building morphology factors and combing information on the surrounding exposure and surface conditions of a zone (Zhou et al., 2022). Therefore, this terminology has a great potential for revealing urban land changes through the portrayal of urban morphological characteristics. In addition, the analysis of multiple megacities using remote sensing images becomes feasible given the Earth Observation (EO)-based approach for classifying LCZ.

Local Climate Zone(LCZ) was designed from the very beginning with the purpose of providing an objective protocol for measuring the magnitude of the urban heat island effect in any city. With this term, it is not necessary to restrict the discussion of UHI(Urban Heat Island) to a distinction between urban and non-urban with just two options (Stewart & Oke, 2012). Standard set is divided into built types (1–10),and land cover types (A-G). Built types include categories 1–10, mainly distinguished by building height and impact. land cover types are also used as a means of differentiation, including paved and hard-packed (concrete, steel, stone, brick, tile, and concrete construction materials.), and permeable land cover types (low plants, scattered trees). Land cover types include categories A-G, which mainly distinguish between land cover landscape types, where the type of vegetation (height, species) is subdivided (Stewart, 2012).

There are four primary approaches to LCZ mapping, including the manual approach, GIS-based approach (Zhao, 2019), Earth Observation (EO)-based approach (Bechtel, 2012; Bechtel, 2015), and hybrid approach (Zhou, 2021). The EO-based WUDAPT method, for example, employs a standard workflow to produce LCZ maps using the same data source (landsat-8) and classifier (random forest), while

DOI: 10.1201/9781003244561-19

generating sample sets based on different regions (Bechtel, 2015). However, the accuracy of this method can be limited. In contrast, Convolutional Neural Networks (CNN) have demonstrated superior performance compared to other supervised deep learning algorithms in remote sensing visual recognition (Zhang, 2016; Ma, 2019). In the field of LCZ mapping, CNN models have been found to outperform traditional methods, such as random forests (Yoo, 2019).

This chapter delves into the potential of Convolutional Neural Networks (CNNs) in mapping Local Climate Zones (LCZs), providing a comparative study against traditional mapping methodologies. We introduce a novel data processing technique involving a 32*32 moving window and demonstrate its efficacy over conventional approaches. Leveraging data augmentation strategies for balanced LCZ representation, we evaluate the performance of several established CNN models in the context of LCZ mapping. With an overall accuracy of 80% to 90%, our findings highlight the promise of CNNs for future advancements in urban climate studies.

METHODOLOGY

There are four main approaches for LCZ mapping, the manual approach, GIS-based approach (C. Zhao, Jensen, Weng, Currit, & Weaver, 2019), EO-based (Earth Observation) approach (Bechtel et al., 2015; Bechtel & Daneke, 2012) and hybrid approach (Zhou et al., 2021). The manual method requires a lot of human and financial resources, not suitable for large scale mapping needs, it is even less possible to analyze multiple cities.

GIS-Based Approach

The basis for the GIS-based approach to map LCZ is based on values of geometric and surface cover properties, including sky view factor(SKF), aspect ratio(AR), building surface fraction(BSF), impervious surface fraction(ISF), pervious surface fraction(PSF), height of roughness elements(Height), and terrain roughness class (Stewart & Oke, 2012). Different research scholars have described these indices based on different data sources methods. Basic input data includes satellite and aerial imagery, land cover datasets, data with height information, road/street maps, building footprint maps and others (Oliveira et al., 2020; C. Zhao et al., 2019). The most used methods are the transformation based on existing land use datasets (Oliveira et al., 2020; C. Zhao et al., 2019) and other standard datasets for reclassification (Hidalgo et al., 2019), and rule-based fuzzy preliminary classification using linear interpolation scores or ranks based on a representative set of metrics at the site, grid, subdistrict or block level (Kotharkar & Bagade, 2018; Lelovics et al., 2014; Quan, 2019). GIS-based methods are often inseparable from data with elevation information (e.g., NDSM or 3D city models) because elevation data are directly or indirectly related to SKF, ISF, BSF, PSF, and height of roughness elements (C. Zhao et al., 2019). The open source DSM and DEM data are both too coarse in spatial resolution, and additionally cannot be analyzed over multiple times

(Zhou et al., 2021). Precise point clouds can be obtained from LIDAR data to obtain NDSM elevation information, but due to its expensive equipment and cumbersome processing procedures, it is often used for small-scale measurements and usage. Although some cities or tracts have obtained airborne LIDAR range data, it is still difficult to obtain for the rest of the cities, and it is even more difficult to obtain at different times (C. Zhao et al., 2019). The disadvantage of using the LULC data for conversion to LCZ is that the definition of land use data varies from place to place, and there is no way to develop a uniform method for conversion (C. Zhao et al., 2019).

EO-Based Approaches

EO-based approaches currently use machine learning methods for classification, which are classified as Object-based approach and pixel-based approach. Object-based classification requires a lot of operations, more detailed information, and contextual knowledge (Ma et al., 2021a; Ma et al., 2021b), while the pixel-based approach can ignore some details and classify use pixel blocks, which is widely used in LCZ classification (Bechtel & Daneke, 2012). The approach involves selecting and solving the feature vector (to describe the most accurate and comprehensive information) in advance based on the features of the EO observations (to classify the LCZ), and then using classifiers to judge the feature vector, the main classifiers being random forests, support vector machines, and so on (Bechtel et al., 2015; Bechtel & Daneke, 2012). The method of selecting the different characteristics of Earth observation data and their best fit with LCZ observations has been studied by research scholars (Bechtel & Daneke, 2012). The WUDAPT method has developed a standard workflow for producing LCZ maps, which uses the same data source (landsat-8) and uses the same classifier (random forest) but can produce sample sets based on different regions, thus allowing scholars to produce LCZ maps that meet their needs based on their own study area (Bechtel et al., 2015). In contrast to deep learning, the WUDAPT method of shallow learning is used by a wide range of researchers because it is computationally convenient, takes less time and is less complex. However, this method is often not very accurate (R. Wang et al., 2018). Hence, many researchers have attempted to map the LCZ using multi-source Earth observation data sources, such as sentinel-1 (La, Bagan, & Yamagata, 2020) and sentinel-2 data (Shi & Ling, 2021), NTL data (visible light emerging from the Earth at night) (La et al., 2020), and so on. The multi-source data make the accuracy improved, but the ease of data acquisition raises the difficulty.

Hybrid Models

The hybrid model combines the advantages and disadvantages of both GIS-based and EO-based approaches. It has been verified by scholars that the accuracy of the LCZ maps obtained by the GIS-based method is generally higher than that of the WUDAPT method (R. Wang et al., 2018). The main reason is that the GIS-based approach is able to operate strictly according to the rules of the LCZ land cover attributes. And because of the different data sources, the LCZ blocks are represented in different forms (e.g., vector data and grid areas), which enables to obtain sufficient size and boundaries for each LCZ class (Kotharkar & Bagade, 2018; Oliveira et al., 2020). The biggest drawback of the GIS-based approach is that data sources vary in quality and format, and even the methods of conversion are developed based on specific data sources (Quan, 2019). Some of the data are not global and data acquisition becomes difficult if we change the study area. Although the MUDAPT method is very simple and convenient for more research scholars to be used in many research areas, the disadvantage is that the method of WUDAPT is a supervised learning method, which requires manual judgment of the class of training samples in advance, and the quality of label data very much affect the final results. In a grid block, different types of land use and land cover types are often mixed due to the dense characteristics of the city, and the criteria for labeling by different scholars are also different (Quan, 2019).Some scholars, in order to synthesize the advantages and disadvantages of the above two methods, proposed to use EO observation data for LCZ classification while combining GIS data such as OSM data to help segment LCZ blocks, which has greatly improved the final classification effect (Fonte et al., 2019; Shi & Ling, 2021; R. Wang et al., 2018; Zhou et al., 2021).

Convolutional Neural Networks (CNN) Models

Convolutional neural networks (CNN) have been shown to outperform most deep learning supervised algorithms in the field of remote sensing visual recognition (Ma et al., 2019; Zhang et al., 2016), such as scene classification (Hu et al., 2015; Scott et al., 2017; Yuan, Han, Gu, & Yan, 2021), target detection (Shin et al., 2016), semantic segmentation (Cao et al., 2019). With its powerful ability to extract highly abstracted, generalized features, CNN models have been tried in the field of LCZ mapping and proven to be better than methods such as random forests (Yoo et al., 2019). So2LCZ42 provides a standardized dataset of 42 city clusters worldwide (plus 10 smaller regions), resulting in approximately 500,000 pairs of Sentinel-1 and Sentinel-2 image block data, and a carefully designed evaluation process resulting in an overall confidence level of 85% (Zhu et al., 2020). Several studies have shown that training with the So2Sat LCZ42 dataset can be trained to obtain better classification models (Chunping, Schmitt, Lichao, & Xiaoxiang; Qiu et al., 2019; Qiu et al., 2020; Zhou et al., 2021).

Most of the CNN models used for LCZ classification are adaptations of classical models, such as VGG, ResNet, etc. (X. Huang, Liu, & Li, 2021; S. Liu & Shi, 2020; Qiu et al., 2019; Qiu et al., 2020; Yoo et al., 2019; Zhou et al., 2021), while most of the models are trained from the initialized parameters. It is undeniable that none of us can prove the superiority of the models designed by the authors.

Some scholars are beginning to recognize that completely designing and training CNN models from scratch is very challenging because it requires large amounts of data and takes a long time (Yuan et al., 2021), and that CNNs pre-trained on large datasets have general feature extraction capabilities that can be transferred to regions with smaller amounts of data (Shin et al., 2016). Many scholars have applied migratory learning for scene classification, and they have explored the effects of different depth levels of features on migratory learning, as well as the effects of different learning modes (parameter freezing, fine-tuning, initialization from random) on migratory learning results (Hu et al., 2015; Pires De Lima & Marfurt, 2019; H. Zhao, Liu, Zhang, & Liang, 2019). However, no scholars have tested in detail the best patterns and classifiers for migration learning for LCZ classification.

In summary, the CNNs-based approach is able to obtain a more generalized LCZ classification model by training. This will be far different from previous GIS-based and shallow learning-based methods, as GIS-based methods usually have a small number of fixed-time data sources and shallow learning-based methods usually have a small number of training samples. In particular, CNNs-based methods based on migration learning can lead to a very generalized model, which makes it possible to study the temporal variation of LCZs at the urban scale. However, the CNNs-based approach still suffers from low accuracy and a discrete distribution of classification results.

In this paper, a variety of CNNs models are tested using a migration learning approach in order to obtain a more generalized model that can enable the study of the temporal variation of LCZs at the urban scale. At the same time, experiments are carried out to obtain high accuracy CNNs models.

Data Processing and CNN Models Training

In our scheme, we considered the surrounding areas around the focal pixel, using a moving window of size 32*32 at 32-pixel intervals. This approach resulted in better results than the RF-based scheme and other CNN-based schemes (e.g., 10*10 size feature extraction block) (Yoo, 2019). To ensure an equivalent number of polygons and patches of different LCZ types and independent LCZ types from each other, we utilized data augmentation in each polygon (i.e., by rotating 90°, 180°, and 270°). Polygons refer to manually annotated polygonal regions that are homogeneous in nature, and batches represent a 32*32-pixel block. A list of the training and test datasets of LCZ types for each city is provided in Table 15.1.

This study testes the traditional CNN models, including VGG (Simonyan, 2014), ResNet (He, 2016), DenseNet (Huang, 2017), and GoogLeNet (Szegedy, 2014) to obtain accurate LCZ maps of the four megacities and synchronously perform the massive LCZ classification more swiftly. All initial parameters were sourced from the official pre-training parameters, the learning rates were set to 0.001, and the entire training lasted 100 epochs.

This study reclassified the LCZ types when calculating the overall accuracy (OA) to obtain highly precise results in classifying the three urbanization processes. OA_{ex}, OA_{high}, and $OA_{compact}$ were calculated to indicate the OA of the three urbanization processes, namely, new urbanization, intensified height, and intensified compactness, respectively.

$$OA = \begin{cases} OA_{ex} = N_{ex}/N_m \\ OA_{high} = N_{high}/N_m \\ OA_{compact} = N_{compact}/N_m \end{cases} \quad (15.1)$$

where N_{ex}, N_{high}, and $N_{compact}$ indicate the numbers of accurately judged samples for LCZ type of new urbanization, intensified height, and intensified compactness, respectively. N_m expresses the total number of all samples.

Since its introduction, convolutional neural networks (CNNs) have been widely used in image classification due to their powerful ability in extracting abstract features (Zhang et al., 2016), and many related models have been proposed and performed well in the standard dataset ImageNet, including VGG, ResNet, SqueezeNet, DenseNet, and GoogleNet. Each model has its own characteristics.

TABLE 15.1

Training and Test Dataset of Each LCZ Type by City

LCZ	Beijing			
	2020		2000	
	Training	Test	Training	Test
1	2* (456)	2* (234)	–	
2	21 (1344)	6 (696)	19 (918)	7 (372)
3	7 (498)	2 (138)	13 (1488)	1 (564)
4	33 (2184)	22 (768)	12 (270)	5 (132)
5	3* (48)	3* (24)	4 (228)	2 (48)
6	13 (834)	5 (402)	3* (96)	3* (72)
7	13 (684)	11 (270)	22 (1134)	6 (492)
8	2* (30)	2* (18)	–	
9	–		–	
10	13 (1638)	4 (1260)	5 (156)	1 (132)
A	3 (2058)	6 (840)	6 (1650)	1 (600)
B	3 (60)	1 (30)	–	
C	–		–	
D	18 (3132)	12 (1278)	16 (3060)	6 (1482)
E	3 (2592)	3 (1440)	3* (726)	3* (330)
F	–		7 (582)	3 (228)
G	7 (834)	3 (276)	8 (474)	3 (258)

* (The red-star classes) has been used to refer to situations with only a few reference polygons of the LCZ classes (no more than 3) (Yoo, 2019). The numbers in the table represent the number of polygons, whereas the figures in parentheses represent the number of patches.

VGG

VGG is one of the most classic neural network models, using small convolutional kernels (3*3) to extract more complex information than the previous large convolutional kernels (5*5 and 7*7). VGG consists of multiple blocks, each consisting of two or three convolutional layers, and a down sampling or pooling layer, which allows features of different depths to be extracted. The feature extraction layers are followed by three fully connected layers to classify the final result (Simonyan & Zisserman, 2014). In this paper, the input layers (3*225*225 to 7*32*32) are modified for the application. VGG is mainly divided into VGG13, VGG16, and VGG19 according to the depth of the network.

ResNet

ResNet uses a residual module to enable the network to be trained more deeply. Through the VGG model scholars found that deeper networks can extract deeper features and information, but if too much training will also appear overfitting phenomenon, in order to avoid this phenomenon, ResNet adds a residual module so that the gradient can be passed between the front and back layers of the network (He et al., 2016). In this paper, the input channels are changed to seven channels. ResNet is divided into ResNet18, ResNet34, ResNet50, ResNet101, and ResNet154.

SqueezeNet

SqueezeNet uses the Fire module to encapsulate multiple convolutional and down sampling layers, consisting of Squeeze (1*1 convolutional kernel) and Expand (1*1 convolutional kernel and 3*3 convolutional kernel) (Iandola et al., 2016). In this paper, the input channels are changed to seven channels. SqueezeNet includes SqueezeNet 1.0 and SqueezeNet 1.1 versions.

DenseNet

As mentioned earlier, in CNNs, the gradient disappearance and overfitting problems become more and more obvious as the depth of the network deepens. ResNet proposes a residual module that enables the fusion of the front and back gradients within the module, but cannot fuse features that span large distances. DenseNet continues the idea of feature fusion by fusing the features of all modules and trying to ensure that layer-to-layer DenseNet continues the idea of feature fusion by fusing the features of all modules and trying to ensure the maximum continuity of information between layers (G. Huang, Liu, Van Der Maaten, & Weinberger, 2017). In this paper, the input channels are changed to seven channels. DenseNet is mainly divided into DenseNet121, DenseNet161, DenseNet169,and DenseNet201.

GoogLeNet

GoogLeNet is composed of several inception modules, and the inception modules internally fuse features from multiple convolutional kernels (1*1, 3*3, 5*5), and this approach enables the fusion of different scale features (Szegedy et al., 2014). At the same time, the deeper the network is, the larger the proportion of 3*3 and 5*5 is, the more abstract features can be extracted. In this paper, the input channels are changed to seven channels.

LCZ MAPPING RESULTS

The performance evaluation of various Convolutional Neural Network (CNN) models for Beijing is presented in Table 15.2, case of LCZ classification results for the Beijing region are shown in Figure 15.1. The models include DenseNet121,

TABLE 15.2

The Performance Evaluation of the CNN Models for Beijing, Demonstrating High Overall Accuracy (OA) Ranging from 80% to 90%

Country	Year	CNNs	OA_{all}	OA_{ex}	OA_{high}	$OA_{compact}$
		DenseNet121	0.8243	0.9906	0.9450	0.9252
		GoogleNet	0.8603	0.9867	0.9432	0.9134
		ResNet50	0.8420	0.9886	0.9289	0.9216
		VGG16	0.8450	0.9864	0.9297	0.9097
		VGG19	0.8324	0.9853	0.9222	0.9038
		DenseNet121	0.8529	0.9630	0.9550	0.9059
		GoogleNet	0.8548	0.9655	0.9618	0.9277
		ResNet50	0.7208	0.9558	0.9459	0.8960
		VGG16	0.8047	0.9675	0.9448	0.9302
		VGG19	0.7295	0.9456	0.9463	0.8714

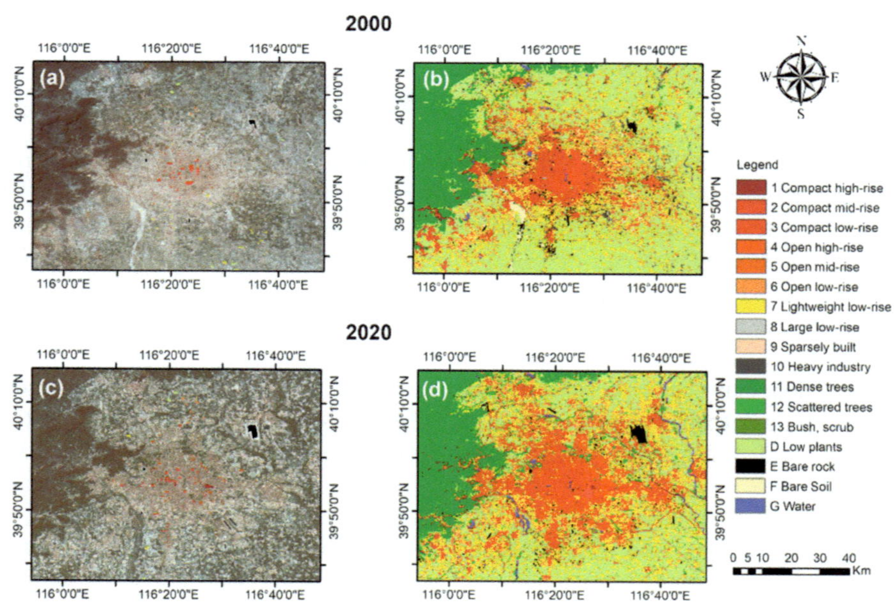

FIGURE 15.1 LCZ classification results for Beijing. (a) and (c) indicate the references of 2000 and 2020, respectively; (b) and (d) indicate the LCZ classification results of 2000 and 2020, respectively.

GoogleNet, ResNet50, VGG16, and VGG19. These models demonstrate high overall accuracy (OA), with values ranging from 80% to 90%. The DenseNet121 model achieved an accuracy of 82.43% in the first instance and improved to 85.29% in the second instance. GoogleNet's performance was slightly better, with accuracy scores of 86.03% and 85.48% in the first and second instances, respectively. The ResNet50 model's performance was consistent, with accuracy scores of 84.20% and 72.08% in the first and second instances, respectively. The VGG16 model showed a decrease in performance from the first instance (84.50% accuracy) to the second instance (80.47% accuracy). Similarly, the VGG19 model's performance decreased from 83.24% accuracy in the first instance to 72.95% accuracy in the second instance. The performance results across the models suggest that GoogleNet and DenseNet121 provided the most reliable and consistent results in this assessment.

DISCUSSION AND CONCLUSION

In our research, we examined the performance of various classical convolutional neural network (CNN) models, including VGG, ResNet, DenseNet, and GoogLeNet, in local climate zone (LCZ) classification tasks. In our experiments, we found that DenseNet and GoogLeNet provided the most reliable and stable results in our tasks. This could be because these models effectively integrate features of various scales and ensure the maximum continuity of information between the layers of the network as much as possible. Particularly, DenseNet, by integrating the features of all modules and trying to ensure the maximum continuity of information between the network layers, solves the problems of gradient disappearance and overfitting in deep networks.

Compared to that, although VGG and ResNet also showed high accuracy in some situations, their performance varied between the two instances, suggesting that they may not have the stability of DenseNet and GoogLeNet (Huang, 2021; Li, 2021). However, it is important to note that although these models perform well in our tasks, their performance may vary in other tasks (Yu, 2022). Therefore, we recommend researchers to choose the most suitable model based on their specific tasks and data.

In summary, our research shows that by using suitable CNN models, we can effectively classify and map local climate zones (LCZ), and further understand the process of urbanization. However, to optimize performance and accuracy, it is crucial to choose the most suitable model and tune parameters. Future work might explore more models and methods to further improve the accuracy and efficiency of LCZ classification (Liang, 2023).

REFERENCES

Bechtel, B., Alexander, P., Böhner, J., Ching, J., Conrad, O., Feddema, J., … Stewart, I. (2015). Mapping local climate zones for a worldwide database of the form and function of cities. *ISPRS International Journal of Geo-Information*, 4(1), 199–219. doi:10.3390/ijgi4010199

Bechtel, B., & Daneke, C. (2012). Classification of local climate zones based on multiple earth observation data. *IEEE Journal of Selected Topics in Applied Earth Observations and Remote Sensing*, 5(4), 1191–1202. doi:10.1109/jstars.2012.2189873

Cao, Z., Fu, K., Lu, X., Diao, W., Sun, H., Yan, M., ... & Sun, X. (2019). End-to-end DSM fusion networks for semantic segmentation in high-resolution aerial images. *IEEE Geoscience and Remote Sensing Letters*, 16(11), 1766–1770.

Fonte, C. C., Lopes, P., See, L., & Bechtel, B. (2019). Using OpenStreetMap (OSM) to enhance the classification of local climate zones in the framework of WUDAPT. *Urban Climate*, 28, 100456. doi:10.1016/j.uclim.2019.100456.

He, K., Zhang, X., Ren, S., & Sun, J. (2016). Deep residual learning for image recognition. Paper presented at the Proceedings of the IEEE conference on computer vision and pattern recognition.

Hidalgo, J., Dumas, G., Masson, V., Petit, G., Bechtel, B., Bocher, E., … Mills, G. (2019). Comparison between local climate zones maps derived from administrative datasets and satellite observations. *Urban Climate*, 27, 64–89. doi:10.1016/j.uclim.2018.10.004.

Hu, F., Xia, G.-S., Hu, J., & Zhang, L. (2015). Transferring Deep Convolutional Neural Networks for the Scene Classification of High-Resolution Remote Sensing Imagery. *Remote Sensing*, 7(11), 14680–14707. doi:10.3390/rs71114680.

Huang, G., Liu, Z., Van Der Maaten, L., & Weinberger, K. Q. (2017). Densely connected convolutional networks. Paper presented at the Proceedings of the IEEE conference on computer vision and pattern recognition.

Huang, X., Chen, W., & Yang, W. (2021). Improved algorithm based on the deep integration of googlenet and residual neural network. Paper presented at the Journal of Physics: Conference Series.

Huang, X., Liu, A., & Li, J. (2021). Mapping and analyzing the local climate zones in China's 32 major cities using Landsat imagery based on a novel convolutional neural network. *Geo-spatial Information Science*, 24(4), 528–557. doi:10.1080/10095020.2021.1892459.

Iandola, F. N., Han, S., Moskewicz, M. W., Ashraf, K., Dally, W. J., & Keutzer, K. (2016). SqueezeNet: AlexNet-level accuracy with 50x fewer parameters and <0.5 MB model size. *arXiv preprint arXiv:1602.07360*

Kotharkar, R., & Bagade, A. (2018). Local Climate Zone classification for Indian cities: A case study of Nagpur. *Urban Climate*, 24, 369–392. doi:10.1016/j.uclim.2017.03.003.

La, Y., Bagan, H., & Yamagata, Y. (2020). Urban land cover mapping under the Local Climate Zone scheme using Sentinel-2 and PALSAR-2 data. *Urban Climate*, 33, 100661. doi:10.1016/j.uclim.2020.100661.

Lelovics, E., Unger, J., Gál, T., & Gál, C. (2014). Design of an urban monitoring network based on Local Climate Zone mapping and temperature pattern modelling. *Climate Research*, 60(1), 51–62. doi:10.3354/cr01220.

Li, Q., Yang, Y., Guo, Y., Li, W., Liu, Y., Liu, H., & Kang, Y. J. I. A. (2021). Performance Evaluation of Deep Learning Classification Network for Image Features. 9, 9318–9333.

Liang, Y., Song, W., Cao, S., & Du, M. J. R. S. (2023). Local Climate Zone Classification Using Daytime Zhuhai-1 Hyperspectral Imagery and Nighttime Light Data. 15(13), 3351.

Liu, S., & Shi, Q. (2020). Local climate zone mapping as remote sensing scene classification using deep learning: A case study of metropolitan China. *ISPRS Journal of Photogrammetry and Remote Sensing*, 164, 229–242. doi:10.1016/j.isprsjprs.2020.04.008.

Ma, L., Liu, Y., Zhang, X., Ye, Y., Yin, G., & Johnson, B. A. (2019). Deep learning in remote sensing applications: A meta-analysis and review. *ISPRS Journal of Photogrammetry and Remote Sensing*, 152, 166–177. doi:10.1016/j.isprsjprs.2019.04.015

Ma, L., Yang, Z., Zhou, L., Lu, H., & Yin, G. (2021a). Local climate zones mapping using object-based image analysis and validation of its effectiveness through urban surface temperature analysis in China. *Building and Environment*, 206, 108348. doi:10.1016/j.buildenv.2021.108348.

Ma, L., Zhu, X., Qiu, C., Blaschke, T., & Li, M. (2021b). Advances of Local Climate Zone Mapping and Its Practice Using Object -Based Image Analysis. *Atmosphere*, 12(9), 1146. doi:10.3390/atmos12091146.

Oliveira, A., Lopes, A., & Niza, S. (2020). Local climate zones in five southern European cities: An improved GIS-based classification method based on Copernicus data. *Urban Climate*, 33, 100631. doi:10.1016/j.uclim.2020.100631.

Pires De Lima, R., & Marfurt, K. (2019). Convolutional Neural Network for Remote-Sensing Scene Classification: Transfer Learning Analysis. *Remote Sensing*, 12(1), 86. doi:10.3390/rs12010086.

Qiu, C., Mou, L., Schmitt, M., & Zhu, X. X. (2019). Local climate zone-based urban land cover classification from multi-season alSentinel-2 images with a recurrent residual network. *ISPRS Journal of Photogrammetry and Remote Sensing*, 154, 151–162. doi:10.1016/j.isprsjprs.2019.05.004.

Qiu, C., Tong, X., Schmitt, M., Bechtel, B., & Zhu, X. X. (2020). Multilevel Feature Fusion-Based CNN for Local Climate Zone Classification From Sentinel-2 Images: Benchmark Results on the So2Sat LCZ42 Dataset. *IEEE Journal of Selected Topics in Applied Earth Observations and Remote Sensing*, 13, 2793–2806. doi:10.1109/jstars.2020.2995711.

Quan, J. (2019). Enhanced geographic information system-based mapping of local climate zones in Beijing, China. *Science China Technological Sciences*, 62(12), 2243–2260. doi:10.1007/s11431-018-9417-6.

Scott, G. J., England, M. R., Starms, W. A., Marcum, R. A., & Davis, C. H. (2017). Training Deep Convolutional Neural Networks for Land–Cover Classification of High-Resolution Imagery. *IEEE Geoscience and Remote Sensing Letters*, 14(4), 549–553. doi:10.1109/lgrs.2017.2657778.

Shi, L., & Ling, F. (2021). Local Climate Zone Mapping Using Multi-Source Free Available Datasets on Google Earth Engine Platform. *Land*, 10(5), 454. doi:10.3390/land 10050454.

Shin, H.-C., Roth, H. R., Gao, M., Lu, L., Xu, Z., Nogues, I., . . . Summers, R. M. (2016). Deep Convolutional Neural Networks for Computer-Aided Detection: CNN Architectures, Dataset Characteristics and Transfer Learning. *IEEE Transactions on Medical Imaging*, 35(5), 1285–1298. doi:10.1109/tmi.2016.2528162.

Simonyan, K., & Zisserman, A. (2014). Very deep convolutional networks for large-scale image recognition. arXiv preprint arXiv:1409.1556.

Stewart, I. D., & Oke, T. R. (2012). Local climate zones for urban temperature studies. *Bulletin of the American Meteorological Society*, 93(12), 1879–1900. doi:10.1175/bams-d-11-00019.1

Szegedy, C., Liu, W., Jia, Y., Sermanet, P., Reed, S., Anguelov, D., …Rabinovich, A. (2014). Going Deeper with Convolutions. arXiv pre-print server. doi: Nonearxiv:1409.4842

Wang, R., Ren, C., Xu, Y., Lau, K. K.-L., & Shi, Y. (2018). Mapping the local climate zones of urban areas by GIS-based and WUDAPT methods: A case study of Hong Kong. *Urban Climate*, 24, 567–576. doi:10.1016/j.uclim.2017.10.001.

Xiang, Y., Cen, Q., Peng, C., Huang, C., Wu, C., Teng, M., & Zhou, Z. (2023). Surface urban heat island mitigation network construction utilizing source-sink theory and local climate zones. *Building and Environment*, 243, 110717.

Yoo, C., Han, D., Im, J., & Bechtel, B. (2019). Comparison between convolutional neural networks and random forest for local climate zone classification in mega urban areas using Landsat images. *ISPRS Journal of Photogrammetry and Remote Sensing*, 157, 155–170. doi:10.1016/j.isprsjprs.2019.09.009

Yu, Z., Dong, Y., Cheng, J., Sun, M., Su, F. J. S., & Networks, C. (2022). Research on Face Recognition Classification Based on Improved GoogleNet. 2022, 1–6.

Yuan, B., Han, L., Gu, X., & Yan, H. (2021). Multi-deep features fusion for high-resolution remote sensing image scene classification. *Neural Computing and Applications*, 33(6), 2047–2063. doi:10.1007/s00521-020-05071-7.

Zhang, L., Zhang, L., & Du, B. (2016). Deep learning for remote sensing data: A technical tutorial on the state of the art. *IEEE Geoscience and Remote Sensing Magazine*, 4(2), 22–40. doi:10.1109/mgrs.2016.2540798

Zhao, C., Jensen, J., Weng, Q., Currit, N., & Weaver, R. (2019). Application of airborne remote sensing data on mapping local climate zones: Cases of three metropolitan areas of Texas, U.S. *Computers, Environment and Urban Systems*, 74, 175–193. doi:10.1016/j.compenvurbsys.2018.11.002

Zhao, H., Liu, F., Zhang, H., & Liang, Z. (2019). Convolutional neural network based heterogeneous transfer learning for remote-sensing scene classification. *International Journal of Remote Sensing*, 40(22), 8506–8527. doi:10.1080/01431161.2019.1615652.

Zhou, L., Yuan, B., Hu, F., Wei, C., Dang, X., & Sun, D. (2022). Understanding the effects of 2D/3D urban morphology on land surface temperature based on local climate zones. *Building and Environment*, 208, 108578.

Zhou, Y., Wei, T., Zhu, X., & Collin, M. (2021). A parcel-based deep-learning classification to map local climate zones from sentinel-2 images. *IEEE Journal of Selected Topics in Applied Earth Observations and Remote Sensing*, 14, 4194–4204. doi:10.1109/jstars.2021.3071577

Zhu, X. X., Hu, J., Qiu, C., Shi, Y., Kang, J., Mou, L., . . . Wang, Y. (2020). So2Sat LC Z42: A Benchmark Data Set for the Classification of Global Local Climate Zones [Software and Data Sets]. *IEEE Geoscience and Remote Sensing Magazine*, 8(3), 76–89. doi:10.1109/mgrs.2020.2964708.

16 The Potential Diffusion and Economic Impacts of Private Autonomous Vehicles

A Spatial Agent-Based Model

Chengxiang Zhuge and
Justin Hayse Chiwing G. Tang

INTRODUCTION

The emergence of Autonomous Vehicles (AVs) has been deemed a promising solution for radically improving transportation and environmental issues, for instance, traffic congestion and the associated pollution issues. AVs can improve overall traffic, increase road capacities, offer in-vehicle activities during commutes, enable automated valet parking, provide non-drivers with access, and reduce energy consumption. In addition to these potential benefits, the diffusion of AVs is expected to dramatically re-shape travel behaviors and patterns, as well as living conditions, and urban and suburban landscapes in the near future. The adoption of private AVs (PAVs) could assist in the progress of flexible mobility and accessibility, not only in urban areas but also in suburban areas. Consequently, as indicated by Krueger et al. (2019), the initial urban inhabitants may contemplate relocating further away from metropolitan regions, thereby triggering residential migration, urban expansion, and suburbanization.

Understanding the potential diffusion of AVs and its impacts would be advantageous for various stakeholders involved in the AV industry, including vehicle manufacturers, policymakers, and infrastructure companies. More specifically, automakers would be keen on determining the potential rate of AV penetration in the future, enabling them to devise effective production strategies. For policymakers, they need to gain a comprehensive understanding of how AVs might diffuse over time and across space, as well as the potential impacts, such as economic benefits, to facilitate the development of AV policies. For infrastructure companies, similarly, would be interested in assessing the infrastructural

DOI: 10.1201/9781003244561-20

implications of AV diffusion, such as the possible removal of parking spaces and increased road capacity.

To cater to the various stakeholders in the AV industry, this study aims to investigate the diffusion of private AVs over time and space, and further how the diffusion would bring economic impacts. Specifically, the study will examine the potential savings in housing expenses resulting from the residential relocation of AV adopters, as well as their inclination toward in-vehicle activities (e.g., work). To achieve this, a spatial agent-based model will be devised to simulate the diffusion of AVs at individual level, encompassing the effects of social influence, including global, neighbor, and friendship effects. In order to be behaviorally sound, the model incorporates the core behavioral rules of consumer agents, based on empirical evidence gathered through a questionnaire survey conducted in Beijing in 2020.

LITERATURE REVIEW

Analysis and Modeling of AV Adoption

Influential Factors to the AV Adoption

As an emerging technology, the diffusion of AVs may be influenced not only by sociodemographic attributes of potential adopters, but also by the high upfront cost of AVs and social influence.

In terms of sociodemographic attributes, many studies have shown that gender, age, education level and income are statistically associated with the adoption of AVs. Specifically, males have shown more interest in AVs than females, as females may require a higher level of trust in AVs (Wien, 2019). In terms of age group, younger generations are also more likely to adopt AVs, with young adults being the most probable early adopters (Wang et al., 2020). Further, individuals with a higher education level are more likely to adopt AVs, as evidenced by a positive correlation between AV adoption and education level (Barbour et al., 2019). People with a higher income might also intend to use AVs, as they tend to spend more money to satisfy their needs. However, according to the findings by Nair et al. (2018), higher-income households may be resistant to using AVs due to privacy concerns.

The high upfront cost of AVs is a crucial factor to consider in AV adoption. AVs generally require numerous sophisticated apparatuses, including diverse and redundant sensors (e.g., radar and laser), automated controls (e.g., braking, signals and steering), microprocessors, GPS navigation, wireless networks, and transmissions (Shah et al., 2018). These components can be expensive and unaffordable for lower-income groups (Kato et al., 2015). As a result, the high upfront cost could potentially limit the usage of AVs among citizens. However, the production of AVs on a large scale could offer an alternative to expensive pricing.

Social influence refers to the impact of comments from consumers on the likelihood and evaluation of adoption. To conduct a systematic study on social cognitive factors, this broad term can be classified into three types, including friendship, neighbor and global influences (Zhuge & Shao, 2019). In terms of friendship influence, a consumer's intention to adopt new technology can be directly influenced by social circles, namely friends and acquaintances. Neighbor influence stems from the adoption of AVs by a consumer's neighbors, which can impress and influence the decision to purchase an AV. Global influence mainly refers to influential information from social media (e.g., news and advertisements) provided by relevant stakeholders, such as governments and vehicle manufacturers.

Modeling of the AV Diffusion

Various modeling approaches have been used to investigate the diffusion of AVs at both macro and micro levels. Microscopic studies have typically explored individual AV adoption behavior using discrete choice models. For example, Chen et al. (2019) employed a nested logit model to differentiate vehicle choices for the optimization problem of purchasing subsidies for human-driven electric vehicles (HDEVs) and autonomous electric vehicles (AEVs). Agent-Based Modeling (ABM), a typical microscopic approach for simulating the diffusion of new technologies, has also been used to explore AV diffusion at the individual level. For example, Talebian and Mishra (2018) developed an ABM to predict the adoption of CAVs. The ABM design is particularly noteworthy because it allows for communication among agents, resulting in a more complex and dynamic perspective of CAV adoption. While macroscopic studies have examined the AV market without explicit consideration of individual behaviors. For example, Luo et al. (2019) explored AV subsidy issues by developing a diffusion of innovations (DOI) model to reveal the evolution of AV's market penetration and share as well as variations of AV's efficiency benefits toward AV's market share.

Impacts of AV Adoption

The potential diffusion of AVs could impact those connected elements, including transportation, land use, economy, energy and environment systems, as reviewed by Zhuge and Wang (2021). This paper focuses on two specific types of economic impact, namely housing expenses saved through residential relocation and in-vehicle work.

(1) The Value of Travel Time Savings (VTTS) and Residential Relocation

Residential relocation refers to the act of moving from one place of residence to another. The advent of AVs could potentially cause people living in urban areas to relocate to suburban areas due to the perception of cheaper housing prices and reduced perceived commuting times (Llorca et al., 2022). The Value of Travel Time Savings (VTTS) is a specific term used to describe the cost of changing the time allocated to travel. Steck et al. (2018) and Kolarova et al. (2019) both investigated commuters' VTTS and their choices for commuting by autonomous driving. Their results indicated substantial decreases in VTTS for autonomous driving (both private and shared AVs) compared to manual driving, with reductions of 31% and 41%, respectively. Conversely, Singleton (2019) argued that the reduction in VTTS impact on driverless vehicles might be modest compared to what the general public expects, and that the reduction in valuation of time could create "positive utility," which suggests psychological well-being due to the relief of anxiety associated with manual driving.

(2) In-Vehicle Activities

Idealistically, former drivers and passengers using AVs are able to engage in non-driving, more leisurely and productive activities, resulting in a higher level of comfort and convenience. They can make good use of their travel time to work, sleep, read, eat, and relax. Malokin et al. (2019) found that multitasking in AVs tended to decrease the disutility of commute/travel time. Pudāne et al. (2019) qualitatively assessed the expectations of 27 commuters regarding in-vehicle activities. Several participants stated that they would engage in simple activities during their travel time that did not require additional facilities in AVs. Respondents under time pressure might re-arrange their activities, including inflexible high-priority tasks, to create a more efficient daily schedule using AVs. Correia et al. (2019) provided insight into the VTTS for working and leisure activities in AVs. Specifically, AVs with office interiors were found to have a lower VTTS for working activities compared to conventional vehicles, while VTTS for leisure activities was found to be non-decreasing.

Research Gaps and Aims

As reviewed earlier, the uptake of AVs may be influenced by various factors, including sociodemographic characteristics, AV price, and social influence. Both microscopic and macroscopic models have been developed to explore the diffusion of AVs using influential factors as independent variables. This paper attempts to develop a spatial agent-based model to simulate the diffusion of private AVs at the individual level, and further quantify the economic impacts of the AV diffusion. Specifically, the contributions of this study are twofold:

- First, the proposed agent-based AV diffusion model will be spatially explicit, and consider three levels of social influence, namely global influence, neighbor effect, and friendship effect, which are particularly important for the diffusion of new technologies. To ensure behavioral realism, we will use empirical findings from Beijing to define behavioral rules of consumer agents in their decision-making on vehicle choices.
- Second, this paper will assess the potential economic impacts of the AV diffusion, including the potential reduction in housing expenses through residential relocation and in-vehicle work, which have received less attention in previous AV studies.

A SPATIAL AGENT-BASED AUTONOMOUS VEHICLE (AV) DIFFUSION MODEL

Model Framework

Given a synthetic population (Sun & Erath, 2015), which contains person and household agents, as well as their sociodemographic attributes, residential locations, and social networks (Zhuge, Shao, & Wei, 2018), the agent-based AV diffusion model simulates how consumer agents choose between AV and Conventional

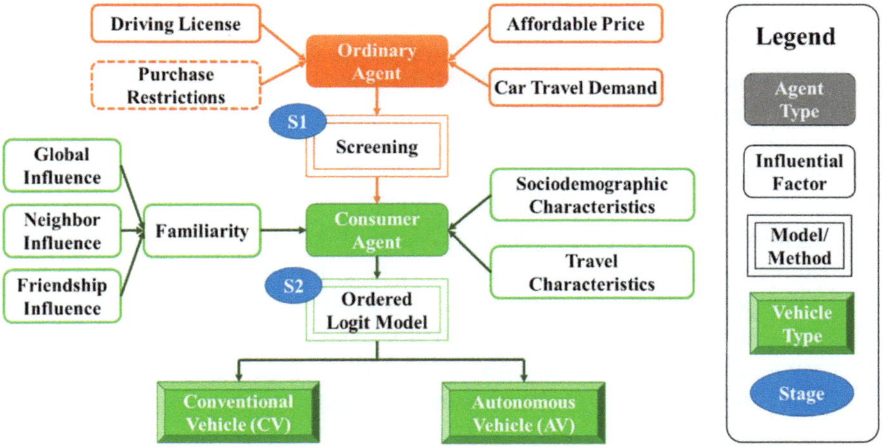

FIGURE 16.1 Framework of the agent-based AV diffusion model.

Vehicle (CV) with no automation function. Figure 16.1 shows the two-stage decision-making process of an agent in the AV market model, which is developed based on a vehicle market model by Zhuge et al. (2019). In the first stage, four conditions are applied to screen all agents in the population. Specifically, only those agents who hold a driving license, can afford a vehicle, have a demand for car travel, and possess a vehicle purchase permit will become consumer agents entering the vehicle market. In the second stage, consumer agents must decide which vehicle type to purchase (i.e., AV or CV), and they might interact with each other through their neighbors and friends who have adopted AVs when making their decisions.

In general, ABM is useful for capturing the interactions between agents, resulting in emerging phenomena. The agent-based AV diffusion model can help to capture important interactions in the complex vehicle market as follows: First, consumer agents need to compete for limited vehicle purchase permits, due to vehicle purchase restrictions, such as the license plate lottery policy in Beijing (Zhu et al., 2022). Second, agents can communicate and share the information on AV adoption via social media, their neighbors (based on their residential locations) and their friends through social networks.

STAGE 1 – SCREENING: GENERATING CONSUMER AGENTS

As aforementioned, the agent-based AV diffusion model will screen agents in the population, resulting in a list of consumer agents who will enter the vehicle market:

- **Criterion 1 – Driving License:** All agents with a driving license are assumed to be interested in purchasing a vehicle. This criterion has also been used in other vehicle choice models (Zhuge et al., 2019).
- **Criterion 2 – Affordable Price:** Each household agent in the population is evaluated to determine whether they can afford a vehicle. Household income is used as an indicator of their capacity, and it is defined that the

ratio of household income to vehicle price should be greater than a specific threshold $\varnothing_{Income2VehPrice}$ that can be specified by model users.

- **Criterion 3 – Car Travel Demand:** In general, people tend to use car for long trips. For example, the average total daily car trip distance for a car owner in Beijing in 2019 was 40.8 km. Therefore, it is defined that only those person agents with a total daily trip distance longer than a threshold $\varnothing_{TripDist}$ would have a demand for a vehicle.
- **Criterion 4 – Purchase Restrictions.** In some megacities (e.g., Beijing and Shanghai), vehicle purchase restrictions are in place. Therefore, the model also includes purchase restrictions as an optional selection criterion. For example, in Beijing, the so-called license plate lottery policy (Zhuge et al., 2020) limits the number of vehicle purchase permits issued each year. Hence, potential vehicle purchasers need to compete for a limited number of permits.

STAGE 2 – Decision-Making: Choose a Vehicle Type

After screening, a list of consumer agents who will eventually purchase a vehicle which can be either conventional or autonomous, is obtained. An ordered logit model (Ben-Akiva & Lerman, 1985) is used here to determine the likelihood of a consumer agent purchasing an AV (i.e., the probability of purchasing AV, P_{AV}). The model incorporates three groups of variables, namely sociodemographic characteristics V_{Social} (e.g., age), travel characteristics V_{Travel} (e.g., transport mode), and familiarity with AVs $V_{Familiarity}$, as shown by Equation (16.1). The ordered logit model is estimated with survey data collected in Beijing.

$$P_{AV} = f\left(V_{Social}, V_{Travel}, V_{Familiarity}\right) \tag{16.1}$$

In particular, a consumer agent's familiarity with AVs was quantified as a sum of global, neighbor and friendship influences as follows (Liu et al., 2022):

- **Global Influence:** It refers to the influence from social media, such as news and advertisements. AV stakeholders, e.g., governments and vehicle manufacturers, may promote the uptake of AVs by disseminating positive information about AVs through social media.
- **Neighbor Influence:** This refers to the influence from those neighbors who have adopted AVs and live close to the potential consumer agent. Neighbor influence has been found to be an important factor in the uptake of new technologies, such as EV (Liu et al., 2017).
- **Friendship Influence:** It refers to the influence from friends who have adopted AVs (Axsen et al., 2013). Specifically, those AV adopters might encourage their friends to try AVs as well. In the study, we used an agent-based spatial social network model by Zhuge et al. (2018) to link person agents in the population according to their similarity in sociodemographic attributes, and spatial closeness.

We used binary variables B_{Global}, $B_{Neighbour}$, and $B_{Friendship}$ to describe whether a consumer agent has been influenced by social media, neighbors or friends, respectively. If yes, the variable will be 1; otherwise, it will be 0. Therefore, familiarity with AVs ($B_{Familarity}$) is defined as the sum of these three types of social influence, which can be mathematically described by Equation (16.2).

$$B_{Familarity} = B_{Global} + B_{Neighbour} + B_{Friendship} \tag{16.2}$$

A CASE STUDY OF BEIJING, CHINA

Study Area: Beijing

Our study area is the capital of China, Beijing, which covers an area of 16,410.54 square kilometers. As of 2020, Beijing's population was capped at 20.5 million, with a Gross Domestic Product (GDP) of approximately 3.61 trillion RMB. Since 2017, the central government of China has launched a long-term national development plan for autonomous industry in Chinese cities, including Beijing, Shanghai, and Guangzhou, to achieve the extensive production of AVs by 2025 (BMCT, 2018). In particular, Beijing was chosen as the first city to conduct a series of traffic tasks for AVs, and has eagerly undertaken autonomous driving road tasks for consecutive years. For example, the tech giant Baidu completed approximately 140,000 kilometers of autonomous driving distance in Beijing in 2018, which accounted for approximately 91% of the total autonomous driving distance traveled by the eight licensed transportation companies in Beijing (Liao, 2019).

Description of AV Diffusion Scenarios

Generating a Virtual Beijing for Model Initialization

To initialize the agent-based AV diffusion model, we used a virtual Beijing in 2020, which was generated by an agent-based land use and transport model for Beijing, as introduced by Zhuge et al. (2018). The 2020 virtual Beijing contains a synthetic population and physical environment, comprising both transport and activity facilities, which serve as inputs for the AV diffusion model. To save computing time, we used a scaling factor of 4%, meaning only 4% of the whole population was used in the simulation. After the simulation, the results were proportionally scaled up. The synthetic population consists of both person and household agents, along with their sociodemographic characteristics (e.g., age, income, and car ownership). Furthermore, each person agent has a daily plan consisting of activity and travel information. Person agents are interconnected through social networks at the individual level, using a utility function that considers the similarity and spatial closeness of agents (Zhuge, Shao, & Wei, 2018). We compared the generated and observed distributions of key sociodemographic characteristics, including age, gender, level of education, family size, the number of private cars, and population size. The overall Mean Absolute Percentage Error (MAPE) for all these distributions was 15.01%, suggesting that the 2020 virtual Beijing can represent the real world to

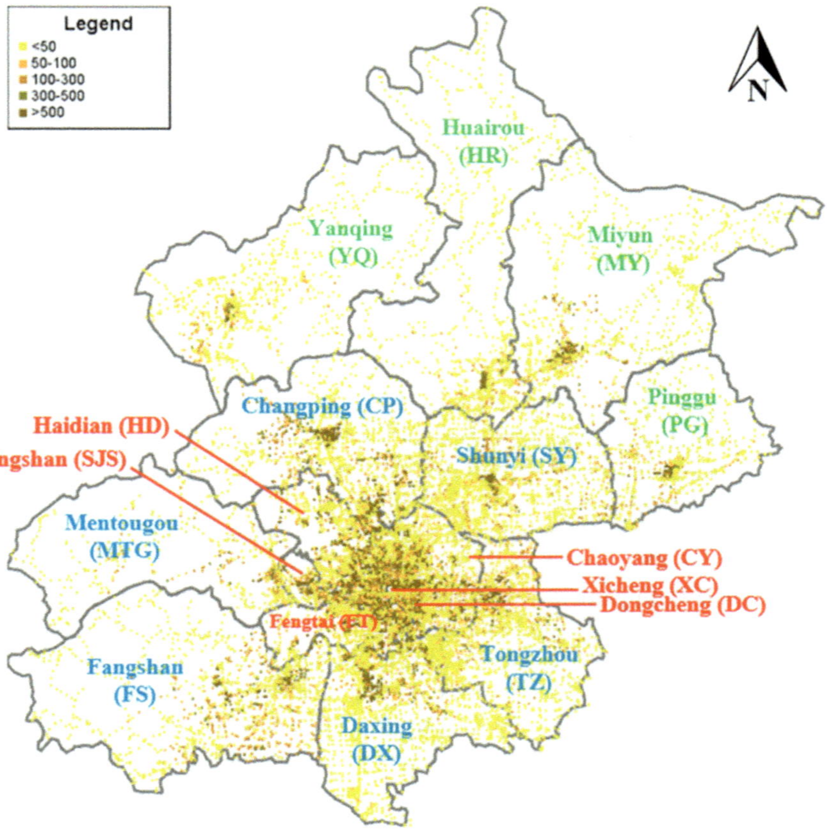

FIGURE 16.2 Spatial distribution of the vehicle owners in the Beijing Synthetic Population. (Note: The Central Area of Beijing is composed of CY, DC, FT, HD, SJS, and XC; the suburb of Beijing is composed of CP, DX, FS, MTG, SY, and TZ; the outer suburb of Beijing is composed of HR, MY, PG, and YQ).

some extent, and can be further used to simulate the AV diffusion. Figure 16.2 shows the spatial distribution of the vehicle owners in the synthetic population based on their residential location, indicating that the majority reside in the central areas.

To establish more accurate behavioral rules for potential consumers and AV adopters in the agent-based diffusion model, we used a survey dataset from Beijing. The dataset contained the preferences of 402 participants regarding AVs, as well as their sociodemographic and travel characteristics (e.g., age and transport mode). The data were collected through a face-to-face survey conducted in the six central districts of Beijing from 10 to 12 January 2020. In this study, we used information from relevant survey questions, namely AV adoption, willingness to accept a longer trip duration with AVs, and willingness to engage in in-vehicle work with AVs. We found that around 74.6% of the respondents expressed interest in AVs, and around half of them were willing to accept a longer trip duration for their main daily activities if they had an AV. Furthermore, 33.1% of the respondents were willing to work in their AVs during their journeys. These data were used to develop Ordered Logit (OL) and Binary Logit (BL) models, which are two typical types of discrete choice model (Ben-Akiva & Lerman, 1985), to relate individual adoption and use behaviors of AVs to their sociodemographic and travel characteristics, facilitating the simulation of AV-related decision makings in the agent-based AV diffusion model.

To more accurately estimate the potential housing expenses saved by AV adopters through residential relocation, we particularly optimized the housing prices in the virtual Beijing using a real estate price dataset. The dataset was obtained from Lianjia, one of the largest real estate agents in China, and included the selling prices of 7,399 communities in February 2021. Specifically, we linked the residential buildings in the virtual Beijing to their nearest real communities in the dataset, and then used the real selling prices from the dataset instead of the generated ones.

Reference and "What-If" Scenarios

In the reference scenario (i.e., baseline), we run the agent-based EV diffusion model from 2021 to 2030 with an assumption that the vehicle market in Beijing would evolve as before. For each year, there are 92,400 vehicle permits available, which is determined according to the current license plate lottery policy in Beijing. The prices of CV and AV are 0.1 million and 2.5 million RMB, respectively, and the prices are assumed to remain the same over this period. Yet, since the license plate lottery policy and high upfront cost of AVs can greatly influence the diffusion of AVs, we set up two groups of "what-if" scenarios to test how the lottery policy and sale price would affect the adoption of AVs in comparison to the reference scenario.

Policy Scenarios: Scenarios for the License Plate Lottery Policy

In the Policy scenarios, we explore how different numbers of vehicle purchase permits would influence the uptake of AVs through several scenarios, namely Lottery_0.5, Lottery_1.5, Lottery_2.0, Lottery_2.5, and Lottery_3.0. These scenarios allocate 50%, 150%, 200%, 250%, and 300% of the current number of permits in the Reference scenario, respectively.

Price Scenarios: Scenarios for AV Prices

When AVs are first introduced into the vehicle market, their upfront cost is likely to be high. However, as AV technologies improve over time, the cost of AVs is expected to decrease, making them more affordable to the general public. To explore how a decreasing AV price would influence AV diffusion, we set up several scenarios, namely AV Price_0.05, AV Price_0.10, AV Price_0.15, AV Price_0.20, and AV Price_0.25. In each scenario, we apply yearly decreasing rates of 5%, 10%, 15%, 20%, and 25% to AV prices, respectively.

The Potential Diffusion of AVs

Policy Scenarios: Scenarios for the License Plate Lottery Policy

In the Policy scenario, we explored how the number of vehicle purchase permits (see Figure 16.3(a)) would influence the AV diffusion in comparison to the

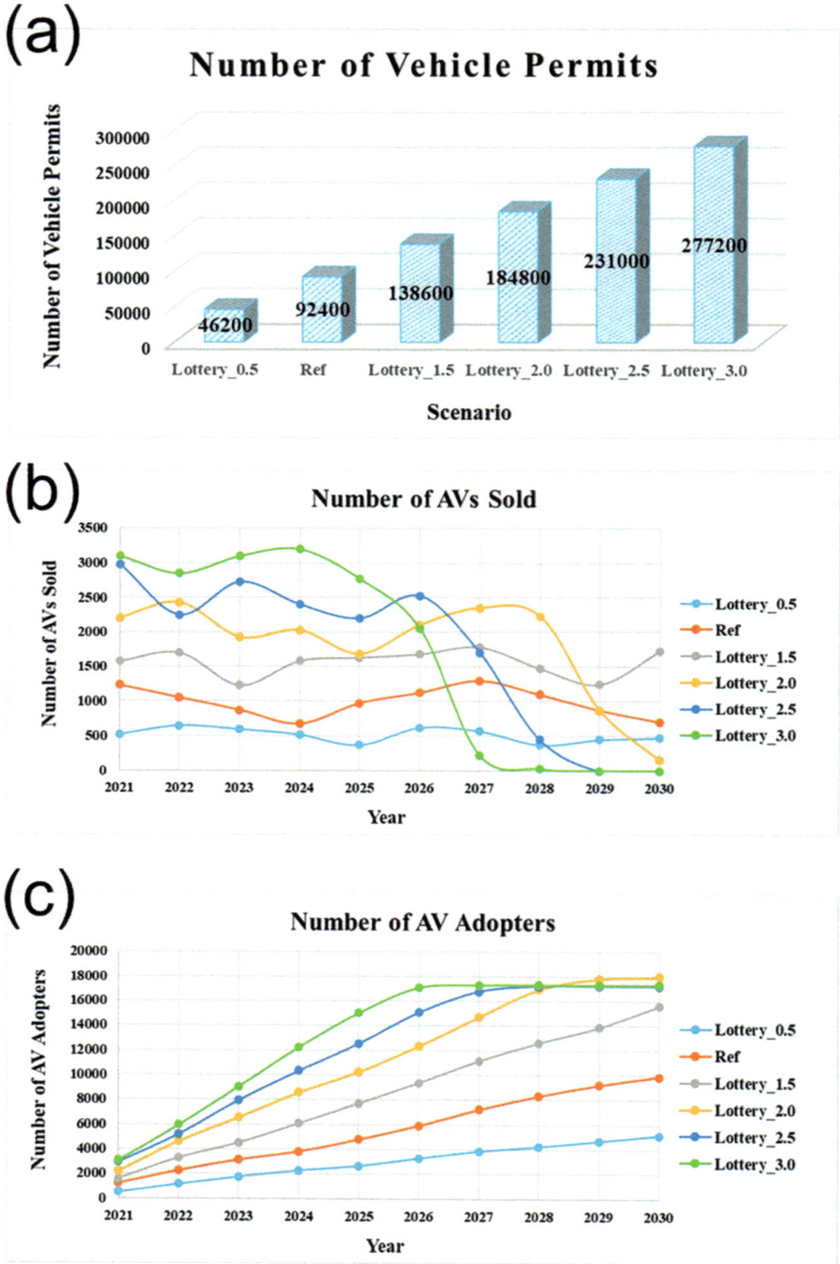

FIGURE 16.3 Results about the AV diffusion in the reference and policy scenarios. (a) The numbers of vehicle permits in the reference and policy scenarios; (b) the numbers of AVs sold in the reference and policy scenarios; (c) The numbers of AV adopters in the reference and policy scenarios. (Note: there are no housing expenses saved in 2029 and 2020 in the Lottery_2.5 and Lottery_3.0 scenarios, as there are no AV adopters in these two years, as shown in Figure 16.3(b)).

Reference (Ref) scenario. As shown in Figure 16.3(b) and (c), the number of AV adopters increases in the first few simulation years (from 2021 to 2026) as the number of vehicle permits increased. For example, the Lottery_3.0 scenario had 3,100 AV adopters in 2021, ranking first among all the scenarios. However, for scenarios with a higher number of vehicle purchase permits (e.g., Lottery_2.0, Lottery_2.5, and Lottery_3.0), the numbers of AV adopters went down in the last few years (i.e., over the period from 2027 to 2030). Also, it can be found from Figure 16.3(c) that these three scenarios had nearly the same number of AV adopters in 2030 (around 17,500 adopters). This is likely because AV prices remained constant throughout the period, and the rest of potential vehicle consumers in these scenarios could not afford an AV.

Price Scenarios: Scenarios for AV Prices

As indicated by the Policy scenarios, if the AV price remains high over the period, the number of AV adopters would level off in the scenarios with more vehicle purchase permits (e.g., Lottery_2.0, Lottery_2.5, and Lottery_3.0), after all the consumer agents who can afford an AV had purchased one. Therefore, we set up several "what-if" scenarios to further test how a decreasing AV price would influence the diffusion of AVs, as shown in Figure 16.4(a). Essentially, a lower AV price would greatly promote the adoption of AVs. Specifically, the total number of AV adopters was on the rise, as the AV price decreased at a specific rate (see Figure 16.4(b) and (c)). The lower the AV price, the higher the number of AV adopters. For instance, in the AV Price_0.25 scenario, the number of AV adopters in 2030 was 134,075, which was almost 14 times that of the Reference scenario. However, it is worth noting that for the scenarios with a higher decreasing rate (i.e., AV Price_0.15, AV Price_0.20, and AV Price_0.25 scenarios), the number of AV adopters went up at a higher rate for the first few years, and then leveled off (see Figure 16.4(c)). This was likely because these scenarios had more consumer agents purchasing an AV in the initial years, resulting in fewer consumer agents interested in and able to afford an AV over the remaining simulation years.

Potential Economic Impacts of the AV Diffusion

As aforementioned, the diffusion of AVs would impact many connected elements in cities. This study attempts to quantify the economic impacts of AV diffusion, with a focus on the potential housing expenses saved through residential relocation of AV adopters and the economic benefits of in-vehicle work. We further explored these two types of economic impacts in the AV diffusion scenarios discussed earlier.

Impacts on Housing Expenses Saved through Residential Relocation

According to the Beijing survey, around 50% of participants would be willing to accept a longer travel time for their main activities, such as work. This would enable citizens to choose a new housing property that is farther away from their main activity location (e.g., workplace) but has a lower selling price than their current property. By doing so, these AV adopters could further reduce their housing

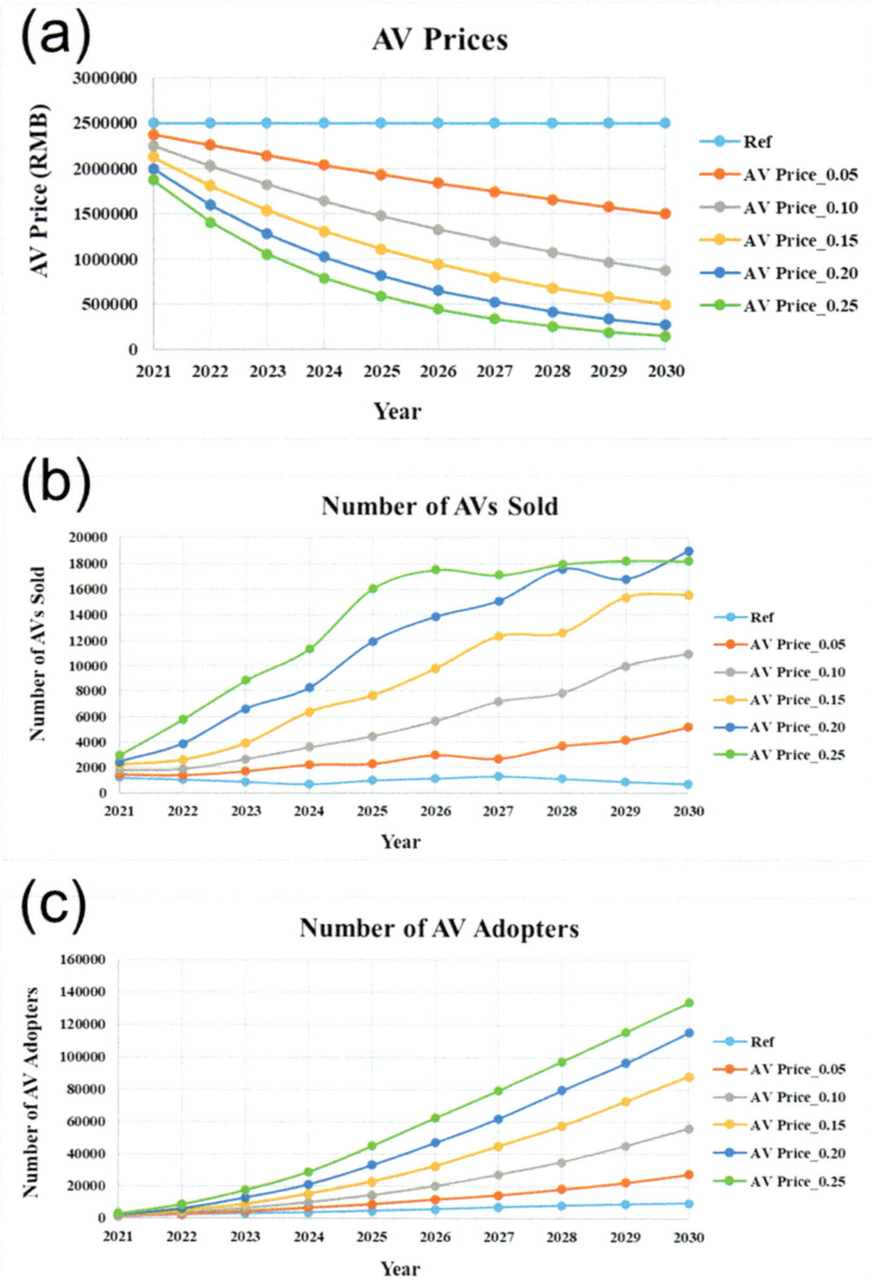

FIGURE 16.4 Results about the AV diffusion in the reference and price scenarios. (a) AV prices in the reference and price scenarios (Note: the AV price in 2030 in the AV Price_0.25 Scenario is around 140,000 RMB, which is still higher than the sale price of CV, i.e., 100,000 RMB); (b) the numbers of AVs sold in the reference and price scenarios; (c) the numbers of AV adopters in the reference and price scenarios

expenses. Here, we estimated the potential housing expense savings for this group of AV owners. We assumed that they would randomly choose a new housing property from a list of candidate properties with a lower selling price but still within an acceptable travel distance for their main activities. The outcomes would provide insights into the extent to which AVs could potentially help adopters in saving housing expenses, on average. The estimation does not consider any other important factors in residential relocation, such as access to educational facilities.

Policy Scenarios: Scenarios for the License Plate Lottery Policy

In the Policy scenarios, AV adopters who are willing to extend their travel time for their main activities (such as work) could save an average of around 5 million RMB on housing expenses if they choose a new residential location with a lower selling price but still within an acceptable travel distance. The average housing expense saved is significant, as the AV adopters tend to be those rich people who also spend much money on housing. However, there is no significant difference between the Reference and Policy scenarios in the total savings on housing expenses through the adoption of AVs (see Table 16.1). Figure 16.5 shows that the savings on housing expenses vary over time. This is likely because the AV adopters in different simulation years may live in housing properties with different selling prices, and

TABLE 16.1
Average Housing Expenses Saved for AV Adopters in Both Reference and Policy Scenarios (Unit: RMB)

Scenario	Lottery_0.5	Ref	Lottery_1.5	Lottery_2.0	Lottery_2.5	Lottery_3.0
Average Housing Expenses Saved	5,197,262	5,140,403	4,763,259	4,872,666	4,930,807	5,118,485

FIGURE 16.5 Housing expenses saved for AV adopters in each simulation year in both reference and policy scenarios (Unit: RMB).

they may have varying preferences for residential relocation, which are associated with their sociodemographic characteristics.

From a spatial perspective, most of the AV adopters who would move to a new Residential Location (RL) with a lower selling price are currently living in central areas in both the Reference and Policy scenarios. The majority of them can find a new RL within the districts where they currently reside. For example, most AV adopters from Chaoyang (CY) and Haidian (HD) in both Reference and Policy scenarios still relocate within their respective districts (see Figure 16.6(a) and (b)). Furthermore, some of them can find a new RL in suburb or neighboring districts. For instance, a few AV adopters from Chaoyang (CY) district may move to Shunyi (SY) and Tongzhou (TZ) districts, where the average housing prices tend to be much lower. It can also be found from Figure 16.6(c) that, on average, the number of AV adopters

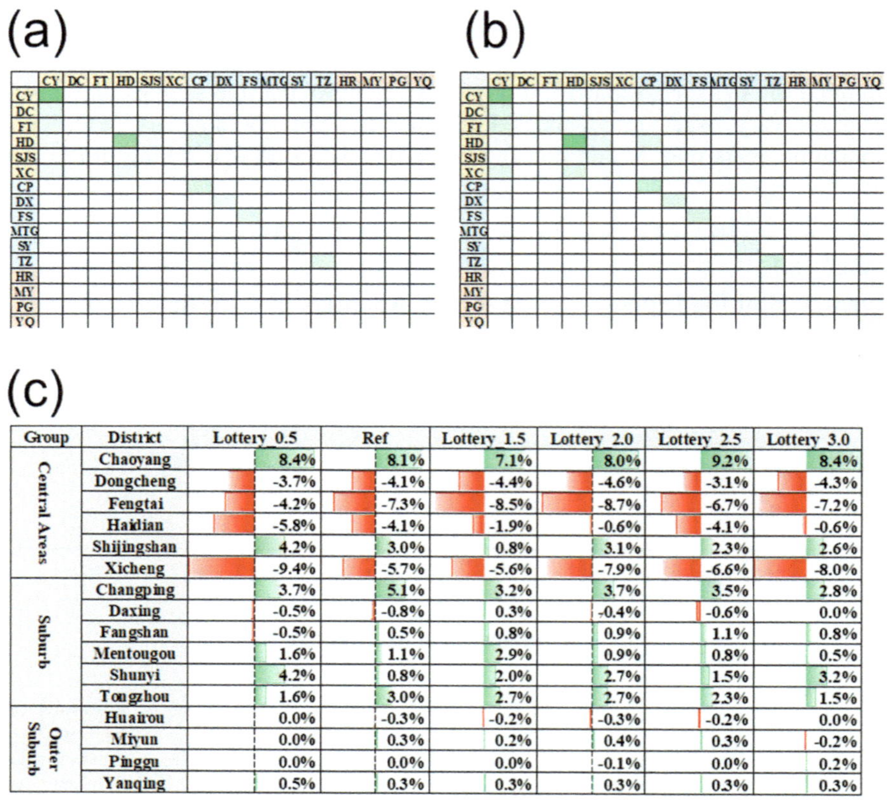

(c)

Group	District	Lottery 0.5	Ref	Lottery 1.5	Lottery 2.0	Lottery 2.5	Lottery 3.0
Central Areas	Chaoyang	8.4%	8.1%	7.1%	8.0%	9.2%	8.4%
	Dongcheng	-3.7%	-4.1%	-4.4%	-4.6%	-3.1%	-4.3%
	Fengtai	-4.2%	-7.3%	-8.5%	-8.7%	-6.7%	-7.2%
	Haidian	-5.8%	-4.1%	-1.9%	-0.6%	-4.1%	-0.6%
	Shijingshan	4.2%	3.0%	0.8%	3.1%	2.3%	2.6%
	Xicheng	-9.4%	-5.7%	-5.6%	-7.9%	-6.6%	-8.0%
Suburb	Changping	3.7%	5.1%	3.2%	3.7%	3.5%	2.8%
	Daxing	-0.5%	-0.8%	0.3%	-0.4%	-0.6%	0.0%
	Fangshan	-0.5%	0.5%	0.8%	0.9%	1.1%	0.8%
	Mentougou	1.6%	1.1%	2.9%	0.9%	0.8%	0.5%
	Shunyi	4.2%	0.8%	2.0%	2.7%	1.5%	3.2%
	Tongzhou	1.6%	3.0%	2.7%	2.7%	2.3%	1.5%
Outer Suburb	Huairou	0.0%	-0.3%	-0.2%	-0.3%	-0.2%	0.0%
	Miyun	0.0%	0.3%	0.2%	0.4%	0.3%	-0.2%
	Pinggu	0.0%	0.0%	0.0%	-0.1%	0.0%	0.2%
	Yanqing	0.5%	0.3%	0.3%	0.3%	0.3%	0.3%

FIGURE 16.6 Spatial residential relocation patterns of AV adopters in reference and policy scenarios. (a) Matrix for intra-district and inter-district relocation in reference; (b) matrix for intra-district and inter-district relocation in lottery_3.0; (c) the difference between before and after residential relocation in the number of AV adopters in each district. (Note: We first calculated the difference in the number of AV Adopters before and after residential relocation for each district and then divided the difference by the total number of AV adopters with potential residential relocation).

in central areas decreases significantly after residential relocation. Some AV adopters, such as those from Dongcheng (DC), Haidian (HD), and Xicheng (XC) districts, relocate to their neighboring districts (such as Chaoyang (CY)), where housing properties with a lower selling price are available. For instance, the average housing prices in DC, HD, and XC in 2021 were 93,418, 80,214, and 103,361 RMB per m^2, respectively, while the average price in CY was only 62,939 RMB per m^2. In addition, there is no significant spatial differences between Reference and Policy scenarios in the relocation flow of AV adopters.

AV Price Scenarios: Scenarios for AV Prices

A significant difference between Reference and Price scenarios in housing expenses saved can be found (see Table 16.2 and Figure 16.7). Essentially, as the AV price decreases at a higher rate, the average housing expenses saved decrease. For example, the housing expenses saved decreased from 5,140,403 RMB in the Reference scenario to 4,011,882 RMB in the AV Price_0.25 scenario. This is likely because AVs become more affordable in those scenarios with a

TABLE 16.2

Average Housing Expenses Saved for AV Adopters in Both Reference and Price Scenarios (Unit: RMB)

Scenario	Ref	AV Price_0.05	AV Price_0.10	AV Price_0.15	AV Price_0.20	AV Price_0.25
Average Housing Expenses Saved	5,140,403	5,054,452	4,486,699	4,172,965	4,132,020	4,011,882

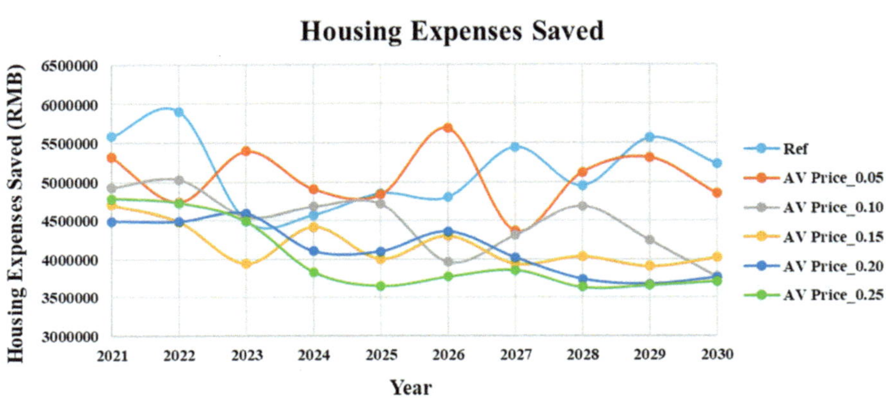

FIGURE 16.7 Housing expenses saved for AV adopters in each simulation year in both reference and price scenarios (Unit: RMB).

higher decreasing rate. As a result, more household agents with a lower household income can afford AVs. These households tend to live in housing properties with a lower selling price, and it would become difficult for them to find new housing properties with a much lower selling price.

In the Price scenarios, we can observe similar spatial relocation patterns among AV adopters who would potentially choose new RLs with a lower selling price. As shown by the district matrices (see Figure 16.8(a) and (b)), most of the AV adopters, such as those from Chaoyang (CY) and Haidian (HD) districts, can find a new RL with a lower housing price in the same districts within their current districts. However, there is no significant difference between Reference and Price

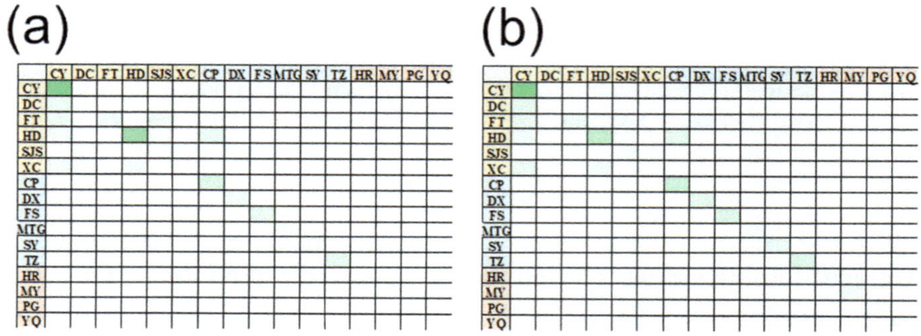

(c)

Group	District	Ref	AV Price_0.05	AV Price_0.10	AV Price_0.15	AV Price_0.20	AV Price_0.25
Central Areas	Chaoyang	8.1%	8.6%	9.7%	8.5%	7.6%	8.0%
	Dongcheng	-4.1%	-4.9%	-4.6%	-4.6%	-4.1%	-4.0%
	Fengtai	-7.3%	-8.2%	-8.0%	-7.8%	-6.9%	-6.7%
	Haidian	-4.1%	-0.7%	-4.4%	-3.2%	-2.2%	-3.4%
	Shijingshan	3.0%	2.0%	2.7%	2.1%	1.5%	1.7%
	Xicheng	-5.7%	-7.5%	-5.5%	-6.0%	-6.2%	-6.0%
Suburb	Changping	5.1%	3.2%	4.0%	4.2%	3.7%	4.0%
	Daxing	-0.8%	-0.1%	0.1%	0.2%	-0.3%	-0.6%
	Fangshan	0.5%	0.4%	0.3%	0.2%	0.5%	0.5%
	Mentougou	1.1%	0.7%	1.3%	1.4%	1.4%	1.2%
	Shunyi	0.8%	3.1%	2.6%	2.7%	2.6%	2.8%
	Tongzhou	3.0%	2.7%	1.3%	1.8%	2.1%	1.9%
Outer Suburb	Huairou	-0.3%	-0.6%	-0.7%	-0.5%	-0.8%	-0.8%
	Miyun	0.3%	0.7%	0.8%	0.6%	0.9%	0.9%
	Pinggu	0.0%	0.1%	0.1%	0.1%	0.0%	0.1%
	Yanqing	0.3%	0.4%	0.3%	0.2%	0.2%	0.4%

FIGURE 16.8 Spatial relocation patterns of AV adopters in reference and price scenarios. (a) Matrix for intra-district and inter-district relocation in reference; (b) matrix for intra-district and inter-district relocation in AV Price_0.25; (c) the difference between before and after residential relocation in the number of AV adopters in each district. (Note: we first calculated the difference in the number of AV Adopters before and after residential relocation for each district and then divided the difference by the total number of AV adopters with potential residential relocation).

scenarios in the spatial patterns of residential relocation. Also, the numbers of AV adopters in several central districts (e.g., Xicheng (XC) district) decreased significantly, but the number in suburb districts increased (see Figure 16.8(c)). This is likely because a few of AV adopters relocate to suburbs, where more housing properties with lower prices are available.

Impacts on In-Vehicle Work

AVs allow vehicle owners to engage in in-vehicle activities. Here, we simulated whether AV adopters in the Beijing scenario would work while traveling within AVs using the estimated binary logit model. If they do work, we further estimated their in-vehicle work duration, which is associated with the potential economic benefits brought by AVs.

Policy Scenarios: Scenarios for the License Plate Lottery Policy

We used two indicators to assess the impacts of AV diffusion on in-vehicle activities, namely the percentage of AV owners working in their AVs and the average duration of in-vehicle work for those AV adopters willing to work in their AVs. As shown in Table 16.3 and Figure 16.9, the indicators in both Reference and Policy scenarios displayed different values in the first few years but started to level off after 2027. Furthermore, the indicators appear to converge as the simulation progresses. The percentage of AV owners working in their AVs in both the Reference and Policy scenarios was around 40% (we found that 33.1% of participants in the Beijing survey would work in AVs), and the average duration of in-vehicle work was around 40 minutes.

From a spatial perspective, AV adopters willing to work in their AVs tend to live and work in central districts in both the Reference and Policy scenarios (see Figure 16.10(c) and (d)). Furthermore, the numbers of AV adopters living and working in the central areas are roughly equal, accounting for around 70% of the total number. According to the district matrices (see Figure 16.10(a) and (b)), most of the AV adopters need to commute from one district to another, though some of

TABLE 16.3

Statistical Indicators about In-Vehicle Work of AV Adopters in Both Reference and Policy Scenarios

Scenario	Lottery_0.5	Ref	Lottery_1.5	Lottery_2.0	Lottery_2.5	Lottery_3.0
Percentage of AV Owners Working in Vehicle	39.1%	40.2%	40.1%	38.4%	39.6%	38.4%
Average In-Vehicle Work Duration (Minute)	40.8	35.8	43.2	41.0	40.3	41.5

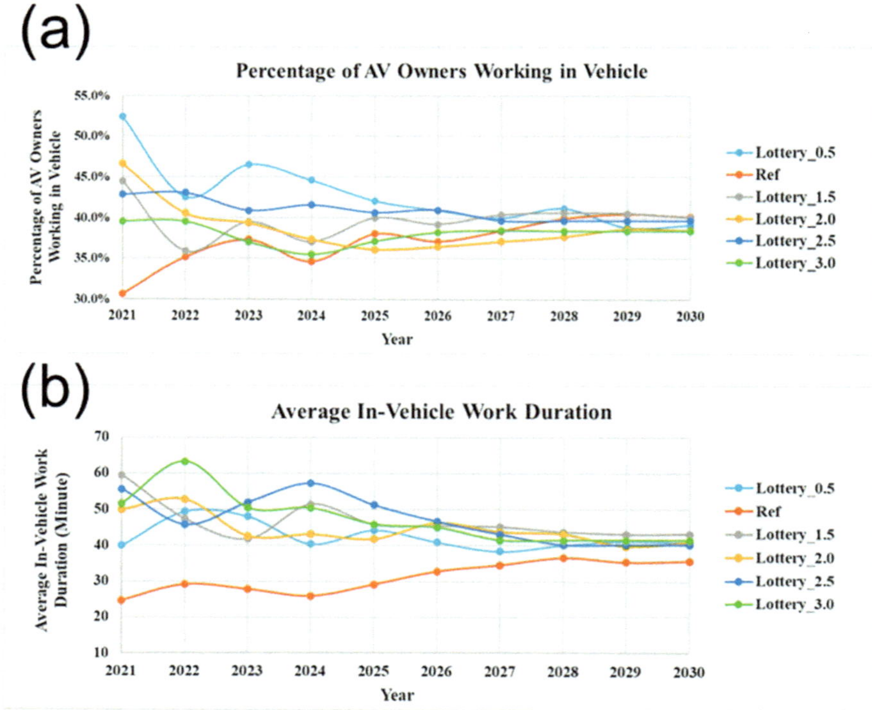

FIGURE 16.9 Results about in-vehicle work of AV adopters in both reference and policy scenarios. (a) Percentage of AV owners working in their AVs in each simulation year; (b) average duration of in-vehicle work of AV adopters in each simulation year.

them live and work in the same districts. Furthermore, there is a slight spatial difference between Reference and Policy scenarios: The scenarios with more vehicle permits (e.g., Lottery_3.0) tend to have more AV adopters commuting between central and (outer) suburban districts.

Price Scenarios: Scenarios for AV Prices

Table 16.4 and Figure 16.11 show the statistical indicators related to the in-vehicle work of AV adopters in Reference and Price scenarios. In terms of the percentage of AV owners working in their AVs, we observe similar convergence phenomena in the Price scenarios, with around 39% of AV owners willing to work in their AVs, slightly lower than that in the Policy scenarios (i.e., 40%; see Table 16.3). However, the average in-vehicle work duration in the Price scenarios tends to be higher than that in the Reference or Policy scenarios. Essentially, the scenarios with a higher decreasing rate (i.e., more AV adopters) tend to have a longer in-vehicle work duration. This indicates that these scenarios tend to have more AV adopters with longer trip durations, on average.

Both the Reference and Price scenarios have more AV adopters living and working in the central areas (see Figure 16.12(c) and (d)), accounting for around 60% of

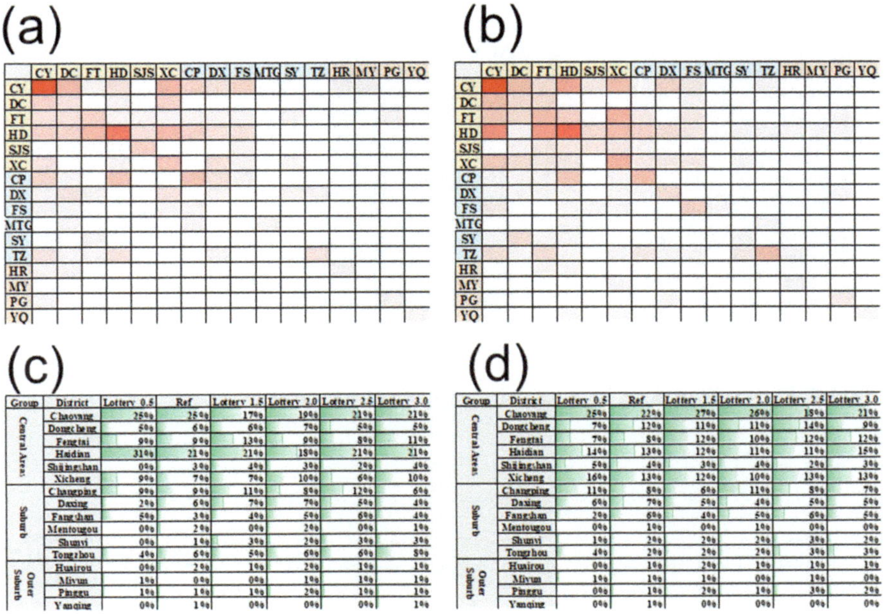

FIGURE 16.10 Spatial Commuting Patterns of those AV Adopters Working in their AVs in both Reference and Policy Scenarios. (a) Commuting Origin-Destination (OD) Matrix in Ref at the District Level; (b) Commuting OD Matrix in Lottery_3.0 at the District Level; (c) Residential Locations of the AV Adopters in each District; (d) Workplaces of the AV Adopters in each District.

the total number. We can find the differences between the Reference and Price scenarios (see Figure 16.12(a) and (b)): those scenarios with a lower AV price (e.g., AV Price_0.25) tend to have more AV adopters who live and work in the same districts, with fewer inter-district commuting trips. Furthermore, these scenarios also have more inter-district commuting trips among suburban districts.

TABLE 16.4

Statistical Indicators about In-Vehicle Work of AV Adopters in Both Reference and Price Scenarios

Scenario	Ref	AV Price_0.05	AV Price_0.10	AV Price_0.15	AV Price_0.20	AV Price_0.25
Percentage of AV Owners Working in Vehicle	40.2%	38.7%	39.9%	38.7%	38.9%	37.9%
Average In-Vehicle Work Duration (Minute)	35.8	43.0	48.4	52.7	54.9	56.1

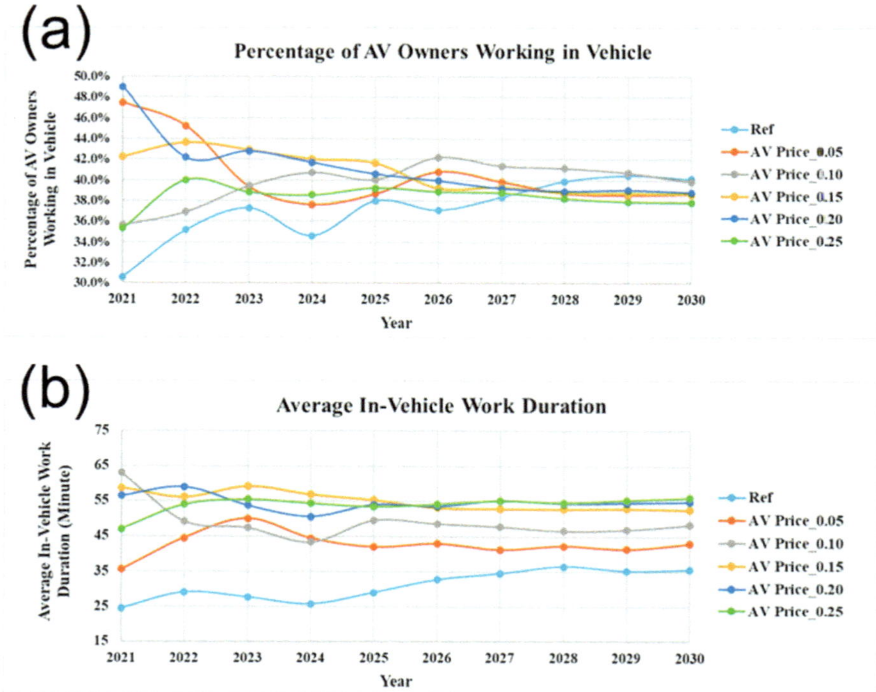

FIGURE 16.11 Results about In-Vehicle Work of AV Adopters in both Reference and Price Scenarios. (a) Percentage of AV Owners Working in their AVs in each Simulation Year; (b) Average Duration of In-Vehicle Work of AV Adopters in each Simulation Year.

CONCLUSION

This paper presented an agent-based Autonomous Vehicle (AV) diffusion model that simulated how private AVs would diffuse over time and space at the micro scale. In particular, the model incorporated three levels of social influence, i.e., global influence, neighbor effect and friendship influence, to capture the interactions and communications between agents (e.g., AV adopters and non-adopters) in the diffusion of AVs. With the model, we explored how AVs would potentially diffuse in Reference, Policy and Price scenarios, and how AV diffusion would influence housing expenses saved (through residential relocation with AVs) and in-vehicle work, which are two typical types of economic impact.

The results suggested that both the license plate lottery policy and upfront cost of AVs could heavily influence the uptake of AVs at both the micro and macro levels, and further the housing expenses saved and in-vehicle work. Specifically, an increase in the number of vehicle permits can promote the adoption of AVs to some extent. However, the number of AV adopters would remain at a specific level due to the high upfront cost of AVs. In the Price scenario, the number of AV adopters increased dramatically. For example, in the scenario with a yearly price decreasing rate of 25%, the number of AV adopters in 2030 was 134,075, which was 14 times

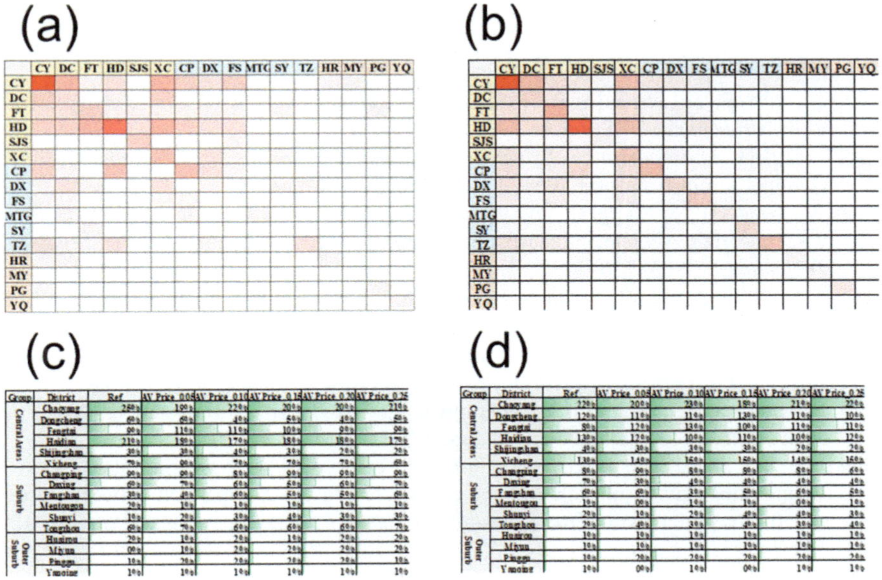

FIGURE 16.12 Spatial Commuting Patterns of those AV Adopters Working in their AVs in both Reference and Price Scenarios. (a) Commuting Origin-Destination (OD) Matrix in Ref at the District Level; (b) Commuting OD Matrix in AV Price_0.25 at the District Level; (c) Residential Locations of the AV Adopters in each District; (d) Workplaces of the AV Adopters in each District.

that of the Reference scenario. AVs could help adopters save, on average, around 5 million and 4 million housing expenses in Policy and Price scenarios, respectively. In addition, around 40% of AV adopters would perform in-vehicle work in 2030 in both Policy and Price scenarios, with in-vehicle work durations of 40 and 56 minutes, respectively.

Future work will focus on improving the agent-based AV diffusion model and its applications. To improve the model, it would be useful to extend it by incorporating other agent types, such as government agents (responsible for AV-related policy making) and vehicle manufacturer agents (responsible for producing and pricing AVs), so as to capture their interactions in the vehicle market. In terms of model applications, as the potential adoption of AVs would influence many connected urban systems, such as transport, land use, economy, energy, environment, and population systems, we could set up a series of "what-if" scenarios to further explore the potential impacts on AVs using the proposed agent-based AV diffusion model.

REFERENCES

Axsen, J., Orlebar, C., & Skippon, S. (2013). Social influence and consumer preference formation for pro-environmental technology: The case of a UK workplace electric-vehicle study. *Ecological Economics*, 95, 96–107.

Barbour, N., Menon, N., Zhang, Y., & Mannering, F. (2019). Shared automated vehicles: A statistical analysis of consumer use likelihoods and concerns. *Transport Policy*, 80, 86–93.

Ben-Akiva, M. E., & Lerman, S. R. (1985). *Discrete choice analysis: Theory and application to travel demand.* MIT Press.

BMCT. (2018). *Beijing autonomous vehicle road testing report in 2018.* Beijing Municipal Commission of Transport (BMCT). (In Simplified Chinese).

Chen, S., Wang, H., & Meng, Q. (2019). Designing autonomous vehicle incentive program with uncertain vehicle purchase price. *Transportation Research Part C: Emerging Technologies*, 103, 226–245.

Correia, G. H. de A., Looff, E., van Cranenburgh, S., Snelder, M., & van Arem, B. (2019). On the impact of vehicle automation on the value of travel time while performing work and leisure activities in a car: Theoretical insights and results from a stated preference survey. *Transportation Research Part A: Policy and Practice*, 119, 359–382.

Kato, S., Takeuchi, E., Ishiguro, Y., Ninomiya, Y., Takeda, K., & Hamada, T. (2015). An open approach to autonomous vehicles. *IEEE Micro*, 35(6), 60–68.

Kolarova, V., Steck, F., & Bahamonde-Birke, F. J. (2019). Assessing the effect of autonomous driving on value of travel time savings: A comparison between current and future preferences. *Transportation Research Part A: Policy and Practice*, 129, 155–169.

Krueger, R., Rashidi, T. H., & Dixit, V. V. (2019). Autonomous driving and residential location preferences: Evidence from a stated choice survey. *Transportation Research Part C: Emerging Technologies*, 108, 255–268.

Liao, R. (2019). *Search giant Baidu has driven the most autonomous miles in Beijing.* TechCrunch. https://techcrunch.com/2019/04/02/baidu-self-driving-2018/

Liu, J., Zhuge, C., Tang, J. H. C. G., Meng, M., & Zhang, J. (2022). A spatial agent-based joint model of electric vehicle and vehicle-to-grid adoption: A case of Beijing. *Applied Energy*, 310, 118581.

Liu, X., Roberts, M. C., & Sioshansi, R. (2017). Spatial effects on hybrid electric vehicle adoption. *Transportation Research Part D: Transport and Environment*, 52, 85–97.

Llorca, C., Moreno, A., Ammar, G., & Moeckel, R. (2022). Impact of autonomous vehicles on household relocation: An agent-based simulation. *Cities*, 126, 103692.

Luo, Q., Saigal, R., Chen, Z., & Yin, Y. (2019). Accelerating the adoption of automated vehicles by subsidies: A dynamic games approach. *Transportation Research Part B: Methodological*, 129, 226–243.

Malokin, A., Circella, G., & Mokhtarian, P. L. (2019). How do activities conducted while commuting influence mode choice? Using revealed preference models to inform public transportation advantage and autonomous vehicle scenarios. *Transportation Research Part A: Policy and Practice*, 124, 82–114.

Nair, G. S., Astroza, S., Bhat, C. R., Khoeini, S., & Pendyala, R. M. (2018). An application of a rank ordered probit modeling approach to understanding level of interest in autonomous vehicles. *Transportation*, 45(6), 1623–1637.

Pudāne, B., Rataj, M., Molin, E. J. E., Mouter, N., van Cranenburgh, S., & Chorus, C. G. (2019). How will automated vehicles shape users' daily activities? Insights from focus groups with commuters in the Netherlands. *Transportation Research Part D: Transport and Environment*, 71, 222–235.

Shah, S., Dey, D., Lovett, C., & Kapoor, A. (2018). Airsim: High-fidelity visual and physical simulation for autonomous vehicles. In Field and Service Robotics: Results of the 11th International Conference Springer International Publishing. (pp. 621–635).

Singleton, P. A. (2019). Discussing the "positive utilities" of autonomous vehicles: Will travellers really use their time productively? *Transport Reviews*, 39(1), 50–65.

Steck, F., Kolarova, V., Bahamonde-Birke, F., Trommer, S., & Lenz, B. (2018). How autonomous driving may affect the value of travel time savings for commuting. *Transportation Research Record*, 2672(46), 11–20.

Sun, L., & Erath, A. (2015). A Bayesian network approach for population synthesis. *Transportation Research Part C: Emerging Technologies*, 61, 49–62.

Talebian, A., & Mishra, S. (2018). Predicting the adoption of connected autonomous vehicles: A new approach based on the theory of diffusion of innovations. *Transportation Research Part C: Emerging Technologies*, 95, 363–380.

Wang, S., Jiang, Z., Noland, R. B., & Mondschein, A. S. (2020). Attitudes towards privately-owned and shared autonomous vehicles. *Transportation Research Part F: Traffic Psychology and Behaviour*, 72, 297–306.

Wien, J. (2019). *An assessment of the willingness to choose a self-driving bus for an urban trip: A public transport user's perspective*. Delft University of Technology (TU Delft).

Zhu, L., Wang, J., Farnoosh, A., & Pan, X. (2022). A game-theory analysis of electric vehicle adoption in Beijing under license plate control policy. *Energy*, 244, 122628.

Zhuge, C., & Shao, C. (2019). Investigating the factors influencing the uptake of electric vehicles in Beijing, China: Statistical and spatial perspectives. *Journal of Cleaner Production*, 213, 199–216.

Zhuge, C., Shao, C., Wang, S., & Hu, Y. (2018). An agent- and GIS-based virtual city creator: A case study of Beijing, China. *Journal of Transport and Land Use*, 11(1), Article 1.

Zhuge, C., Shao, C., & Wei, B. (2018). An agent-based spatial urban social network generator: A case study of Beijing, China. *Journal of Computational Science*, 29, 46–58.

Zhuge, C., & Wang, C. (2021). Integrated modelling of autonomous electric vehicle diffusion: From review to conceptual design. *Transportation Research Part D: Transport and Environment*, 91, 102679.

Zhuge, C., Wei, B., Dong, C., Shao, C., & Shan, Y. (2019). Exploring the future electric vehicle market and its impacts with an agent-based spatial integrated framework: A case study of Beijing, China. *Journal of Cleaner Production*, 221, 710–737.

Zhuge, C., Wei, B., Shao, C., Shan, Y., & Dong, C. (2020). The role of the license plate lottery policy in the adoption of Electric Vehicles: A case study of Beijing. *Energy Policy*, 139, 111328.

Index

Note: Locators in *italics* represent *figures* and **bold** indicate **tables** in the text.